T0141816

Studies in Computational Intelligence

Volume 872

Series editor
Janusz Kacprzyk, Polish Academy of Sciences, Warsaw, Poland

The series "Studies in Computational Intelligence" (SCI) publishes new developments and advances in the various areas of computational intelligence—quickly and with a high quality. The intent is to cover the theory, applications, and design methods of computational intelligence, as embedded in the fields of engineering, computer science, physics and life sciences, as well as the methodologies behind them. The series contains monographs, lecture notes and edited volumes in computational intelligence spanning the areas of neural networks, connectionist systems, genetic algorithms, evolutionary computation, artificial intelligence, cellular automata, self-organizing systems, soft computing, fuzzy systems, and hybrid intelligent systems. Of particular value to both the contributors and the readership are the short publication timeframe and the world-wide distribution, which enable both wide and rapid dissemination of research output.

The books of this series are submitted to indexing to Web of Science, EI-Compendex, DBLP, SCOPUS, Google Scholar and Springerlink.

More information about this series at http://www.springer.com/series/7092

Orestes Llanes Santiago • Carlos Cruz Corona
Antônio José Silva Neto • José Luis Verdegay
Editors

Computational Intelligence in Emerging Technologies for Engineering Applications

 Springer

Editors

Orestes Llanes Santiago (iD)
CUJAE
Universidad Tecnológica de La Habana José
Antonio Echeverría
Marianao, La Habana, Cuba

Antônio José Silva Neto (iD)
Instituto Politécnico-Universidade do
Estado do Rio de Janeiro
Nova Friburgo, Brazil

Carlos Cruz Corona (iD)
University of Granada
Granada, Spain

José Luis Verdegay (iD)
University of Granada
Granada, Spain

ISSN 1860-949X   ISSN 1860-9503   (electronic)
Studies in Computational Intelligence
ISBN 978-3-030-34411-5   ISBN 978-3-030-34409-2   (eBook)
https://doi.org/10.1007/978-3-030-34409-2

Mathematics Subject Classification: 15A29, 80A20, 93C42, 74P99

© Springer Nature Switzerland AG 2020
This work is subject to copyright. All rights are reserved by the Publisher, whether the whole or part of
the material is concerned, specifically the rights of translation, reprinting, reuse of illustrations, recitation,
broadcasting, reproduction on microfilms or in any other physical way, and transmission or information
storage and retrieval, electronic adaptation, computer software, or by similar or dissimilar methodology
now known or hereafter developed.
The use of general descriptive names, registered names, trademarks, service marks, etc. in this publication
does not imply, even in the absence of a specific statement, that such names are exempt from the relevant
protective laws and regulations and therefore free for general use.
The publisher, the authors, and the editors are safe to assume that the advice and information in this book
are believed to be true and accurate at the date of publication. Neither the publisher nor the authors or
the editors give a warranty, expressed or implied, with respect to the material contained herein or for any
errors or omissions that may have been made. The publisher remains neutral with regard to jurisdictional
claims in published maps and institutional affiliations.

This Springer imprint is published by the registered company Springer Nature Switzerland AG.
The registered company address is: Gewerbestrasse 11, 6330 Cham, Switzerland

To our families

Preface

Although it seems a paradox, the world is already in the next technological revolution and worldwide it has transformative effects on the way we live, work, and develop our economies. New opportunities for entrepreneurs and businesses and the society in general are being created that were unthinkable in the recent past. Digitalization is transforming jobs across all sectors and economies, new types of jobs and employment are arising, and the nature and conditions of work are also changing. Global production of ICT goods and services now amounts to an estimated 6.5% of global gross domestic product (GDP), and some 100 million people are employed in the ICT services sector alone.

The United Nations Conference on Trade and Development (UNCTAD) Information Economy Report 2017 highlighted that the digital economy is growing rapidly underpinned by the introduction of new and emergent technologies, such as cloud computing, big data analytics, the Internet of Things, robotics, 3D printing, and artificial intelligence.

The early detection of promising new and emergent technological areas, wherever they come from, is essential in order to identify and seize opportunities of long-term benefit for the society. The problems associated with these areas require the integration of new methodological approaches, tools, and techniques capable of modeling some kind of data ignorance, dynamism, noise, etc. Then, it is here where Computational Intelligence methodologies could play an important role in addressing these challenges.

Committed to this idea, the aim of this book is to offer a comprehensive and up-to-date portfolio of solutions based on Computational Intelligence to different problems related to new and emerging technologies.

Thus, there are fifteen chapters that present fundamental concepts and analysis of different computational techniques solving problems of acoustic levitation, solar panels, automotive batteries, and UAV Autonomous Navigation, to mention but a few.

Each chapter is briefly introduced below following the order of the index:

Fran Sérgio Lobato, Geisa Arruda Zuffi, Aldemir Ap. Cavalini Jr., and Valder Steffen Jr. present in their paper **Uncertainty Analysis of a Near-Field Acoustic**

Levitation System (Chap. 1) a proposal to evaluate the influence of uncertainties during the design of engineering systems. They used two different methodologies to evaluate uncertainties affecting the maximum force necessary to acoustically levitate a given object. Specifically, they proposed Robust Design (RD) by using the Effective Mean Concept (EMC) and Reliability-Based Design (RBD) by using the Inverse Reliability Analysis (IRA). Related to this, two strategies are presented: MODE+EMC to solve multi-objective problems in the robustness context by using the EMC strategy, and MODE+IRA to solve multi-objective problems in the reliability context by using the IRA strategy.

The analysis of the uncertainty and robustness of infinite dimensional objects such as curves and surfaces is presented by Mohamed Bassi, Emmanuel Pagnacco, and Eduardo Souza de Cursi in the chapter **Uncertainty Quantification and Statistics of Curves and Surfaces** (Chap. 2). Two basic approaches were considered: on the one hand, the use of Hilbert basis to reduce the problem to the analysis of probabilities on spaces formed of sequences of real numbers and, on the other hand, approaches based on the variational characterization of the mean. The authors have shown that both are effective to calculate and that a mixed approach can produce gains analogous to those of uncertainty quantification of finite-dimensional objects.

The roof measurement in solar panels installations is usually expensive and risky. Luis Diago, Junichi Shinoda, and Ichiro Hagiwara try to solve it using a new methodology to automatically create three-dimensional (3D) house model using its elevation views (i.e., north, south, east, and west views) in their proposal **Meta-heuristic Approaches for Automatic Roof Measurement in Solar Panels Installations** (Chap. 3). They used a polygon matching algorithm based on image processing algorithms in combination with Genetic Algorithms with niching methods to search for the best matching. After the best match is found, the scale given by the satellite image is used to rectify the proportions of the 3D house model and to estimate the space available for a solar installation.

The radiative transfer analysis is studied by a lot of authors because of its application to different areas of interest. Lucas Correia da Silva Jardim, Diego Campos Knupp, Wagner Figueiredo Sacco, and Antônio José Silva Neto formulated in the chapter **Solution of a Coupled Conduction-Radiation Inverse Heat Transfer Problem with the Topographical Global Optimization Method** (Chap. 4) an inverse problem combining conduction and radiation used for the determination of thermal and radiative properties. To minimize residuals between predictions yielded by a mathematical/computational model and experimental measurements, the authors used a clustering optimization technique based on the topographic information of the objective function, Topographical Global Optimization (TGO), combined with the Nelder–Mead method as local optimization method.

Soumya Banerjee, Valentina E. Balas, Abhishek Pandey, and Samia Bouzefrane investigated various levels of Cyber-Physical Systems (CPS) formulation driven by machine learning and evolutionary algorithms with their strategic similarities in their proposal **Towards Intelligent Optimization of Design Strategies of Cyber-Physical Systems: Measuring Efficacy Through Evolutionary Computations** (Chap. 5). The work was focused on the analytical aspects of the design paradigm of CPS, but it was also observed that there is a significant trend of multi-objective optimization in terms of resource utilization, scheduling, and even learning the dynamic attributes of design.

Leakage detection and location is a fundamental task that must be performed to guarantee an efficient operation in Water Distribution Networks. In this line, Maibeth Sánchez-Rivero, Marcos Quiñones-Grueiro, Alejandro Rosete Suárez, and Orestes Llanes Santiago propose **A Novel Approach for Leak Localization in Water Distribution Networks Using Computational Intelligence** (Chap. 6). The approach does not depend on the sensitivity matrix neither the labeling method for the nodes, and it solves the inverse problem by using metaheuristic optimization algorithms such as differential evolution, particle swarm optimization, and simulated annealing.

Lucas Camargos Borges, Eduarda Cristina de Matos Camargo, João Jorge Ribeiro Damasceno, Fabio de Oliveira Arouca, and Fran Sérgio Lobato in their chapter **Determination of Nano-aerosol Size Distribution Using Differential Evolution** (Chap. 7) formulated and solved an inverse problem to characterize the relation between the monodispersed and polydispersed aerosol stream measured by electric mobility in a differential mobility analyzer. A Differential Evolution algorithm was used as the optimization tool in which the objective function consists of determining the transfer functions that minimize the sum of difference between predicted and experimental concentrations of sodium chloride.

The inverse heat conduction problems of estimating timewise and/or spacewise varying functions have become an emerging area of research and development, among other things, due to the huge number of innovative applications that they can enable in engineering and medicine. Wellington B. da Silva, Julio C. S. Dutra, Diego C. Knupp, Luiz A. S. Abreu, and Antônio José Silva Neto in their chapter **Estimation of Timewise Varying Boundary Heat Flux via Bayesian Filters and Markov Chain Monte Carlo Method** (Chap. 8) formulated the inverse problem through the Bayesian framework and proposed its solution using Markov Chain Monte Carlo (MCMC) methods, the Sampling Importance Resampling (SIR). Also, a combination of these methodologies is used, consisting of employing the SIR filter solution as the initial state for the MCMC method.

Incremental Capacity Analysis (ICA) relates battery degradations to changes in the derivative of the charge stored in the battery with respect to the voltage at its terminals. Related to this, Luciano Sánchez, José Otero, Inés Couso, and David Anseán in their chapter **Health Monitoring of Automotive Batteries in Fast-Charging Conditions Through a Fuzzy Model of the Incremental Capacity** (Chap. 9) proposed a method for approximating ICA curves in fast-charging conditions. It is based on a dynamic fuzzy model of the derivative of the stored charge with respect to the voltage that can be fitted to data from fast charges and discharges. The model contains a fuzzy knowledge base, where the antecedents of the rules match the extrema of the ICA curve and the consequents correspond to the heights of the same curve.

A bioinspired, complex-adaptive modeling methodology that allows modeling single and multiple faults on smart grid devices using Probabilistic Boolean

Networks (PBN) is presented by Pedro J. Rivera-Torres and Orestes Llanes Santiago in the chapter **Fault Detection and Isolation in Smart Grid Devices Using Probabilistic Boolean Networks** (Chap. 10). The proposal is based on a PBN model of Intelligent Power Router (IPR), in which each of the IPR's faults are modeled using reliability analysis to detect and isolate single and multiple faults.

Automated Machine Learning (AutoML) is one of the most successful approaches to select the most appropriate machine learning (ML) algorithm and its best hyper-parameter setting given the characteristics of the problem at hand (Model Selection Problem). Juan S. Angarita-Zapata, Antonio D. Masegosa, and Isaac Triguero explored the benefits of AutoML for Traffic Forecasting supervised regression problems in the chapter **Evaluating Automated Machine Learning on Supervised Regression Traffic Forecasting Problems** (Chap. 11). Auto-WEKA results were compared with some state-of-the-art ML algorithms for the prediction of traffic at scales of predictions focused on the point and the road segment levels within freeway and urban environments.

The formation of groups of robots to handle and execute the tasks simultaneously and efficiently can be modeled by the multi-robot coalition formation (MRCF) problem. Amit Rauniyar and Pranab K. Muhuri in their chapter titled **Multi-Robot Coalition Formation and Task Allocation Using Immigrant-Based Adaptive Genetic Algorithms** (Chap. 12) developed different variants of genetic algorithms (GA) as a solution technique for this problem. They incorporated immigrants-based schemes into standard GA (SGA) and develop RIGA (random immigrants GA), EIGA (elitism-based immigrants GA), and also integrated adaptive settings of genetic operations.

Images taken from an active sensor called Light Detection and Ranging (LiDAR) allow the flight of the Unmanned Aerial Vehicle (UAV) over water-covered areas and under low or no light conditions. Related to this, José Renato G. Braga, Haroldo F. de Campos Velho, and Elcio H. Shiguemori in their chapter **Lidar and Non-extensive Particle Filter for UAV Autonomous Navigation** (Chap. 13) proposed to estimate the aircraft position by applying data fusion from two positioning techniques: computer vision and visual odometry. Computer vision system (CVS) correlates a geo-referenced image, and an image without geographic coordinate information, to incorporate geographic location to each pixel of the second image. Visual odometry determines the position of the aircraft by processing two subsequent images of the same scene. A novel and interesting approach based on non-extensive particle filter was used for data fusion applied to the UAV positioning.

The Time-Dependent Traveling Salesman Problem (TD TSP) is one of the most realistic extensions under real traffic conditions of the TSP. Ruba Almahasneh, Boldizsar Tuu-Szabo, Peter Foldesi, and Laszlo T. Koczy propose a novel Intuitionistic Fuzzy Time-Dependent Traveling Salesman Problem (IFTD TSP) in their chapter **Quasi-optimization of the Time-Dependent Traveling Salesman Problem by Intuitionistic Fuzzy Model and Memetic Algorithm** (Chap. 14). This proposal introduces an even more real-life model of TSP using intuitionistic fuzzy sets for the definition of uncertain costs, time, and space of the rush hour traffic jam region affecting graph sections. A memetic version of the bacterial evolutionary algorithm (DBMEA) was used to compute the experiments. It is a combination of the bacterial evolutionary algorithm as global search, and 2-opt and 3-opt (with some enhanced techniques) as local search.

Finally, Taymi Ceruto, Orenia Lapeira, and Alejandro Rosete in the chapter **Analyzing Information and Communications Technology National Indices by Using Fuzzy Datamining Techniques** (Chap. 15) present an experimental study in which data are analyzed from a fuzzy point of view in order to discover similarities among apparently not related original data. Thus, different data mining techniques (fuzzy clusters, fuzzy predicates, graphs, correlations) are applied to several indices characterizing the information and communications technology (ICT) in different countries.

In short, the book can be seen as confirmation of the great potential of Computational Intelligence techniques to obtain robust, sustainable, and low-cost solutions for the problems that arise in emerging and new technological areas.

We hope that students, researchers, engineers, and practitioners have in this book a comprehensive and up-to-date portfolio of solutions to different problems, on the one hand, and a starting point to develop new research lines related to future and emerging technologies, on the other.

To conclude and highlight in this way what we want to say, editors wish to remark our deep gratitude to Prof. Janusz Kacprzyk for all the help he has given us and for his constant encouragement for this book to be published.

Marianao, Cuba   Orestes Llanes Santiago
Granada, Spain   Carlos Cruz Corona
Nova Friburgo, Brazil   Antônio José Silva Neto
Granada, Spain   José Luis Verdegay

Acknowledgments

The editors acknowledge the decisive support provided by CAPES—Foundation for the Coordination and Improvement of Higher Level Education Personnel, through the project "Computational Modelling for Applications in Engineering and Environment," Program for Institutional Internationalization CAPES PrInt 41/2017, Processo No. 88887.311757/2018-00 and the support in part obtained by the project TIN2017-86647-P (Spanish Ministry of Economy and Competitiveness, includes FEDER funds from the European Union). Our gratitude is also due to the Brazilian Society of Computational and Applied Mathematics (SBMAC), to the Springer representation in Brazil, to CNPq—National Council for Scientific and Technological Development to FAPERJ—Foundation Carlos Chagas Filho for Research Support of the State of Rio de Janeiro, as well as to the Cuban Ministry of Higher Education (MES) for the publication of this book.

Acknowledgments are also due to the authors and reviewers, whose expertise was fundamental in the preparation of the book.

Contents

Contributors

Luiz A. S. Abreu Mechanical Engineering and Energy Department, IPRJ-UERJ, Nova Friburgo, RJ, Brazil

Ruba Almahasneh Telecommunications and Media Informatics, Budapest University of Technology and Economics, Budapest, Hungary

Juan S. Angarita-Zapata DeustoTech, Faculty of Engineering, University of Deusto, Bilbao, Spain

David Anseán Departamento de Ingeniería Eléctrica, Electrónica, C. y S., Universidad de Oviedo, Gijón, Spain

Valentina E. Balas Aurel Vlaicu University of Arad, Arad, Romania

Soumya Banerjee CEDRIC Lab, Conservatoire National des Arts et Metiers, Paris Cedex 03, France

Mohamed Bassi LMN, INSA Rouen Normandie, Normandie Université, Saint-Étienne-du-Rouvray, France

Lucas Camargos Borges NUCAPS - Laboratory of Separation Processes, School of Chemical Engineering, Federal University of Uberlândia, Uberlândia, Brazil

Samia Bouzefrane CNAM-CEDRIC Lab, Conservatoire National des Arts et Metiers, Paris Cedex 03, France

José Renato G. Braga Instituto Nacional de Pesquisas Espaciais (INPE), São José dos Campos, SP, Brazil

Aldemir Ap. Cavalini Jr. LMEst, Laboratory of Mechanics and Structures, School of Mechanical Engineering, Federal University of Uberlândia, Uberlândia, Brazil

Taymi Ceruto Grupo de Investigación SCoDA (SoftComputing and Data Analysis), Facultad de Informática, Universidad Tecnológica de La Habana José Antonio Echeverría, Cujae, Marianao, Cuba

Inés Couso Departamento de Estadística e I. O., Universidad de Oviedo, Gijón, Spain

João Jorge Ribeiro Damasceno NUCAPS - Laboratory of Separation Processes, School of Chemical Engineering, Federal University of Uberlândia, Uberlândia, Brazil

Wellington B. da Silva Chemical Engineering Program, CCAE-UFES, Alegre, ES, Brazil

Lucas Correia da Silva Jardim Rio de Janeiro State University, Polytechnic Institute, Nova Friburgo, RJ, Brazil

Antônio José Silva Neto Rio de Janeiro State University, Polytechnic Institute, Nova Friburgo, RJ, Brazil

Haroldo F. de Campos Velho Instituto Nacional de Pesquisas Espaciais (INPE), São José dos Campos, SP, Brazil

Eduardo Souza de Cursi LMN, INSA Rouen Normandie, Normandie Université, Saint-Étienne-du-Rouvray, France

Eduarda Cristina de Matos Camargo NUCAPS - Laboratory of Separation Processes, School of Chemical Engineering, Federal University of Uberlândia, Uberlândia, Brazil

Fabio de Oliveira Arouca NUCAPS - Laboratory of Separation Processes, School of Chemical Engineering, Federal University of Uberlândia, Uberlândia, Brazil

Luis Diago Interlocus Inc, Yokohama, Japan
Meiji University, Nakano, Tokyo, Japan

Julio C. S. Dutra Chemical Engineering Program, CCAE-UFES, Alegre, ES, Brazil

Peter Foldesi Department of Logistics Technology, Széchenyi István University, Gyor, Hungary

Ichiro Hagiwara Meiji University, Nakano, Tokyo, Japan

Diego Campos Knupp Rio de Janeiro State University, Polytechnic Institute, Nova Friburgo, RJ, Brazil

Laszlo T. Koczy Telecommunications and Media Informatics, Budapest University of Technology and Economics, Budapest, Hungary
Department of Information Technology, Széchenyi István University, Gyor, Hungary

Orenia Lapeira Departamento de Ingeniería de Software, Facultad de Informática, Universidad Tecnológica de La Habana José Antonio Echeverría, Cujae, Marianao, Cuba

Orestes Llanes Santiago Automation and Computing Department, Universidad Tecnológica de La Habana José Antonio Echeverría (CUJAE), La Habana, Cuba

Fran Sérgio Lobato NUCOP, Laboratory of Modeling, Simulation, Control and Optimization of Processes, School of Chemical Engineering, Federal University of Uberlândia, Uberlândia, Brazil

Antonio D. Masegosa DeustoTech, Faculty of Engineering, University of Deusto, Bilbao, Spain
IKERBASQUE, Basque Foundation for Science, Bilbao, Spain

Pranab K. Muhuri Department of Computer Science, South Asian University, New Delhi, India

José Otero Departamento de Informática, Universidad de Oviedo, Gijón, Spain

Emmanuel Pagnacco LMN, INSA Rouen Normandie, Normandie Université, Saint-Étienne-du-Rouvray, France

Abhishek Pandey University of Petroleum and Energy Studies, Dehradun, India

Marcos Quiñones-Grueiro Automation and Computing Department, Universidad Tecnológica de La Habana José Antonio Echeverría (CUJAE), La Habana, Cuba

Amit Rauniyar Department of Computer Science, South Asian University, New Delhi, India

Pedro J. Rivera-Torres Department of Computer Science-School of Natural Sciences, University of Puerto Rico at Río Piedras, San Juan, PR, USA

Wagner Figueiredo Sacco Federal University of Western Pará, Institute of Engineering and Geosciences, Santarém, Brazil

Luciano Sánchez Departamento de Informática, Universidad de Oviedo, Gijón, Spain

Maibeth Sánchez-Rivero Automation and Computing Department, Universidad Tecnológica de La Habana José Antonio Echeverría (CUJAE), La Habana, Cuba

Elcio H. Shiguemori Instituto de Estudos Avançados (IEAv), Departamento de Ciência e Tecnologia Aeroespacial (DCTA), Trevo Coronel Aviador José Alberto Albano do Amarante 01 - Putim, São José dos Campos, SP, Brazil

Junichi Shinoda Interlocus Inc, Yokohama, Japan

Valder Steffen Jr. LMEst, Laboratory of Mechanics and Structures, School of Mechanical Engineering, Federal University of Uberlândia, Uberlândia, Brazil

Alejandro Rosete Suárez Artificial Intelligence Department, Universidad Tecnológica de La Habana José Antonio Echeverría (CUJAE), La Habana, Cuba

Isaac Triguero Computational Optimisation and Learning (COL) Lab, School of Computer Science, University of Nottingham, Nottingham, UK

Boldizsar Tuu-Szabo Department of Information Technology, Széchenyi István University, Gyor, Hungary

Geisa Arruda Zuffi LMEst, Laboratory of Mechanics and Structures, School of Mechanical Engineering, Federal University of Uberlândia, Uberlândia, Brazil

Chapter 1
Uncertainty Analysis of a Near-Field Acoustic Levitation System

Fran Sérgio Lobato, Geisa Arruda Zuffi, Aldemir Ap. Cavalini Jr., and Valder Steffen Jr.

Abstract Near-field acoustic levitation is a physical phenomenon that occurs when a planar object is placed in the proximity of a vibrating surface. Consequently, a thin layer of ambient gas, commonly referred to as squeeze film, is trapped in the clearance between a vibrating surface and an adjacent planar object performing its levitation. Mathematically, this phenomenon is described by using the Reynolds equation, which is derived from the Navier–Stokes momentum and continuity equations. The equation of motion that represents the dynamic behavior of the levitated object is also considered. However, the performance of the near-field acoustic levitation can be significantly affected by uncertainties on its geometrical parameters and operating conditions. Thus, the present contribution aims to evaluate the influence of uncertain parameters on the resulting levitation force. For this purpose, the differential evolution algorithm is associated with two strategies (inverse reliability analysis and effective mean concept). This multi-objective optimization problem considers the maximization of the levitation force associated with the maximization of both the reliability and robustness coefficients. Numerical simulations demonstrated the sensitivity of each uncertain parameter associated to the obtained levitation forces. As expected, it was verified that the levitation force decreases as the considered reliability and robustness coefficients increases.

Keywords Uncertainties · Sensitivity · Near-field acoustic levitation · Multi-objective optimization · Differential evolution

F. S. Lobato (✉)
NUCOP, Laboratory of Modeling, Simulation, Control and Optimization of Processes, School of Chemical Engineering, Federal University of Uberlândia, Uberlândia, Brazil
e-mail: fslobato@ufu.br

G. A. Zuffi · A. A. Cavalini Jr. · V. Steffen Jr.
LMEst, Laboratory of Mechanics and Structures, School of Mechanical Engineering, Federal University of Uberlândia, Uberlândia, Brazil
e-mail: geisazuffi@ufu.br; aacjunior@ufu.br; vsteffen@ufu.br

© Springer Nature Switzerland AG 2020
O. Llanes Santiago et al. (eds.), *Computational Intelligence in Emerging Technologies for Engineering Applications*, Studies in Computational Intelligence 872, https://doi.org/10.1007/978-3-030-34409-2_1

1.1 Introduction

The study of uncertainties characterizes an important research field in engineering due to a number of realistic possible applications that can be performed. The present interest on this area is due to the difference between the results obtained from simulation, control, and optimization of engineering design and their practical implementations. Uncertainties can be associated with design variables, mathematical models, loading, geometry, boundary conditions, material properties, manufacturing processes, and operational environment [1].

To evaluate the influence of uncertainties during the design of engineering systems, strategies based on reliability and robustness analyses have been proposed [2–9]. Reliability-based design (RBD) emphasizes reliability in the design by ensuring the probabilistic achievement of constraints at predefined levels [10]. Additionally, RBD can be understood as the ability of a system or component to perform its required functions under certain conditions, i.e., RBD is based on the probability of the desired system performance failing [11]. On the other hand, robust design (RD) determines solutions presenting small sensitivity to changes in the design variables [5]. It is important to mention that in both approaches, either the Monte Carlo simulation (MCS) or the Latin hypercube design (LHD) method is used to perform analyses around the deterministic solution [12].

Both RBD and RD can be used in any engineering systems design. For this purpose, for each kind of optimization problem a particular formulation should be considered. Thus, new studies can be performed to demonstrate the ability of each methodology. For this aim, in the present contribution, the phenomenon of near-field acoustic levitation is analyzed in the uncertainty context. According to Vandaele [13], contactless handling can present numerous advantages: (1) the reduction of friction which enables devices to operate at high speed and accuracy along its motion; (2) the shedding of contamination by handling contact; (3) the manipulation of fragile, sensitive, micron-sized, and non-rigid components; and (4) the measurement of some physical properties of very small solid and liquid samples. Among the known levitation techniques (magnetic, electric, optical, and aerodynamic), the acoustic levitation is the most advantageous because it does not present restrictions to the composition of the object to be levitated, neither to its shape, has a compact driving system, and can be applied in any environment [13].

The acoustic levitation principle can be applied to various systems [14–17]. These applications are typically based on one of two existing levitation modes. The most simple one is called standing waves acoustic levitation. This technique deploys an ultrasonic transducer and a reflector, which can be flat or curved [18]. The ultrasonic transducer is responsible to generate acoustic waves with high frequency (generally, 20 kHz) that travel in the direction of the reflector. Thus, it will be reflected in the direction of the transducer, colliding with the waves generated by the transducer, forming a standing wave with areas of high-pressure and low-pressure. Consequently, when objects are placed between the transducer and the reflector among the created flat waves, they will move from high-pressure to low-pressure areas and there they will remain [19]. This type of acoustic levitation is usually employed for suspension of small particles, since it can measure only half of the wavelength used to generate the wave which needs to be higher than 20 kHz.

Another acoustic levitation type can be used when the levitating object is large enough. It is called near-field acoustic levitation and occurs when a flat object is placed next to a vibrating surface, called driving surface. In this case, a sinusoidal movement is induced over a frequency of 20 kHz compressing the air trapped into the gap between the driving surface and the flat object. Therefore, the gas layer reaches an average pressure value higher than the ambient pressure, resulting in a force that enables the levitation of the object above the vibrating surface [18].

However, the parameters related to ultrasonic actuators (operation frequency and vibration amplitude), the parameters related to air properties (initial pressure and dynamic viscosity), and the geometric parameters (initial distance between the driving surface and the object to be levitated and sizes of these surfaces) can suffer deviations and affect the levitation force and, consequently, the load capacity of the system [13, 20]. In this case, it is necessary to evaluate the sensitivity of the system to changes in the values of the parameters during the simulation, optimization, and control of this type of process.

Therefore, two non-dimensional parameters are analyzed in the present contribution, namely excursion ratio and squeeze number. For this purpose, two different strategies are considered: RD by using the effective mean concept (EMC) and RBD by using the inverse reliability analysis (IRA). For each approach, a multi-objective optimization problem is formulated. In the RD approach, the proposed multi-objective optimization problem considers the maximization of the force necessary to levitate the object and the maximization of robustness coefficient. In the RBD context, the multi-objective optimization considers both the maximization of the force and the reliability coefficient. In both problems, the multi-objective optimization differential evolution (MODE) algorithm, proposed by Lobato and Steffen [21], is considered as the optimization tool. This chapter is organized as follows. Section 1.2 presents the mathematical modeling of near-field acoustic levitation. Sections 1.3.1 and 1.3.2 provide a brief review of the RD and RBD approaches, respectively. Section 1.4 presents a brief description of the MODE strategy. The proposed methodology is discussed in Sect. 1.5. The numerical results are presented in Sect. 1.6. Finally, the conclusions are outlined in Sect. 1.7.

1.2 Near-field Acoustic Levitation: Mathematical Modeling

The problem to be evaluated consists in two discs with a distance $(h(t))$ from each other, immersed in the air at atmospheric pressure, as shown in Fig. 1.1. The lower disc, called driving surface, is responsible for creating a pressurized gas film, from an ultrasonic vibration induced on its surface. As a consequence of the pressure field generated in the gas film, there will be a force capable of levitating the upper disc (object to be levitated). To simplify the analyses of the pressure field generated in

Fig. 1.1 Near-field acoustic levitation system [18]

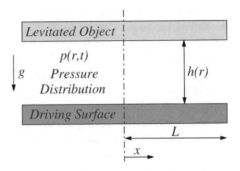

this system, the object to be levitated will be considered fixed while the driving surface will oscillate (sinusoidal behavior) with amplitude δh and frequency ω. Thus, the distance between the object to be levitated and the driving surface varies around h_0, according to Eq. (1.1). In addition, the system is considered symmetric and can be solved by using polar coordinates.

$$h = h_0 + \delta h \sin(\omega t) \tag{1.1}$$

According to Zhao [22], the behavior of the air film can be represented by the Reynolds equation in polar coordinates, as shown in Eq. (1.2). In this case, the air gap between the surfaces of the lower and upper discs is assumed to be small as compared with the other dimensions of the system. In addition, a piston-like movement is assumed for the driving surface. Additionally, no angular movement and temperature changes are considered.

$$\frac{\partial}{\partial r}\left(ph^3\frac{\partial p}{\partial r}\right) = 12\mu\frac{\partial(ph)}{\partial t} \tag{1.2}$$

where p represents the pressure distribution that varies with the distance r and the time t, h represents the gap between the driving surface and the object to be levitated, and μ is the dynamic viscosity of the air.

The Reynolds equation in its non-dimensional form can be obtained by using the following relations [22]:

$$P = \frac{p}{p_0} \tag{1.3a}$$

$$H = \frac{h}{h_0} \tag{1.3b}$$

$$R = \frac{r}{r_0} \tag{1.3c}$$

$$T = \omega t \tag{1.3d}$$

$$\epsilon = \frac{\delta h}{h_0} \tag{1.3e}$$

$$\sigma = \frac{12\omega\mu L^2}{p_0 h_0^2} \tag{1.3f}$$

where σ is called *squeeze number*, L is the characteristic length of the discs, p_0 is the atmospheric pressure, and ϵ is the dimensionless vibration amplitude of the driving surface. Replacing Eqs. (1.3a)–(1.3f) in Eqs. (1.1) and (1.2), the following non-dimensional equations are determined:

$$H = 1 + \epsilon \sin(T) \tag{1.4}$$

$$\frac{\partial}{\partial R}\left(PH^3 \frac{\partial P}{\partial R}\right) = \sigma \frac{\partial(PH)}{\partial T} \tag{1.5}$$

In order to obtain the pressure field along with the compressed air film, the initial pressure in the gap is assumed to be equal to the atmospheric pressure, as well as the pressure at the edges of the discs (Dirichlet boundary condition). The system is considered symmetric with respect to the center of the disc, so that the pressure derivative is zero at this location (Neumann contour condition). The initial and boundary conditions are given by:

$$P(R, T = 0) = 1 \tag{1.6a}$$

$$P(R = 0.5, T) = 1 \tag{1.6b}$$

$$\frac{\partial P(R = 0, T)}{\partial R} = 0 \tag{1.6c}$$

1.3   Robust Optimization and Reliability-Based Design

In order to evaluate uncertainties, in this section the effective mean concept (EMC) and the inverse reliability analysis (IRA) are presented.

1.3.1   Robust Optimization: Effective Mean Concept

Robust optimization configures a research field dedicated to minimizing the influence of small variations in design variables on the system behavior (particularly on the specifications that have been previously prescribed for the system to perform). In general, new constraints and/or objective functions are required to deal with robustness issues in both mono or multi-objective design problems. As an alternative method, Deb and Gupta [4] proposed an extension of the EMC to avoid the introduction of new constraints into the original problem, which is rewritten to consider the mean value of the original objective functions. The robustness measure is based on the following definition:

Definition 1.1 Being f an integrable objective function, the EMC of f associated with the neighborhood δ of the solution x is the function $f^{eff}(x, \delta)$ given by:

$$f^{eff}(\mathbf{x}, \delta) = \frac{1}{|B_\delta(\mathbf{x})|} \int\limits_{\mathbf{y} \in B_\delta(\mathbf{x})} f(\mathbf{y}) d\mathbf{y} \tag{1.7}$$

subject to:

$$g_j^{eff}(\mathbf{x}, \delta) \leq 0 \qquad j = 1, \ldots, k \tag{1.8}$$

$$\mathbf{x}^l \leq \mathbf{x} \leq \mathbf{x}^u \tag{1.9}$$

where g_j represents the j-th inequality constraint ($j=1, \ldots, k$; k is the number of inequality constraints), $|B_\delta(\mathbf{x})|$ is the hypervolume of the neighborhood δ (robustness coefficient), and $\mathbf{x}^l$ and $\mathbf{x}^u$ represent the lower and upper limits of the design variables, respectively. N samples are generated randomly by using the Latin hypercube (LH) approach to estimate this integral.

1.3.2 Reliability-Based Design: Inverse Reliability Analysis

A reliable solution can be interpreted as the one that presents a small probability to produce an infeasible solution [5]. From a given measure of reliability R (desired reliability value), it is necessary to find a solution ensuring the probability of having an infeasible solution given by 1-R. Thus, the RBD problem can be formulated as follows [5]:

$$\min f(\mathbf{x}_d, \mathbf{x}_r) \tag{1.10}$$

$$P(G_j(\mathbf{x}_d, \mathbf{x}_r) \leq 0) \geq R_j \tag{1.11}$$

$$\mathbf{x}_d^l \leq \mathbf{x}_d \leq \mathbf{x}_d^u \tag{1.12}$$

where G_j represents the j-th inequality constraint, $\mathbf{x}_d$ is the vector of design variables, $\mathbf{x}_r$ represents the vector of random variables, and P is the probability of G_j to be less than or equal to zero ($G_j \geq 0$ indicates failure). The vector of design variables contains deterministic values associated with the lower and upper limits $\mathbf{x}_d^l$ and $\mathbf{x}_d^u$, respectively. The probability of failure is defined by the following cumulative distribution function [5]:

$$P(G_j(\mathbf{x}_d, \mathbf{x}_r) \leq 0) = \int \cdots \int_{G_j \leq 0} f_X(\mathbf{x}_r) d\mathbf{x}_r \qquad (1.13)$$

where f_X is a joint probability density function.

The integral defined by Eq. (1.13) has to be evaluated along with the design space defined by the vector of inequality constraints to determine the probability of failure. Either the analytical or numerical evaluation of Eq. (1.13) is computationally expensive due to the specified domain. Thus, the so-called Rosenblatt transformation is used to rewrite the original problem with random variables $\mathbf{x}_r$ as a similar problem with a new vector of random variables u [23]. Considering the new search space (u-space), the most probable point (MPP) for failure is determined by locating the minimum distance between the origin and a given constraint function, which is defined by the reliability coefficient β. Equation (1.13) can be expressed through an inverse transformation for a given probabilistic constraint as presented by Eq. (1.14).

$$P(G(\mathbf{x}_d, \mathbf{u}) \leq 0) = \Phi(\beta) \qquad (1.14)$$

where Φ is the standard normal distribution function. This transformation is used by different approaches to measure reliability and avoid the evaluation of Eq. (1.13). In the reliability coefficient approach, Φ is used to describe the probabilistic constraint in Eq. (1.13).

To evaluate the uncertainties in the RBD context, the IRA approach proposed by Du et al. [12] is taken into account. This approach is based on the following probability equation for a given constraint:

$$P(G(\mathbf{x}_d, \mathbf{u}) \leq G^p) = R \qquad (1.15)$$

where G_p must be determined through the solution of a typical reliability inverse problem and R is defined by the user.

Equation (1.15) indicates the probability of $G(\mathbf{x}_d, \mathbf{x}_r)$ to be equal to a given reliability measure R if $G(\mathbf{x}_d, u) \leq G_p$, in which G_p is estimated by using the well-known reliability approach FORM (first order reliability method) [24]. FORM is applied by considering $G'(\mathbf{x}_d, u)$ and MPP u^* for $P(G'(\mathbf{x}_d, u) \leq 0) = P(G'(\mathbf{x}_d, u) \leq G_P)$, as given by Eq. (1.16).

$$G'(\mathbf{x}_d, \mathbf{u}) = G(\mathbf{x}_d, \mathbf{u}) - G^p \qquad (1.16)$$

The described procedure can be summarized according to the following steps [12]:

(i) Inform the starting point ($k = 0$, $u_0 = 0$, β, and distribution type);
(ii) Evaluate $G_j(\mathbf{x}_d, \mathbf{u}^k)$ and $\nabla G_j(\mathbf{x}_d, \mathbf{u}^k)$. Compute $\mathbf{a}^k$ as follows:

$$\mathbf{a}^k = \frac{\nabla G_j(\mathbf{x}_d, \mathbf{u}^k)}{\left\| \nabla G_j(\mathbf{x}_d, \mathbf{u}^k) \right\|} \tag{1.17}$$

(iii) Update $\mathbf{u}^{k+1} = -\beta \mathbf{a}^k$ and $k = k + 1$;
(iv) Repeat steps (ii) and (iii) until convergence is achieved.

As can be observed, if the reliability coefficient and the distribution are known, the point associated with the minimum value of $G(\mathbf{x}_d, u)$ can be determined.

1.4 Multi-Objective Optimization Differential Evolution

In this section, the multi-objective optimization differential evolution algorithm (MODE), as proposed by Lobato and Steffen Jr [21], is presented. This strategy is based on the differential evolution (DE) algorithm developed by Storn and Price [25]. DE differs from other evolutionary algorithms in the mutation and recombination phases, and presents the following advantages: simple structure, easiness of use, speed, robustness, and ability to escape from local optima [25]. DE creates new candidate solutions by combining the parent individual with several other individuals of the same population. A candidate replaces the parent only if it has a better fitness value. DE has three control parameters, namely the perturbation rate, crossover rate, population size, and number of generations.

MODE incorporates two operators in the original DE algorithm, namely the mechanisms of rank ordering [26] and exploration of the neighborhood potential solution candidates [27]. The general structure of the proposed algorithm for multi-objective optimization problems by using DE is briefly described as follows. An initial population of size NP is randomly generated. All dominated solutions are removed from the population through the operator fast non-dominated sorting [26]. In this way, the population is sorted into non-dominated fronts F_j (sets of vectors that are non-dominated with respect to each other). This procedure is repeated until each vector becomes member of a front. Three parents are selected randomly in the population. A child is generated from three parents and this process continues until NP children are generated. Starting from population P_1 of size $2NP$, neighbors are generated to each one of the individuals of the population. In this way, the neighbors generated are classified according to the dominance criterion and only the non-dominated neighbors (P_2) will be included in P_1 to form P_3. The population P_3 is then classified according to the dominance criterion. If the number of individuals in the population P_3 is larger than a number defined by the user, it is truncated according to the criterion of the crowding distance [26], i.e., which describes the density of solutions surrounding a vector.

More details about the MODE algorithm can be found in Lobato and Steffen Jr. [21].

1.5 Methodology

In this contribution, two strategies are proposed: (1) MODE+EMC to solve multi-objective problems in the robustness context by using the EMC strategy, and (2) MODE+IRA to solve multi-objective problems in the reliability context by using the IRA strategy. Both strategies start by rewriting the original problem (mono-objective) into an equivalent multi-objective optimization problem with two new objective functions: the maximization of the robustness parameter (max δ) in the robustness context and the maximization of the reliability coefficient (max β) in the reliability context. Mathematically, each problem is defined as follows:

$$P_1(\text{Robust}) \equiv \begin{cases} \max F(\epsilon, \sigma) \\ \max \delta \\ 0.01 \leq \epsilon \leq 0.5 \\ 10 \leq \sigma \leq 20 \\ 0.1 \leq \delta \leq 0.5 \end{cases} \tag{1.18}$$

$$P_2(\text{Reliable}) \equiv \begin{cases} \max F(\mu_\epsilon, \mu_\sigma) \\ \max \beta \\ P(\mu_\epsilon - 0.5 \leq 0) = \Phi(\beta) \\ P(\mu_\sigma - 20 \leq 0) = \Phi(\beta) \\ 0.01 \leq \mu_\epsilon \\ 10 \leq \mu_\sigma \\ 0.1 \leq \beta \leq 5 \end{cases} \tag{1.19}$$

where μ_ϵ and μ_σ are the mean values of the design variables ϵ and σ, respectively, and F is the force necessary to levitate the object, as given by Eq. (1.20).

$$F(\epsilon, \sigma) = \int_0^{2\pi} \int_0^{0.5} R\left(P(\epsilon, \sigma) - 1\right) dR dT \tag{1.20}$$

Both problems must satisfy the mathematical model that represents the associated acoustic levitation phenomenon (Eqs. (1.4) and (1.5)).

The MODE+EMC strategy is presented in Fig. 1.2a and implemented as follows:

- Initially, the MODE parameters (population size, maximum number of generations, crossover rate, perturbation rate, reduction rate, number of pseudo-curves, and the strategy considered to generate potential candidates to solve the optimization problem) are defined by the user;

- The multi-objective optimization problem is defined by the minimization of the original objective function and the maximization of the robustness coefficient. Thus, the original problem is converted into an equivalent multi-objective function that presents a new objective function (maximization of the robustness coefficient);
- The population of candidates (vector of design variables) is generated using MODE considering the strategy proposed in DE;
- For each candidate generated by DE, N neighbors (perturbations) are obtained considering the range delimited by the robustness parameter δ;
- The values of the design variables are used to evaluate the original objective function defined by EMC;
- The evaluated population is organized according to the Pareto's dominance criterion;
- The iteration process is repeated until convergence is achieved (maximum number of generations).

The MODE+IRA strategy is presented in Fig. 1.2b and implemented as follows:

- Initially, the MODE parameters (population size, maximum number of generations, crossover rate, perturbation rate, reduction rate, number of pseudo-curves and the strategy considered to generate potential candidates to solve the optimization problem) and the IRA parameter (distribution type) are defined by the user;
- The multi-objective optimization problem is defined by the minimization of the original objective function and the maximization of the reliability coefficient. Thus, the original problem is converted into an equivalent multi-objective function that presents a new objective function (maximization of the reliability coefficient);
- The population of candidates (vector of design variables x_d) is generated using MODE considering the strategy proposed in DE;
- For each candidate, IRA is applied to determine the value of the inequality constraint from the vector of random variables (uncertain variables) considering a reliability coefficient β;
- The values of x_d and u (the vector x_r is transformed into an equivalent vector u) are used to evaluate the original objective function and the inequality constraints;
- The evaluated population is organized according to the Pareto's dominance criterion;
- The iteration process is repeated until convergence is achieved (maximum number of generations).

The flowcharts of the proposed methodology are presented in Fig. 1.2.

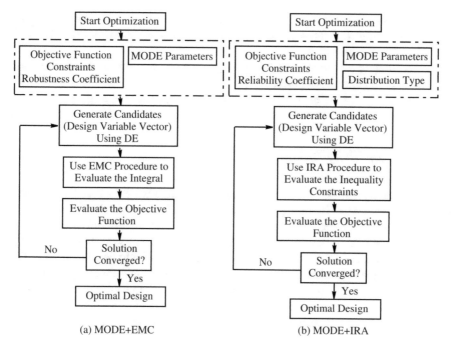

(a) MODE+EMC   (b) MODE+IRA

Fig. 1.2 MODE+EMC and MODE+IRA approaches

1.6   Results and Discussion

To evaluate the MODE+IRA and MODE+EMC strategies, the following conditions
were considered:

- In order to evaluate the influence of the ϵ and σ parameters, a sensitivity analysis
  is performed considering $0.01 \leq \epsilon \leq 0.5$ and $\sigma = 10$, and $\epsilon = 0.25$ and
  $10 \leq \sigma \leq 20$;
- To solve the partial differential equation described in Eq. (1.5), the collocation
  method was used [28]. For this purpose, in each domain (time and space) 50
  points were considered;
- For each problem, the following MODE parameters were considered: population
  with 20 individuals, 100 generations, perturbation rate of 0.8, crossover rate of
  0.8, DE/rand/1/bin strategy to generate potential candidates [25], 10 pseudo-
  curves, and reduction rate of 0.9 [21];
- The number N of samples considered to generate neighbors in EMC is 25;
- The vector of random variables u is considered as being equal to zero in the first
  iteration of IRA;
- The derivatives of the constraints required by IRA were obtained analytically;

- The stopping criterion used to finish IRA is associated with the Euclidean norm of the vector u. If this value along two consecutive iterations is less than 10^{-5}, the process is finished;
- The stopping criterion used to finish optimization algorithms is the maximum number of generations (100);
- All case studies were run 10 times to obtain the upcoming average values.

1.6.1 Sensitivity Analysis

The sensitivity analysis indicates the possibility of improving the near-field acoustic levitation load capacity through changes in the different parameters. In this case, the influence of variations on the operation frequency (ω), initial pressure (p_0), dynamic viscosity (μ), initial distance between the driving surface and the object to be levitated (h_0), sizes of the driving surfaces and object to be levitated (L). The levitation force is obtained by applying variations on the dimensionless parameter σ, as shown in Fig. 1.3. Thus, the load capacity of the system increases exponentially as the value of σ increases. The same behavior is observed by varying the vibration amplitude of the driving surface δ_h, obtained through variations imposed to the dimensionless parameter ϵ (Fig. 1.4).

1.6.2 Robust Design

The proposed robust multi-objective optimization problem (P_1) considers that the design variables ϵ and σ are affected by uncertainties. In this sense, Table 1.1 presents some points belonging to the Pareto's curve (see Fig. 1.5) obtained by using the MODE+EMC approach. Note that the value of the objective function decreases as the robustness parameter increases. It implies that a robust solution is obtained for lower values of F. It is important to mention that all solutions converge to values

Fig. 1.3 Influence of σ on the acoustic levitation design problem

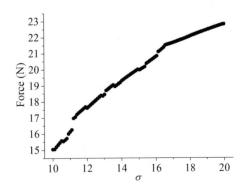

Fig. 1.4 Influence of ϵ on the acoustic levitation design problem

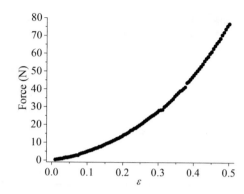

Table 1.1 Robust results associated with the acoustic levitation design problem

Point	ϵ	σ	δ	F (N)
A	0.499616	19.825087	0.005750	86.643546
B	0.496876	19.970418	0.234085	75.861299
C	0.496984	19.547240	0.499570	64.732507

Fig. 1.5 Pareto's curve considering MODE+EMC strategy

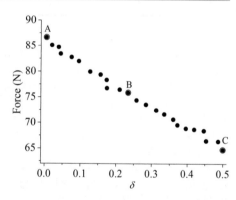

Table 1.2 Statistical data of the random variables used in the acoustic levitation design problem

Symbol	Mean value	Coefficient of variation
ϵ	μ_ϵ	0.1
σ	μ_σ	0.1

close to the associated maximum bounds of each design variable. The average number of evaluations of the objective function required by this methodology for each run was 40,020.

1.6.3 Reliability-Based Design

In this case, only reliability is considered (Problem P_2) and the mean values associated with the design variables (μ_ϵ and μ_σ) are uncertain quantities. The statistical data associated with these variables are presented in Table 1.2.

Table 1.3 Reliable results associated with the acoustic levitation design problem

Point	μ_ϵ	μ_σ	β	F (N)
A	0.498409	19.795086	2.444696	41.323354
B	0.496145	19.186356	2.470856	41.238645
C	0.497855	18.974667	2.398777	44.426098
D	0.499566	18.855578	2.487299	44.278056

Fig. 1.6 Pareto's Curves considering the MODE+IRA strategy

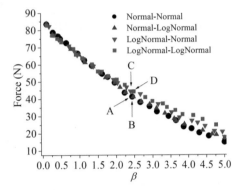

Table 1.3 shows the results obtained by using the MODE+IRA approach by considering different distributions (A = Normal-Normal, B = Normal-LogNormal, C = Log Normal-Normal, and D = LogNormal-LogNormal) for the reliability coefficient equal to 2.5 (approximately). Note that for β smaller than 2, no significant difference can be observed among the considered distributions. On the other hand, for β greater than 2, differences among the considered profiles are verified (see Fig. 1.6).

As observed, the value of the objective function decreases according to β. This result is associated with the considered probabilistic constraints, in which identical values for the design variables may result for different values of the objective function. The average number of evaluations of the objective function required by the proposed methodology for each run was 18,020.

1.7 Conclusions

This chapter presented two different methodologies to evaluate uncertainties affecting the maximum force necessary to acoustically levitate a given object. For this aim, the MODE algorithm was associated with the effective mean concept (EMC) and the inverse reliability analysis (IRA). Therefore, robustness and reliability coefficients could be evaluated. The proposed MODE+EMC and MODE+IRA approaches are based on the definition of a new optimization problem with additional objective functions, which are the maximization of the reliability and robustness coefficients.

Considering the proposed methodologies, the following conclusions can be outlined:

- The proposed methodology consists of the combination of IRA and EMC in a multi-objective optimization context. In this case, the MODE algorithm was used. However, other multi-objective optimization techniques could also be performed successfully;
- Two different probability distributions (normal and lognormal) were chosen to represent the uncertain design variables. However, the proposed methodology is able to handle problems that consider different probability distributions;
- The number of evaluations of the objective function required by the proposed approaches is greater than the one required by the cases in which a deterministic solution is obtained. These results were expected, since there is a need to evaluate the objective function in more points, increasing the value of this parameter;
- The main advantage of these formulations is the possibility of treating the original problem by considering the Pareto's dominance criterion. An optimal solution is found for all the objective functions simultaneously. For each point belonging to the obtained Pareto's curve, it is possible to choose the best configuration in terms of one or more objective functions. Thus, it is possible to choose the most interesting design after obtaining the optimal solution.

Acknowledgements The authors are thankful for the financial support provided to the present research effort by CNPq (574001/2008-5, 304546/2018-8, and 431337/2018-7), FAPEMIG (TEC-APQ-3076-09, TEC-APQ-02284-15, TEC-APQ-00464-16, and PPM-00187-18), and CAPES through the INCT-EIE.

References

1. Choi, S.K., Grandhi, R., Canfield, R.A.: Reliability-Based Structural Design. Springer, London (2006)
2. Carter, D.S.: Mechanical Reliability and Design. Wiley, New York (1997)
3. Melchers, R.E.: Structural Reliability Analysis and Prediction. Wiley, Chichester (1999)
4. Deb, K., Gupta, H.: Introducing robustness in multi-objective optimization. Evol. Comput. **14**(4), 463–494 (2006)
5. Deb, K., Padmanabhan, D., Gupta, S., Mall, A.K.: Handling uncertainties through reliability-based optimization using evolutionary algorithms. IEEE Trans. Evol. Comput. **13**(5), 1054–1074 (2009)
6. Wang, P., Wang, Z., Almaktoom, A.T.: Dynamic reliability-based robust design optimization with time-variant probabilistic constraints. Eng. Optim. **246**(6), 784–809 (2013)
7. Gu, X., Sun, G., Li, G., Mao, L., Li, Q.: A comparative study on multi-objective reliable and robust optimization for crashworthiness design of vehicle structure. Struct. Multidiscip. Optim. **48**(3), 669–684 (2013)
8. Prigent, S., Maréchal, P., Rondepierre, A., Druot. T., Belleville, B.: A robust optimization methodology for preliminary aircraft design. Eng. Optim. **48**(5), 883–899 (2015)
9. Doh, J., Kim, Y., Lee, J.: Reliability-based robust design optimization of gap size of annular nuclear fuels using kriging and inverse distance weighting methods. Eng. Optim. **1**(1), 1–16 (2018)

10. Du, X., Chen. W. Sequential optimization and reliability assessment method for efficient probabilistic design. J. Mech. Des. **126**(2), 225–233 (2004)
11. Poles, S., Lovison, A.: A polynomial chaos approach to robust multiobjective optimization, hybrid and robust Approaches to multiobjective optimization. In: Deb, K., Greco, S., Miettinen, K., Zitzler, E. (eds.) Proceedings of Dagstuhl Seminar, pp. 1–15. Schloss Dagstuhl-Leibniz-Zentrum fuer Informatik, Dagstuhl. (2009)
12. Du, X., Sudjianto, A., Chen, W.: An integrated framework for optimization under uncertainty using inverse reliability strategy. J. Mech. Des. **126**(4), 562–570 (2004)
13. Vandaele, V., Lambert, P., Delchambre, A.: Non-contact handling in microassembly: acoustical levitation. Precis. Eng. **29**(4), 491–505 (2005)
14. Yamazaki, T., Hu, J., Nakamura, K., Ueha, S.: Trial construction of a noncontact ultrasonic motor with an ultrasonically levitated rotor. Jpn. J. Appl. Phys. **35**(5S), 3286 (1996)
15. Peng, T., Yang, Z., Kan, J., Tian, F., Che, X.: Performance investigation on ultrasonic levitation axial bearing for flywheel storage system. Front. Mech. Eng. China **4**(4), 415 (2009)
16. Stolarski, T.A., Xue, Y., Yoshimoto, S.: Air journal bearing utilizing near-field acoustic levitation stationary shaft case. Proc. Inst. Mech. Eng. J J. Eng. Tribol. **225**(3), 120–127 (2011)
17. Thomas, G.P., Andrade, M.A., Adamowski, J.C., Silva, E.C.N.: Development of an acoustic levitation linear transportation system based on a ring-type structure. IEEE Trans. Ultrason. Ferroelectr. Freq. Control **64**(5), 839–846 (2017)
18. Ilssar, D., Bucher, I.: On the slow dynamics of near-field acoustically levitated objects under High excitation frequencies. J. Sound Vib. **354**, 154–166 (2015)
19. Hrka, S.: Acoustic Levitation. University of Ljubljana, Faculty of Mathematics and Physics, Ljubljana (2015)
20. Ilssar, D., Bucher, I.: The effect of acoustically levitated objects on the dynamics of ultrasonic actuators. J. Appl. Phys. **121**(11), 114504 (2017)
21. Lobato, F.S., Steffen Jr., V.: A new multi-objective optimization algorithm based on differential evolution and neighborhood exploring evolution strategy. J. Artif. Intell. Soft Comput. Res. **1**, 1–12 (2011)
22. Zhao, S.: Investigation of Non-contact Bearing Systems Based on Ultrasonic Levitation. PZH, Produktionstechn (2010)
23. Rosenblatt, M.: Remarks on a multivariate transformation. Ann. Math. Stat. **23**, 470–472 (1952)
24. Zhao, Y.G., Ono, T.: A general procedure for first/second-order reliability method (FORM/SORM). Struct. Saf. **21**(1), 95–112 (1999)
25. Storn, R., Price, K.V.: Differential evolution: a simple and efficient adaptive scheme for global optimization over continuous spaces. Int. Comput. Sci. Inst. **12**, 1–16 (1995)
26. Deb. K.: Multi-Objective Optimization using Evolutionary Algorithms. Wiley, Chichester (2001)
27. Hu, X., Coello, C.A.C., Huang, Z.: A new multi-objective evolutionary algorithm: neighborhood exploring evolution strategy. Eng. Optim. **37**(4), 351–379 (2005)
28. Villadsen, J.V., Michelsen, M.L.: Solution of differential equation models by polynomial approximation. Prentice-Hall, New Jersey (1978)

Chapter 2
Uncertainty Quantification and Statistics of Curves and Surfaces

Mohamed Bassi, Emmanuel Pagnacco, and Eduardo Souza de Cursi

Abstract The analysis of the uncertainty and robustness of infinite dimensional objects such as functions, curves and surfaces is an emerging problem, which is requested, for instance, in uncertain multiobjective optimization. Indeed, the determination of statistics—such as the mean or standard deviation—of infinite dimensional objects involves probabilities in infinite dimensional spaces, what introduces operational difficulties. We examine two existing approaches and furnish some comparisons. It is shown that both are effective to calculate and that a mixed approach can produce gains analogous to those of uncertainty quantification of finite-dimensional objects.

Keywords Uncertainties · Multiobjective optimization · Hybrid algorithms

2.1 Introduction

The analysis of the uncertainty and robustness of infinite dimensional objects such as functions, curves and surfaces is an emerging problem found in several practical situations [1–6].

For example, when considering the analysis and control of dynamic systems subject to variability and uncertainty, boundary cycles, periodic orbits and Poincaré sections become random: the characterization of their probability distributions and the evaluation of the statistics of these objects are necessary, in particular for the purposes of risk assessment and reliability improvement.

M. Bassi · E. Pagnacco · E. S. de Cursi (✉)
LMN, INSA Rouen Normandie, Normandie Université, Saint-Étienne-du-Rouvray, France
e-mail: mohamed.bassi@insa-rouen.fr; emmanuel.pagnacco@insa-rouen.fr;
eduardo.souza@insa-rouen.fr

© Springer Nature Switzerland AG 2020
O. Llanes Santiago et al. (eds.), *Computational Intelligence in Emerging Technologies for Engineering Applications*, Studies in Computational Intelligence 872, https://doi.org/10.1007/978-3-030-34409-2_2

A second example arises from uncertain multiobjective optimization. In such a situation, Pareto fronts are uncertain and we are interested, on the one hand, in the determination of their probability distribution and, on the other hand, in the determination of confidence intervals and statistics such as their mean or the variance.

In both the cases above, we are interested in the distributions of probabilities and the associated statistics for manifolds—such as curves and surfaces. For example, a Pareto front in biobjective optimization and a limit cycle of a dynamic system with a single degree of freedom are, in general, curves—which are objects belonging to infinite dimensional spaces. When considering multiobjective optimization with a larger number of objectives or dynamic systems having several degrees of freedom, these same objects become manifolds that are also, in general, infinite dimensional objects.

These situations lead to a main difficulty concerning the definition of probability distributions and statistics in infinite dimensional spaces having functions as elements. For instance, the construction of confidence interval for a limit cycle needs the definition of quantiles in convenient functional space containing all the possible limit cycles—which is an infinite dimensional one (recall that a confidence interval is a region having a given probability, so that confidence intervals are closely connected to quantiles). Analogously, the determination of a confidence interval of a Pareto front requests the definition of probabilities in a functional space containing all the possible Pareto fronts—again, it is an infinite dimensional vector space.

To illustrate the difficulty, let us recall that a curve in $\mathbb{R}^2$ may be described by a vector map or an algebraic equation, i.e., a vector function associating an interval $I \subset \mathbb{R}$ to a set of points in $\mathbb{R}^2$ ($I \ni t \rightarrow \mathbf{x}(t) \in \mathbb{R}^2$) or an algebraic equation $\varphi(\mathbf{x}) = 0, \mathbf{x} \in S$. We are interested in the situation where the curve depends upon a random variable $\mathbf{Z} \in \mathcal{Z} \subset \mathbb{R}^m$: denoting by $\mathbf{z}$ a variate from $\mathbf{Z}$, the equation becomes $\varphi(\mathbf{x}, \mathbf{z}) = 0$, $\mathbf{x} \in S(\mathbf{z})$ and the map reads as $\mathbf{x}(t, \mathbf{z}) : I(\mathbf{z}) \rightarrow \mathbb{R}^2$. In the sequel, we denote $\mathbf{x}_t$ the function $\mathbf{z} \rightarrow \mathbf{x}(t, \mathbf{z})$ and $\mathbf{x}_\mathbf{z}$ the function $t \rightarrow \mathbf{x}(t, \mathbf{z})$.

A first idea consists in considering $\mathbf{x}(t, \mathbf{z}) \in \mathbb{R}^2$ as being a stochastic process $\{\mathbf{x}_t : t \in I\}$ indexed by t and look for the distribution of $\mathbf{x}_t$ for a given t, what furnishes a mean $E(\mathbf{x}_t)$ for each t. We have

$$E(\mathbf{x}_t) = \int_{\mathcal{Z}} \mathbf{x}_t(\mathbf{z}) P(\mathbf{z} \in d\mathbf{z}) = \int_{\mathcal{Z}} \mathbf{x}(t, \mathbf{z}) P(\mathbf{z} \in d\mathbf{z}). \qquad (2.1)$$

When $\mathbf{Z}$ has a probability density $\phi_\mathbf{Z}$, we have $P(\mathbf{z} \in d\mathbf{z}) = \phi_\mathbf{Z}(\mathbf{z})d\mathbf{z}$, so that

$$E(\mathbf{x}_t) = \int_{\mathcal{Z}} \mathbf{x}_t(\mathbf{z})\phi_\mathbf{Z}(\mathbf{z})d\mathbf{z} = \int_{\mathcal{Z}} \mathbf{x}(t, \mathbf{z})\phi_\mathbf{Z}(\mathbf{z})d\mathbf{z}. \qquad (2.2)$$

Unfortunately, this punctual approach is not adapted to the situations under consideration: let us illustrate it by using the simple dynamical system $\ddot{u} = -Z^2 u, u(0) = 1, \dot{u}(0) = 0$, where Z is random, uniformly distributed on $\mathcal{Z} = (-\pi, \pi)$: the density is $\phi(z) = \frac{1}{2\pi}$. Since Z is random, $u, \dot{u}$ and the orbit $\mathbf{x} = (u, \dot{u})$

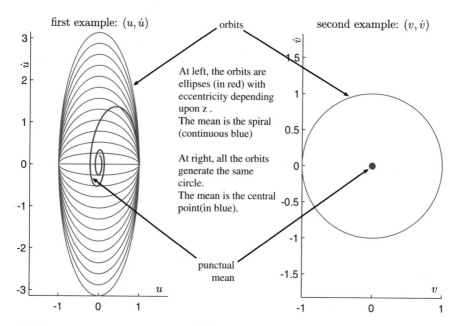

Fig. 2.1 Two simple dynamical systems illustrating that a punctual mean is not adapted to the analysis of infinite dimensional objects. On the left, the orbits are ellipses (in red continuous line) for $z \neq 0$ and reduce to the point $(1, 0)$ for $z = 0$, while the punctual mean furnishes a spiral (in blue dotted line). On the right, all the orbits coincide in a single circle (in red continuous line), but the punctual mean furnishes a point (in blue)

are random too. Indeed, variates from these variables may be obtained by solving the motion equation: we have $u(t, z) = \cos(zt)$, $\dot{u}(t, z) = -z \sin(zt)$. For $z \neq 0$, the variate of the orbit **x** corresponds to an ellipse having equation $u^2 + (\dot{u}/z)^2 = 1$. For $z = 0$, it reduces to a point $u = 1$, $\dot{u} = 0$. Then, for $t > 0$,

$$E(u_t) = \frac{\sin(\pi t)}{\pi t}, \quad E(\dot{u}_t) = \frac{\pi t \cos(\pi t) - \sin(\pi t)}{\pi t^2}. \tag{2.3}$$

As observed in Fig. 2.1, the punctual approach furnishes a spiral as mean, what is not representative of the behaviour of the system. Analogously, let us consider the simple dynamical system $\ddot{v} = -v$, $v(0, z) = \cos(z)$, $\dot{v}(0, z) = -\sin(z)$, where z is a variate from Z uniformly distributed on $\mathcal{Z} = (-\pi, \pi)$. The punctual approach leads to $E(v_t) = E(\dot{v}_t) = 0$, while all the orbits coincide with same circle $v^2 + \dot{v}^2 = 1$ (see Fig. 2.1).

Indeed, the punctual approach does not furnish adequate results in these situations and we must adopt another point of view: we may find in the literature two approaches tending to give suitable responses to these questions.

A first one consists in using Hilbert basis in order to reduce the problem to the analysis of probabilities on spaces formed of sequences of real numbers, such as, for instance ℓ^2 [7, 8]. Such an approach is effective to calculate and suggests also a method for infinite dimensional optimization [9, 10]. However, up to this date, no mathematical result establishes the independence from the basis and guides the choice of the parameterizations. A second approach consists in the use of the variational characterization of the mean and of the median to determine the statistics [11].

In the following, we examine these two approaches and compare them in some significant situations. It is shown that both are effective to calculate and that a mixed approach can produce gains analogous to those of uncertainty quantification of finite-dimensional objects. For the sake of simplicity, we consider curves, but the arguments exposed extend straightly to surfaces or, more generally, manifolds.

2.2   The Hilbertian Approach

The Hilbertian approach assumes that $\mathbf{x}$ belongs to a Hilbert space V having an inner product $(\bullet, \bullet)$ and a Hilbert basis or total family $\Phi = \{\varphi_i\}_{i \in \mathbb{N}^*}$. Then, $\mathbf{x}$ admits an expansion

$$\mathbf{x} = \sum_{i \in \mathbb{N}^*} \mathbf{x}_i \varphi_i. \tag{2.4}$$

In practice, finite sums are considered:

$$\mathbf{x} \approx P_n \mathbf{x} = \sum_{1 \le i \le n} \mathbf{x}_i \varphi_i. \tag{2.5}$$

In this case, the coefficients $\mathbf{X} = (\mathbf{x}_1, \ldots, \mathbf{x}_n)$ are the solutions of the linear system

$$\mathcal{A}\mathbf{X} = \mathcal{B}, \ \mathcal{A}_{ij} = (\varphi_i, \varphi_j), \ \mathcal{B}_i = (\mathbf{x}, \varphi_i). \tag{2.6}$$

In addition,

$$\lim_{n \to \infty} P_n \mathbf{x} = \mathbf{x}. \tag{2.7}$$

If the family is orthonormal, we have $(\varphi_i, \varphi_j) = \delta_{ij}$; $\mathcal{A}$ is the identity matrix and the coefficients of the expansion are given by $\mathbf{x}_i = (\mathbf{x}, \varphi_i)$.

The main difficulties when using this approach are:

1. the choice of the family Φ: in general, for a given functional space V, the choice of the family is not unique. For instance, in dimension one, we may use wavelets, trigonometric functions or polynomial functions. The choice may influence the results.
2. the choice of the parameterization: all the curves must be defined on the same interval I, what brings difficulties when each curve is defined on a different

interval. For instance, the period of the simple system $\ddot{u} = -z^2 u, u(0) = 1, \dot{u}(0) = 0$ is $T(z) = 2\pi/z$, so that a closed orbit corresponding to one period is defined on $I(z) = (0, T(z))$. In order to apply the Hilbertian approach, it is necessary to re-parameterize the curves to get a single interval $\bar{I} = (0, \bar{T})$.

3. The expansion may be affected by changes in the frame of reference used.

In despite of these difficulties, the Hilbertian approach may be used to the evaluation of statistics such as means and covariances. The construction of confidence intervals in this approach is a difficulty which is still with us—instead of confidence intervals, this approach may generate envelopes containing predetermined quantiles.

2.2.1 Classical UQ Approach

The classical UQ approach uses $V = L^2(\Omega, \mu) = \{ U : E(|U|^2) < \infty \}$, where Ω is the universe and μ is a probability measure on Ω. The inner product is $(U, V) = E(U.V)$. It assumes that $Z \in V$ and, for any t, $x_t \in V$. Considering $\Phi = \{ \varphi_i(z) \}_{i \in \mathbb{N}^*}$, the random vector x_t has an expansion given by Eq. (2.4) for each t. Thus, we have (see [8, 12, 13])

$$\mathbf{x}(t, \mathbf{z}) = \sum_{i \in \mathbb{N}^*} \mathbf{x}_i(t) \varphi_i(\mathbf{z}) ; \quad P_n \mathbf{x}(t, \mathbf{z}) = \sum_{1 \leq i \leq n} \mathbf{x}_i(t) \varphi_i(\mathbf{z}). \tag{2.8}$$

In this case, $\mathbf{X}(t) = (\mathbf{x}_1(t), \dots, \mathbf{x}_n(t))$ and the linear system reads as $\mathcal{A}\mathbf{X}(t) = \mathcal{B}(t)$, where $\mathcal{A}_{ij} = (\varphi_i, \varphi_j)$ and $\mathcal{B}_i(t) = (\mathbf{x}(t), \varphi_i)$. We estimate the mean as

$$E(\mathbf{x}(t, \mathbf{Z})) = \sum_{i \in \mathbb{N}^*} \mathbf{x}_i(t) E(\varphi_i(\mathbf{Z})) \approx \sum_{1 \leq i \leq n} \mathbf{x}_i(t) E(\varphi_i(\mathbf{Z})) = E(P_n \mathbf{x}(t, \mathbf{Z})) . \tag{2.9}$$

Other statistics may be defined in an analogous way. It is expected that $E(P_n \mathbf{x}(t, \mathbf{z})) \longrightarrow E(\mathbf{x}(t, \mathbf{z}))$ for $n \to \infty$. Envelopes may be generated by solving the optimization problem

$$\text{Find } \mathbf{y}^* = (y_1^*, \dots, y_n^*) = \arg\max \left\{ \left\| \sum_{1 \leq i \leq n} \mathbf{x}_i(t) y_i \right\| : y_i \in C_i \right\}. \tag{2.10}$$

where, for each i, C_i is a confidence interval for $\varphi_i(\mathbf{Z})$. Notice that the problem formulated in Eq. (2.10) admits multiple solutions, which must be all determined—for each t—to generate a correct envelope—this is a severe limitation of this approach to generate envelopes when n increases. Nevertheless, means and covariances may be obtained by this way.

2.2.2   Inverted UQ Approach

Croquet and Souza de Cursi [14] proposed a different approach, using an expansion on the variable t and coefficients depending on $\mathbf{z}$, what is equivalent to consider $V = L^2(0, T)$, with the inner product $(\mathbf{u}, \mathbf{v}) = \int_0^T \mathbf{u}(t).\mathbf{v}(t)dt$. Then, $\Phi = \{\varphi_i(t)\}_{i \in \mathbb{N}^*}$ and

$$\mathbf{x}(t, \mathbf{z}) = \sum_{i \in \mathbb{N}^*} \mathbf{x}_i(\mathbf{z})\varphi_i(t). \tag{2.11}$$

In this case, we evaluate the mean as

$$E(\mathbf{x}(t, \mathbf{Z})) = \sum_{i \in \mathbb{N}^*} E(\mathbf{x}_i(\mathbf{Z}))\varphi_i(t) \approx \sum_{1 \leq i \leq n} E(\mathbf{x}_i(\mathbf{Z}))\varphi_i(t) = E(P_n\mathbf{x}(t, \mathbf{z})). \tag{2.12}$$

Here, confidence intervals may be determined for the coefficients $\mathbf{x}_i$, so that we may find envelopes by an analogous way:

$$\text{Find } \mathbf{y}^* = (\mathbf{y}_1^*, \dots, \mathbf{y}_n^*) = \arg\max\left\{ \left\| \sum_{1 \leq i \leq n} \mathbf{y}_i\varphi_i(t) \right\| : \mathbf{y}_i \in C_i \right\}. \tag{2.13}$$

Here, C_i is a confidence interval for $\mathbf{x}_i(\mathbf{Z})$. Analogously to the preceding one, this approach has a severe limitation when the dimensions increase, but remains useful to generate means and covariances.

2.2.3   Mixed Approach

In the mixed approach, we consider expansions on the couple $(t, \mathbf{z})$: in this case, $V = L^2(0, T) \times L^2(\Omega, \mu)$ and

$$\mathbf{x}(t, \mathbf{z}) = \sum_{i \in \mathbb{N}^*} \mathbf{x}_i\varphi_i(t, z). \tag{2.14}$$

The basis Φ may be generated by a tensor product between a basis of $L^2(0, T)$ and a basis of $L^2(\Omega, \mu)$.

2.2.4   Application to the First Dynamical System

Let us consider the simple system $\ddot{u} = -Z^2 u, u(0) = 1, \dot{u}(0) = 0$, where Z is random, uniformly distributed on $\mathcal{Z} = (-\pi, \pi)$: as previously observed, a variate of orbit $\mathbf{x} = (u, \dot{u})$ is a curve $\mathbf{x}(t, z) = (\cos(zt), -z\sin(zt))$ with period

$T(z) = 2\pi/z$. To apply these approaches, the solutions must be re-parameterized on a fixed interval $(0, \bar{T})$. For instance, choosing $\bar{T} = 1$ leads to $t = \bar{t}T(z)$, so that

$$x_1(\bar{t}, z) = \cos(zT(z)\bar{t}), \quad x_2(\bar{t}, z) = -z\sin(zT(z)\bar{t}).$$

i.e.,

$$x_1(\bar{t}, z) = \cos(2\pi\bar{t}), \quad x_2(\bar{t}, z) = -z\sin(2\pi\bar{t}).$$

Observe that

1. for the classical UQ approach, we may use a polynomial basis $\varphi_i(z) = z^{i-1}$, so that $\mathbf{x}_1 = (\cos(2\pi\bar{t}), 0)$, $\mathbf{x}_2 = (0, -\sin(2\pi\bar{t}))$, $\mathbf{x}_i = (0, 0)$, for $i > 2$. Since $E(\varphi_2(Z)) = 0$, we have $E(\mathbf{x}) = \mathbf{x}_1$;
2. for the inverted UQ approach, we may use a trigonometrical Fourier basis $\psi_1(\bar{t}) = 1$, $\psi_{2k}(\bar{t}) = \sin(2k\pi\bar{t})$, $\psi_{2k+1}(\bar{t}) = \cos(2k\pi\bar{t})$, so that $\mathbf{x}_2 = (0, -z)$, $\mathbf{x}_3 = (1, 0)$, $\mathbf{x}_i = (0, 0)$, for $i \notin 2, 3$. Since $E(\mathbf{x}_2) = (0, 0)$, we have $E(\mathbf{x}) = \mathbf{x}_3\psi_3(\bar{t})$;
3. for the mixed approach, we may use the tensor product of the preceding basis, so that the only non-null coefficients are $(1, 0)$ associated with $\varphi_1(z)\psi_3(\bar{t})$ and $(0, 1)$ associated with $\varphi_2(z)\psi_2(\bar{t})$. Since $E(\varphi_1(Z)) = 1$ and $E(\varphi_2(Z)) = 0$, we have $E(\mathbf{x}) = (\psi_3(\bar{t}), 0)$.

Thus, any of the approaches presented leads to

$$E(u(\bar{t}, Z)) = \cos(2\pi\bar{t}), \quad E(\dot{u}(\bar{t}, Z)) = 0,$$

so that the mean is a horizontal segment connecting $(-1, 0)$ to $(1, 0)$. In this case, the mean *is not* a member of the family $\mathcal{F} = \{\mathbf{x}_z : z \in (-\pi, \pi)\}$—in the same way that the mean of one roll of a six faced die is $7/2$, what *is not* one of its faces. In the next section, we introduce an approach allowing the determination of a member of $\mathcal{F}$ which may be considered as the most representative member of the family. We have also

$$E(\mathbf{xx}^t) = \begin{pmatrix} \cos^2(2\pi\bar{t}) & 0 \\ 0 & \frac{1}{3}\pi^2\sin^2(2\pi\bar{t}) \end{pmatrix}, \quad \text{cov}(\mathbf{x}, \mathbf{x}) = \begin{pmatrix} 0 & 0 \\ 0 & \frac{1}{3}\pi^2\sin^2(2\pi\bar{t}) \end{pmatrix}.$$

Envelopes may be generated by considering a confidence interval for $\varphi_2(Z)$. Since $E(\varphi_2(Z)) = 0$, a confidence interval of level $1 - \alpha$ corresponds to a quantile $P(|Z| \le z_\alpha) = 1 - \alpha$. Taking into account the distribution of Z, we have $z_\alpha = (1 - \alpha/2)\pi$ and the extreme points $(u_e, \dot{u}_e)$ verify

$$u_e = \cos(2\pi\bar{t}), \quad |\dot{u}_e| = z_\alpha |\sin(2\pi\bar{t})|,$$

what corresponds to the ellipse generated by z_α. The results are exhibited in Fig. 2.2.

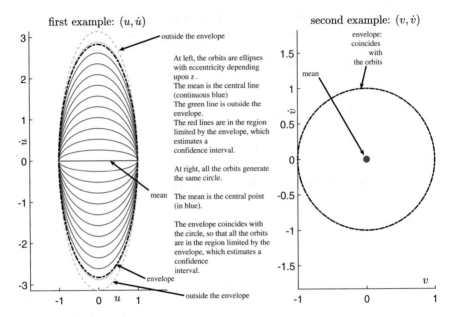

Fig. 2.2 Two simple dynamical systems illustrating the UQ approach to the analysis of infinite dimensional objects. On the left, the orbits are ellipses (in red continuous line) for $z \neq 0$ and reduce to the point $(1, 0)$ for $z = 0$ and mean evaluated by the UQ expansions furnishes a segment (in blue dotted line). On the right, all the orbits coincide in a single circle (in red continuous line) and the mean evaluated by the UQ expansions furnishes a point (in blue). Notice that the envelopes (black line) furnished by the expansions give correct estimates for the confidence intervals

2.2.5 Application to the Second Dynamical System

Let us consider the simple system $\ddot{u} = -u$, $u(0) = \cos(z)$, $\dot{u}(0) = -\sin(z)$, where Z is random, uniformly distributed on $\mathcal{Z} = (-\pi, \pi)$: as previously observed, a variate of orbit $\mathbf{x} = (u, \dot{u})$ is a curve $\mathbf{x}(t, z) = (\cos(z + t), -\sin(z + t))$ with period $T(z) = 2\pi$. Let us choose $\bar{T} = 1$, what corresponds to $t = 2\pi\bar{t}$. Then

$$x_1(\bar{t}, z) = \cos(z + 2\pi\bar{t}), \quad x_2(\bar{t}, z) = -\sin(z + 2\pi\bar{t}),$$

i.e.,

$$x_1(\bar{t}, z) = \cos(z)\cos(2\pi\bar{t}) - \sin(z)\sin(2\pi\bar{t}),$$

$$x_2(\bar{t}, z) = -\sin(z)\cos(2\pi\bar{t}) - \cos(z)\sin(2\pi\bar{t}).$$

This example reveals a difficulty, connected to the phase z: all the orbits correspond to the same circle $\|\mathbf{x}\| = 1$, but they present a phase shift, so that all the preceding approaches lead to

$$E\left(x_1(\bar{t}, Z)\right) = 0, \quad E\left(x_2(\bar{t}, Z)\right) = 0.$$

Analogously to the preceding example, the mean furnished by this approach *is not* a member of the family $\mathcal{F} = \{\mathbf{x}_z : z \in (-\pi, \pi)\}$: the next section will present a method to determine a member of $\mathcal{F}$ which may be considered as the most representative member of the family. We have also

$$E\left(\mathbf{x}\mathbf{x}^t\right) = \begin{pmatrix} \frac{1}{2} & 0 \\ 0 & \frac{1}{2} \end{pmatrix}, \; \mathrm{cov}(\mathbf{x}, \mathbf{x}) = \begin{pmatrix} \frac{1}{2} & 0 \\ 0 & \frac{1}{2} \end{pmatrix}.$$

To generate an envelope, we may consider an interval for the values of Z: $\|Z\| \leq z_\alpha$: all the values of z in this interval generate the orbit, so that the envelope consists in the single circle $\|\mathbf{x}\| = 1$—the real orbits remain in the envelope. The results for $\alpha = 0.1$ are shown in Fig. 2.2.

This example evidentiates the dependence on the parameterization: indeed, we may describe all the orbits by

$$\mathbf{x}(\bar{t}, z) = \left(\cos\left(2\pi\bar{t}\right), -\sin\left(2\pi\bar{t}\right)\right).$$

When using this parameterization, we get

$$E\left(\mathbf{x}(\bar{t}, Z)\right) = \left(\cos\left(2\pi\bar{t}\right), -\sin\left(2\pi\bar{t}\right)\right).$$

Notice that $E\left(\mathbf{x}\mathbf{x}^t\right)$ and $\mathrm{cov}(\mathbf{x}, \mathbf{x})$ coincide to those furnished by the preceding parameterization, analogously to the envelope—all these quantities are the same for both the parameterizations.

2.2.6 Generation of Random Curves and Applications to Infinite Dimensional Optimization

The Hilbertian may be interesting in optimization. For instance, the inverted UQ approach identifies $\mathbf{x}$ to the sequence $\mathbf{X}(\mathbf{z}) = (\mathbf{x}_1(\mathbf{z}), \mathbf{x}_2(\mathbf{z}), \ldots, \mathbf{x}_n(\mathbf{z}), \ldots)$, so that we may define probabilities on V by means of probabilities on the space of sequences $\mathbf{X}$, i.e., the probabilities on V may be defined on the coefficients of the expansion (2.10). Since $V = L^2(0, T)$, the natural space for the sequences $\mathbf{X}$ is ℓ^2—the space of the square summable sequences. If the Hilbert basis is replaced by a total family, the natural space is the set ℓ_0 of the sequences having a finite number on non-null elements. Such an identification furnishes a method for the generation of random functions by generating random sequences from ℓ^2 or ℓ_0 (Cf. [7, 8]), what furnishes a method of optimization in infinite dimensional spaces which does not involve finite-dimensional approximations [10]. Let us illustrate the procedure by a simple example: the brachistochrone, i.e., the problem of finding

$$u = \arg\min\left\{J(y) : y(0) = 0, y(3/2) = -1\right\}, \; J(y) = \int_0^1 \frac{\sqrt{1 + (y')^2}}{y} dt.$$

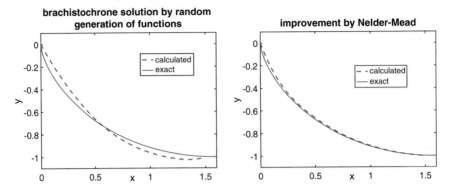

Fig. 2.3 An example of the use of random functions in optimization: the brachistochrone problem is solved by using a representation of the solution as the mean of a random variable. Random polynomials are generated and furnish the estimation shown on the left. On the right, the results improved by Nelder–Mead search using the preceding estimation as starting point

We generate random polynomials p by considering random coefficients $\mathbf{X} \in \ell_0$: the number of coefficients n_c is generated using a Poisson's law having as parameter $\lambda = 10$, the indexes corresponding to these coefficients are generated from a discrete uniform law on $\{1, \ldots, 3n_c\}$ and the values of the coefficients are generated from the normal distribution $N(0, \sigma)$, $\sigma = 4$. Then, we evaluate the estimation of the solution given by: $\mathbf{U} = E\left(\mathbf{X}g(\alpha, J(p))\right)$, $g(\alpha, s) = \exp(-\alpha s)$. We use $\alpha = 10$ and we estimate the mean by using the empirical mean on a random sample of $nr = 10{,}000$ variates. The results for $\alpha = 0.1$ are shown in the left side of Fig. 2.3. This result can be considered as a good starting point for a local optimization algorithm to find an improved solution. Thus, hybridizing this approach with a local optimization algorithm leads to a global optimization method, able to find global solutions with improved accuracy. The right side of Fig. 2.3 displays the result obtained after the execution of the Nelder–Mead local optimization procedure from the result proposed above as a starting point. As expected, it is close to the exact solution with better accuracy.

2.3 The Variational Approach

The difficulties found in Sect. 2.2 encourage to look for alternative solutions, namely solutions which are independent of the parameterization. In [11], a variational approach is proposed: define the mean as the curve that minimizes the distance to the whole family. Indeed, we may observe that, for a random variable A defined on a discrete universe $\Omega = \{\omega_1, \ldots, \omega_k\}$, the mean verifies

$$E(A) = \arg\min \left\{ \frac{1}{k} \sum_{i=1}^{k} (A(\omega_i) - y)^2 : y \in \mathbb{R} \right\}. \tag{2.15}$$

This equality suggests that we may determine the mean by finding the solution of an optimization problem: the mean of A minimizes the distance to all the possible values of A. In terms of curves, we may look for a curve that minimizes the distance to the whole family of curves:

$$E\,(\mathbf{x}) = \arg\min \left\{ \mathrm{dist}\,(\mathbf{x}, \mathbf{y}) : \mathbf{y} : I \to \mathbb{R}^2 \right\}. \tag{2.16}$$

It often occurs that the mean *is not* an element of the family $\mathcal{F} = \{\mathbf{x}(t, \mathbf{z}) : \mathbf{z} \in \mathcal{Z}\}$. If we are interested in determining a member of the family, we may consider

$$\mathrm{med}\,(\mathbf{x}) = \arg\min \{\mathrm{dist}\,(\mathbf{x}, \mathbf{y}) : \mathbf{y} \in \mathcal{F}\}. \tag{2.17}$$

In this case, the membership to $\mathcal{F}$ is a constraint in the optimization procedure: we will determine an element of $\mathcal{F}$ occupying a central position—analogously to a *median*. Even if, in general, this problem may do no admit any solution, approximate solutions may be determined by using samples from $\mathbf{Z}$ (see examples below).

The distance is, in principle, arbitrary. For instance, we may use the distance dn generated by the inner product defined on V:

$$\mathrm{dn}\,(\mathbf{x}, \mathbf{y}) = E\,(\|\mathbf{x_z} - \mathbf{y}\|)\,. \tag{2.18}$$

In this case, the difficulties previously state arise, such as the dependence on the parameterization. In order to get independence of the parameterization, we must consider set distances, such as the Hausdorff distance dh, which is defined as

$$\mathrm{dh}\,(\mathbf{x}, \mathbf{y}) = E\,(\mathrm{d}\,(\mathbf{x_z}, \mathbf{y}))\,. \tag{2.19}$$

Here,

$$\mathrm{d}\,(\mathbf{x_z}, \mathbf{y}) = \max\,\{\sup\,\{\delta\,(\mathbf{x_z}, \mathbf{y}(s)) : s \in I\}, \sup\,\{\delta\,(\mathbf{x_z}(t), \mathbf{y}) : t \in I\}\}, \tag{2.20}$$

where

$$\delta\,(\mathbf{x_z}, \mathbf{y}(s)) = \inf\,\{\|\mathbf{x}(t, \mathbf{z}) - \mathbf{y}(s)\| : t \in I\} \tag{2.21}$$

and

$$\delta\,(\mathbf{x_z}(t), \mathbf{y}) = \inf\,\{\|\mathbf{x}(t, \mathbf{z}) - \mathbf{y}(s)\| : s \in I\}. \tag{2.22}$$

In this approach, we may generate confidence intervals: once $E\,(\mathbf{x})$ or med $(\mathbf{x})$ is determined, we may look for a region including this object and containing a given proportion $1 - \alpha$ of the family: the procedure is illustrated in the examples below.

2.3.1 Application to the First Dynamical System

Let us consider again the system $\ddot{u} = -Z^2 u$, $u(0) = 1$, $\dot{u}(0) = 0$, with Z uniformly distributed on $\mathcal{Z} = (-\pi, \pi)$: we have $\mathbf{x} = (\cos(zt), -z\sin(zt))$, $T(z) = 2\pi/z$, $t = \bar{t}T(z)$,

$$\mathbf{x}(\bar{t}, z) = \left(\cos(2\pi\bar{t}), -z\sin(2\pi\bar{t})\right).$$

Let us consider the distance dn:

$$\|\mathbf{x_z} - \mathbf{y}\|^2 = \int_0^1 \left\|\mathbf{x_z}(\bar{t}) - \mathbf{y}(\bar{t})\right\|^2 d\bar{t}.$$

Thus,

$$\mathrm{dn}^2(\mathbf{x}, \mathbf{y}) = \int_0^1 \left(y_1(\bar{t}) - \cos(2\pi\bar{t})\right)^2 d\bar{t} + \frac{\pi^2}{6} + \int_0^1 y_2^2(\bar{t})d\bar{t}.$$

The minimum of the distance dn is equal to 0 and it is attained at

$$y_1(\bar{t}) = \cos(2\pi\bar{t}), \quad y_2(\bar{t}) = 0.$$

Thus the mean furnished by dn is the segment connecting $(-1, 0)$ to $(1, 0)$, as in Sect. 2.2.4. The Hausdorff distance between this element and the family $\mathcal{F} = \{\mathbf{x_z} : \mathbf{z} \in \mathcal{Z}\}$ may be easily determined by geometrical considerations: for each $\mathbf{x_z}(t)$, the nearest element in this segment is the point $\mathbf{y}(t)$, and the farthest points from $\mathbf{x_z}$ are the upper and lower extremities of the ellipses, so that the distance between $\mathbf{x_z}(t)$ and $\mathbf{y}$ is equal to the distance from the origin $(0, 0)$ to $(0, \pm z)$, i.e., $|z|$ and dh$(\mathbf{x}, \mathbf{y}) = \pi/2$ (See Fig. 2.4).

Let us consider the orbit $\mathbf{x_{z_0}}$, $\mathbf{z_0} \in \mathcal{Z}$. In an analogous way, the distance between $\mathbf{x_z}$ and $\mathbf{x_{z_0}}$ may be evaluated by a simple geometrical argument: the farthest points are the extremities $(0, \pm z)$ and $(0, \pm z_0)$, so that dist$\left(\mathbf{x_z}, \mathbf{x_{z_0}}\right) = \||z| - |z_0|\|$ and dh$\left(\mathbf{x}, \mathbf{x_{z_0}}\right) = \pi/2 - |z_0| + |z_0|^2/\pi$ (See Fig. 2.4). The minimal distance occurs when $|z_0| = \pi/2$, so that, on the one hand, med$(\mathbf{x}) = \mathbf{x_{\pm\pi/2}}$ and, on the other hand, dh$\left(\mathbf{x}, \mathbf{x_{\pm\pi/2}}\right) < \pi/2$: it results that the segment furnished by the Hilbertian approach is not the mean for the variational approach. The median $\mathbf{x_{\pm\pi/2}}$ may be used to determine a confidence interval:

$$P(\mathrm{dh}(\mathbf{x}, \mathrm{med}(\mathbf{x}))) = 1 - \alpha \iff P\left(\left||z| - \frac{\pi}{2}\right| \le d_\alpha\right) = 1 - \alpha.$$

For a bilateral confidence interval, the solution is $z \in (-\pi, -(1-\alpha)\pi) \cup ((1-\alpha)\pi, \pi)$. The results are exhibited in Fig. 2.5.

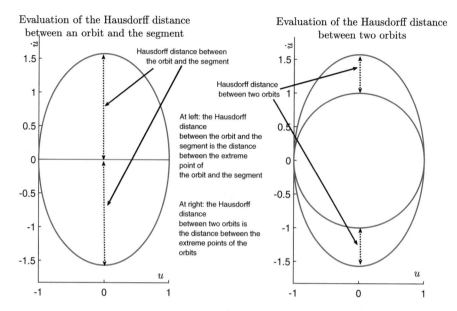

Fig. 2.4 Evaluation of the Hausdorff distance in the example of Sect. 2.3.1

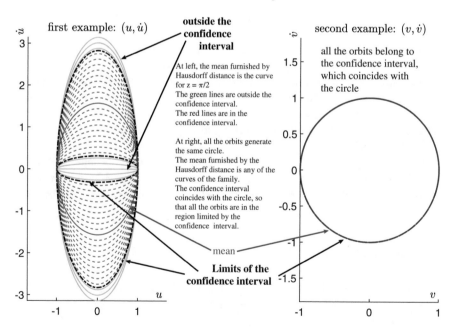

Fig. 2.5 Medians and confidence intervals at the level 90% generated by the Hausdorff distance. On the right, the example of Sect. 2.3.1: the median is a central orbit corresponding to $z = \pi/2$; the orbits in red belong to the confidence interval; the orbits in green are outside the confidence interval; the limits of the confidence interval appear in black. On the left, the example of Sect. 2.3.2: all the orbits generate the same circle, so that the median and the confidence interval are the circle; all the orbits are in the confidence interval

2.3.2 Application to the Second Dynamical System

Let us consider again the system $\ddot{u} = -u$, $u(0, z) = \cos(z)$, $\dot{u}(0, z) = -\sin(z)$, with Z uniformly distributed on $\mathcal{Z} = (-\pi, \pi)$: we have $\mathbf{x} = (\cos(t + z), -\sin(t + z))$, $T(z) = 2\pi$, $t = \bar{t}T(z)$,

$$\mathbf{x}(\bar{t}, z) = \left(\cos(z + 2\pi\bar{t}), -\sin(z + 2\pi\bar{t}) \right).$$

Let us consider the distance dn:

$$\|\mathbf{x_z} - \mathbf{y}\|^2 = \int_0^1 \left\| \mathbf{x_z}(\bar{t}) - \mathbf{y}(\bar{t}) \right\|^2 d\bar{t}.$$

Thus,

$$\mathrm{dn}^2(\mathbf{x}, \mathbf{y}) = \int_0^1 y_1^2(\bar{t}) d\bar{t} + 1 + \int_0^1 y_2^2(\bar{t}) d\bar{t}.$$

The minimum of the distance dn is equal to 0 and it is attained at

$$y_1(\bar{t}) = 0, \quad y_2(\bar{t}) = 0.$$

Thus, the mean furnished by dn is the origin $(0, 0)$, as in Sect. 2.2.5.

Let us consider the orbit $\mathbf{x_{z_0}}$, $\mathbf{z_0} \in \mathcal{Z}$. In an analogous way, the distance between $\mathbf{x_z}$ and $\mathbf{x_{z_0}}$ is equal to zero, since both the curves correspond to the circle of radius 1 and centred at the origin (Fig. 2.6). Thus, the median is the circle, i.e., any orbit) and may be used to generate a confidence interval: since the distance between any arbitrary orbits is zero, all the orbits belong to the confidence interval. The results are shown in Fig. 2.5.

A simple way to improve the results consists in the use of a sample of equally spaced values of Z: $z_i = \pi + 2(i - 1)\pi/n_s$: see results in Fig. 2.7—however, this method cannot be used in general situations where the bounds of Z are unknown.

2.4 Using Samples

In general, only samples from the objects are available in practice. In this case, we must evaluate the means, median envelopes, and confidence intervals by using a finite number of objects: assuming that a sample $S = \{ \mathbf{x_{z_i}} : 1 \leq i \leq n_s \}$ of n_s curves is available, we may estimate the punctual mean as

$$E(\mathbf{x}) \approx \bar{\mathbf{x}}_p(t) = \frac{1}{n_s} \sum_{i=1}^{n_s} \mathbf{x_{z_i}}(t).$$

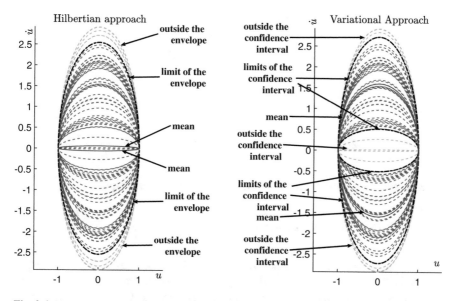

Fig. 2.6 Empirical results for mean and envelope (Hilbertian approach) and for median and confidence interval (variational approach) estimation for a sample of $n_s = 50$ variates from Z

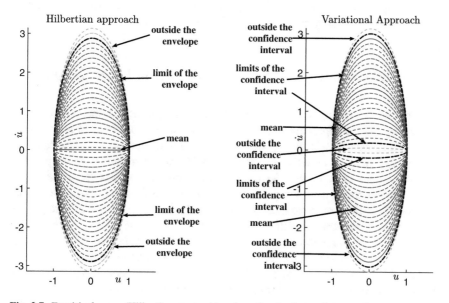

Fig. 2.7 Empirical mean (Hilbertian approach) and median (variational approach) estimation for a sample of $n_s = 50$ *equally spaced* variates from Z

To estimate the Hilbertian mean, we may consider the inverted UQ expansions of each element of the sample and take the means of the coefficients:

$$\mathbf{x}_{\mathbf{z}_\mathbf{i}} = \sum_{k\in\mathbb{N}^*} \mathbf{x}_{i,k}\varphi_k(t) \; ; \; P_n\mathbf{x}_{\mathbf{z}_\mathbf{i}} = \sum_{1\le k \le n} \mathbf{x}_{i,k}\varphi_k(t).$$

Then

$$E(\mathbf{x}) \approx \bar{\mathbf{x}}_h(t) = \sum_{k\in\mathbb{N}^*} \bar{\mathbf{x}}_k\varphi_k(t) \approx \sum_{1\le k \le n} \bar{\mathbf{x}}_k\varphi_k(t) \, , \; \bar{\mathbf{x}}_k = \frac{1}{n_s}\sum_{i=1}^{n_s}\mathbf{x}_{i,k}.$$

To estimate the variational mean, we use the median of the sample

$$\text{med}(\mathbf{x}) \approx \text{med}_\mathcal{S}(\mathbf{x}) = \arg\min\left\{\frac{1}{n_s}\sum_{i=1}^{n_s}d\left(\mathbf{x}_{\mathbf{z}_\mathbf{i}}, \mathbf{y}\right) : \mathbf{y}\in\mathcal{S}\right\}$$

The envelope and the confidence interval are determined by using the sample: we evaluate the distances $dn\left(E(\mathbf{x}), \mathbf{x}_{\mathbf{z}_\mathbf{i}}\right)$ or $dh\left(\text{med}(\mathbf{x}), \mathbf{x}_{\mathbf{z}_\mathbf{i}}\right)$ for $1 \le i \le n_s$. The confidence interval of level $1 - \alpha$ is defined by the $(1 - \alpha)n_s$ curves of $\mathcal{S}$ corresponding to the smaller $(1 - \alpha)n_s$ distances. The envelope is analogously determined: we exclude the largest distances to keep the smaller ones.

For instance, let us consider the dynamic system used as example in Sect. 2.3.1 and a sample from $\mathbf{x}$, corresponding to $n_s = 50$ variates from Z. The results for the Hilbertian and the variational means are shown in Fig. 2.6.

2.4.1 Transformation of Small Sets of Data in Large Samples

One of the interesting features of the Hilbertian approach is the possibility of generating large samples from a few data. When a small set of data is used (i.e., n_s is small), we may use the classical UQ approach to generate a representation of the family $\mathbf{z}$ and, then, use the representation to generate new curves, corresponding to other values of $\mathbf{z}$. In the UQ framework, a procedure exists for the situation where the bounds of $\mathbf{z}$ are unknown or, eventually, $\mathbf{z}$ itself is unknown—the reader may find it in [8]. Large samples may be useful when we look for the probability distribution of random variables connected to the family of curves. They are also useful to estimate probabilities of events connected to $\mathbf{z}$. They may also be used to improve the evaluations of the means, medians and confidence intervals.

Let us illustrate the procedure: assume that we have a sample of $n_s = 10$ orbits from the dynamical systems previously considered. We may apply the classical UQ approach with a polynomial basis $\varphi_i(z) = (z/\pi)^{i-1}$. For each $t_i = (i - 1)\Delta t$, $1 \le i \le 501$, $\Delta t = 0.02$, we determine the coefficients of a polynomial of degree 5 in the variable z by collocation. Then the polynomial is used to generate 101 new curves corresponding to $z_i = (i - 1)\Delta z$, $1 \le i \le 101$, $\Delta z = \pi/100$. The

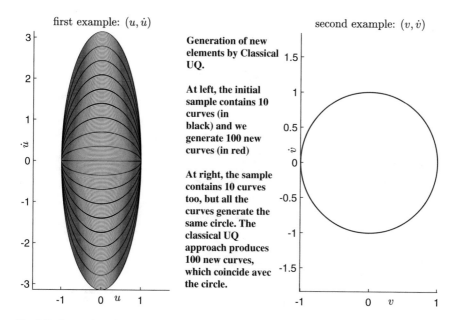

Fig. 2.8 Generation of a large sample from a small one by the Hilbertian approach

curves generated are exhibited in Fig. 2.8. The means furnished by the Hilbertian approach, for each sample are shown in Fig. 2.9: we observe an improvement for the first system—for the second one, the samples coincide into a single circle. We present 101 curves by graphical reasons, but the generation of a larger number of curves consists in the evaluation of the polynomials determined, what involves a reasonable computational cost even for larger samples—we use samples going to 10^6 variates in the sequel.

Let us consider the temporal mean of $\dot{u}^2$, i.e., $Y = \frac{1}{T} \int_0^1 \dot{u}^2 dt$. Since $\dot{u}$ depends on Z, Y is a random variable. We are interested in the probability distribution of Y: the data furnished by the sample of $n_s = 10$ curves is insufficient to obtain the distribution, but we can use the representation determined to generate more data. For instance, we may generate 10,000 variates from Z and use the representation to transform them in a sample of Y, what produces 10,000 variates from Y and allows the determination of the probability distribution. For each z_k, an orbit is determined and generates a sequence of values $\dot{u}_k(t_i)$, $1 \leq i \leq 501$ (points defined above). Then, we obtain a variate Y_k by evaluating the mean value of $\dot{u}_k^2$, so that we generate a sample $\mathcal{Y} = \{Y_k : 1 \leq k \leq 10000\}$. The discrete cumulative function F is evaluated at 51 points y_j, $1 \leq j \leq 51$, equally spaced on the interval $(\min(\mathcal{Y}), \max(\mathcal{Y}))$: $F(y_j)$ obtained by counting the elements of $\mathcal{Y}$ inferior to y_j. The probability density f is determined by the numerical derivation of the discrete cumulative function F by a particle method (see [15]). In Fig. 2.10, we compare the empirical distribution furnished by this procedure and the exact distribution (notice that $Y = z^2/2$, so that the analytical expression of the exact values may be determined).

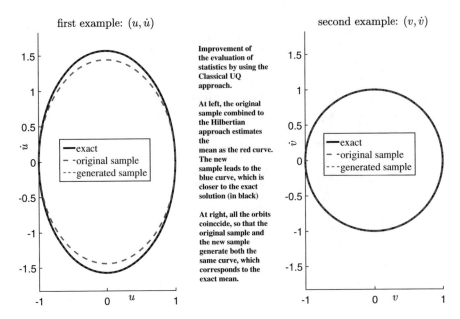

Fig. 2.9 Evaluation of the median using the data generated by the Hilbertian approach

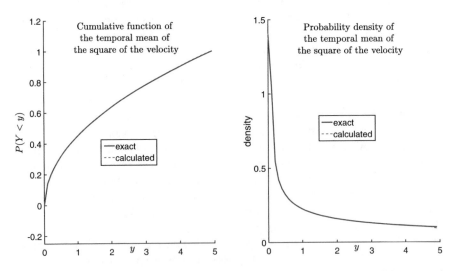

Fig. 2.10 Determination of the probability distribution (cumulative and density) of the temporal mean of the square of the velocity for the first dynamical system $(u, \dot{u})$

Analogously, assume that we are interested in the evaluation of the probability $P(Y > y_\alpha)$. We generate 10^5 variates from Y by an analogous way and we evaluate the empirical probability for $y_\alpha = 4$, what furnishes the value 0.0988—the exact value is 0.0997. For $y_\alpha = 4.9$, we obtain 0.003: the exact value is 0.004. For $y_\alpha = 4.92$, we obtain 0.0011: the exact value is 0.0015.

2.5   Applications to Multiobjective Optimization

As established in the preceding section, the variational approach is effective to furnish the median and, then, to generate a confidence interval of a family of curves, so that we may consider its application to the determination of the median and a confidence interval of Pareto fronts.

For instance, let us consider the Fonseca and Fleming problem [16] involving $\mathbf{Z} = (z_1, z_2, z_3)$, mutually independent and uniformly distributed on $(0, 0.1)$:

$$\underset{\mathbf{x} \in \mathbb{R}^3}{\text{Minimize}} \left\{ \begin{array}{l} f_1(\mathbf{x}) = 1 - \exp\left[-\sum_{i=1}^{3} \left(x_i - \frac{1}{\sqrt{3}} + z_i \right)^2 \right] \\[2mm] f_2(\mathbf{x}) = 1 - \exp\left[-\sum_{i=1}^{3} \left(x_i + \frac{1}{\sqrt{3}} - z_i \right)^2 \right] \end{array} \right.$$

$$\text{such that} : \; -4 \le x_i \le 4 \; ; \; i \in \{1, 2, 3\}$$

This problem may be solved by the variational approach presented in [9] (see also [15])—we use an expansion of x_1 and x_2 as polynomials of degree 6. For a sample of $n_s = 200$ variates from Z, we obtain the results shown in Fig. 2.11.

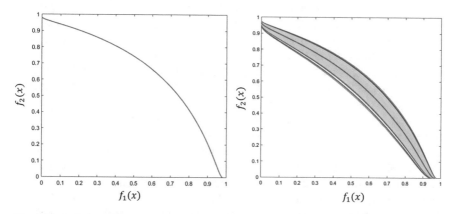

Fig. 2.11 Estimation of the median Pareto front in biobjective optimization (Fonseca–Fleming problem) with a sample of $n_s = 200$ variates from $\mathbf{Z}$. On the left, the Pareto front for the deterministic problem with $\mathbf{Z} = \mathbf{0}$. On the right, the median appears in red and deep blue curves lie outside the confidence interval of level $1 - \alpha$, $\alpha = 0.05$

Fig. 2.12 Generation of 10^5
Pareto fronts by the classical
UQ approach. The median
furnished by the variational
approach appears in white

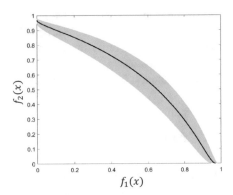

The preceding examples show that the variational approach leads to the construction of the median and of the confidence interval, but does not furnish a representation of the Pareto front. As previously remarked, we may need a larger sample—for instance, if we are interested in the evaluation of very small probabilities. Since each Pareto front results from an optimization procedure, this may lead to a very high computational cost. To reduce the calculations, it is possible to apply the method introduced in Sect. 2.4.1: given a relatively small sample of the Pareto fronts, we may apply the method presented in Sect. 2.4.1. For instance, in the Fonseca–Fleming problem, we generate $n_s = 10^5$ Pareto fronts by this way (See Fig. 2.12).

2.6 Concluding Remarks

Curves and surfaces are infinite dimensional objects, i.e., objects belonging to infinitely dimensional vector spaces. The analysis of the uncertainty and robustness of these objects involves probabilities in infinite dimensional spaces, what introduces operational and conceptual difficulties: for instance, if $\mathbf{x}$ is a family of curves depending upon a random vector $\mathbf{Z}$, giving a signification to $E(\mathbf{x})$ requests the definition of a probability on an infinite dimensional space. To be effective to calculate, two basic approaches may be considered: on the one hand, the use of Hilbert basis to reduce the problem to the analysis of probabilities on spaces formed of sequences of real numbers and, on the other hand, approaches based on the variational characterization of the mean. The first approach is effective to calculate and suggests also a method for infinite dimensional optimization. In addition, it furnishes a method for the generation of large samples: if $\mathbf{x}$ is represented by an expansion in a Hilbert basis, analogous to those considered in Sect. 2.2, we may generate a sample from the random variable $\mathbf{Z}$ and use the representation to transform it in a sample from $\mathbf{x}$. However, the first approach produces results that depend on the parametrization and hard fundamental questions concerning it are still with us, such as the independence from the basis and the best choice of the parameterizations. Nevertheless, this approach is useful to evaluate covariances and generates a method for optimization of functionals in infinitely dimensional vector spaces. The second approach is independent from the parametrization and is also effective to calculate. Among its advantages, it produces the median and a confidence interval—both these quantities may be difficult to obtain by the Hilbertian approach. It may be used to determine confidence intervals for Pareto fronts in multiobjective optimization or orbits of dynamical systems. Future developments will concern robust optimization and reliability.

References

1. Chui, Q.S.H.: J. Braz. Chem. Soc. **18**, 424 (2007)
2. Arunthong, T., Thuenhan, S., Wongsripan, P., Chomkokard, S., Wongkokua, W., Jinuntuya, N.: J. Phys. Conf. Ser. **1144**, 012153 (2018). https://doi.org/10.1088/1742-6596/1144/1/012153
3. Mukherjee, S., Reddy, M.P., Ganguli, R., Gopalakrishnan, S.: Int. J. Comput. Methods Eng. Sci. Mech. **19**(3), 156 (2018). https://doi.org/10.1080/15502287.2018.1431735
4. Králik, J.: Transactions of the VSB: Technical University of Ostrava. Civ. Eng. Ser. **17** (2017). https://doi.org/10.1515/tvsb-2017-0011
5. Phan, B., Salay, R., Czarnecki, K., Abdelzad, V., Denouden, T., Vernekar, S.: Calibrating Uncertainties in Object Localization Task (2018). https://arxiv.org/abs/1811.11210
6. Pfoser, D., Jensen, C.S.: Capturing the Uncertainty of Moving-Object Representations. In: Güting, R.H., Papadias, D., Lochovsky, F. (eds.) Advances in Spatial Databases. SSD 1999. Lecture Notes in Computer Science, vol. 1651. Springer, Berlin, Heidelberg (1999)
7. Souza de Cursi, E.: In: Tarocco, E., de Souza Neto, E.A., Novotny, A.A. (eds.) Variational Formulations in Mechanics: Theory and Applications, pp. 87–106. International Center for Numerical Methods in Engineering (CIMNE) (2007)
8. Souza de Cursi, E., Sampaio, R.: Uncertainty Quantification and Stochastic Modeling with Matlab, 1st edn. Elsevier, Amsterdam (2015)
9. Zidani, H.: Representation de Solutions en Optimisation Continue, Multiobjectif et Applications. Ph.D. Thesis. INSA, Rouen (2013)
10. Zidani, H., Souza de Cursi, E., Ellaia, R.: J. Glob. Optim. **65**(2), 261 (2016). https://doi.org/10.1007/s10898-015-0357-5
11. Bassi, M., Souza de Cursi, E., Pagnacco, E., Ellaia, R.: Lat. Am. J. Solids Struct. **15**(11), 1015 (2018). https://doi.org/10.1007/s10898-015-0357-5. https://www.lajss.org/index.php/LAJSS/article/view/5018
12. Bassi, M., Souza de Cursi, E., Ellaia, R.: In: 3rd International Symposium on Uncertainty Quantification and Stochastic Modeling (2016). http://www.swge.inf.br/PDF/USM-2016-0037_027656.PDF
13. Bassi, M.: Quantification d'incertitudes et Objets en Dimension Infinie. Ph.D. Thesis, INSA Rouen Normandie, Normandie University, Normandie (2019)
14. Croquet, R., Souza de Cursi, E.: In: Topping, B., Adam, J., Pallares, F., Bru, R., Romero, M. (eds.) Proceedings of the Tenth International Conference on Computational Structures Technology, pp. 541–561. Civil-Comp Press, Stirlingshire (2010)
15. Souza de Cursi, E.: Variational Methods for Engineers with Matlab, 1st edn. Wiley, London (2015). https://doi.org/10.1002/9781119230120. e-ISBN: 9781119230120
16. Fonseca, C.M., Fleming, P.J.: Evol. Comput. **3**(1), 1 (1995). https://doi.org/10.1162/evco.1995.3.1.1

Chapter 3
Meta-Heuristic Approaches for Automatic Roof Measurement in Solar Panels Installations

Luis Diago, Junichi Shinoda, and Ichiro Hagiwara

Abstract Optimizing the conversion of solar energy to electricity is central to the world's future energy economy. Solar panels installations require measuring the total area of the roof and any areas on the roof that may account for gaps due to vents, pipes, roof fixtures, or tree shading. There are several options for measuring roofs (e.g., climbing the roof, using drones from the floor, manually measuring with Google Maps). However, these options are usually very expensive, risky, and time-consuming. In this work, we developed a methodology that automatically creates a three-dimensional (3D) model of the house and measures its roof using pictures of its elevation views (north, south, east, and west) and satellite images from Google Maps. After the 3D model is created, a top view projection of the house's ceiling is used for polygonal shape matching with a previously known ceiling from Google images. The matching problem is multimodal, noisy, and highly dimensional since there are several roofs similar to the projection, they are covered by leaves and the search is done with high-dimensional images. In recent years, meta-heuristic algorithms (MHs) have arisen as effective algorithms to solve such multimodal optimization problems. Among the MHs, population based algorithms when applied with a niching strategy, called niching algorithms, have proven to be able to identify such optima. In our previous work, we have used GA to solve the above problem. In this chapter we use genetic algorithms (GA) with niching methods and design a new fitness function to better cope with the multimodality of the problem. After the best

L. Diago (✉)
Interlocus Inc, Yokohama, Japan

Meiji University, Nakano, Tokyo, Japan
e-mail: ldiago@i-locus.com; luis_diago@meiji.ac.jp

J. Shinoda
Interlocus Inc, Yokohama, Japan
e-mail: jshinoda@i-locus.com

I. Hagiwara
Meiji University, Nakano, Tokyo, Japan
e-mail: ihagi@meiji.ac.jp

© Springer Nature Switzerland AG 2020
O. Llanes Santiago et al. (eds.), *Computational Intelligence in Emerging Technologies for Engineering Applications*, Studies in Computational Intelligence 872, https://doi.org/10.1007/978-3-030-34409-2_3

matching polygon is chosen from the set of possible solutions, the scale given by
the satellite image is used to rectify the 3D house-model proportions and to estimate
the space available for a solar installation.

Keywords Solar panels installations · Roof measurement · Polygonal shape
matching · Multi-objective optimization · Meta-heuristic algorithms

3.1 Introduction

Solar energy is the most clean renewable energy source and has good prospects for
future sustainable development. Flat photovoltaic panels are commonly deployed
in residential and commercial rooftop installations without sun tracking systems
and using simple installation guidelines to optimize solar energy collection. So,
installation of solar photovoltaic (PV) systems on building rooftops has been the
most widely applied method for using solar energy resources. However, solar panels
installations require measuring the total area of the roof and any areas on the roof
that may account for gaps due to vents, pipes, roof fixtures, or tree shading. This
task is sometimes very complicated and time-consuming.

3.1.1 Rooftop Measurement

Several alternatives for measuring roofs are presented: climbing on top, using
drones, and manually approximating by having remote sensing images as reference.
Figure 3.1 shows two alternatives frequently used in practice. It is clear that climbing
the roof increases the cost of the procedure, not only because it increases the
level of risk, but also because it requires a lot of time. Nowadays cameras placed
in a high altitude bar (see, e.g., EXFR100KTSET in Fig. 3.1a) measure roofs of
houses that are very close to each other. However, sometimes it is difficult to
observe the details of complex ceilings because of not being able to locate the
camera in a place with an adequate field of vision (FoV). The recent trend is to
use drones or satellite images to make the calculations without even having to go
to the measurement site. From the measurements, the potential for generating PV
energy in large areas can be estimated. For example, Song et al. [1] developed
an approach to simulate the monthly and annual solar radiation on rooftops at an
hourly time step to estimate the solar PV potential, based on rooftop feature retrieval
from remote sensing images. The rooftop features included 2D rooftop outlines and
3D rooftop parameters retrieved from high-resolution remote sensing image data
(obtained from Google Maps) and digital surface model (DSM, generated from the
Pleiades satellite), respectively. The parameters of the PV modules derived from
the building features were then combined with solar radiation data to evaluate solar
photovoltaic potential. In this chapter, images taken in the field are also combined
with satellite images to create 3D building models for solar panels installations.

Fig. 3.1 Alternatives for measuring roofs: (**a**) climbing on top using a high altitude bar (e.g., EXFR100KTSET), (**b**) satellite image from Google Maps

3.1.2 Building Models

3D building modeling is an important approach to obtain the 3D structure information of the buildings and has been widely applied in the fields of telecommunication, urban planning, environmental simulation, cartography, tourism, and mobile navigation systems. It is also a very active research topic in photogrammetry, remote sensing, computer vision, pattern recognition, surveying, and mapping.

In this chapter, a new methodology to create three-dimensional (3D) house model using its elevation views (i.e., north, south, east, and west views) is proposed. Following the proposed methodology, the user is able to reconstruct the 3D shape of the external walls and roofs by clicking in only two of the elevation views. The points introduced by the user are used to generate a 3D polygon. The user can create any number of polygons until the whole structure is complete. The projection of the top view of the roof is a polygon in 3D that is obtained from the combination of the polygons created by the user. The methodology works with the original drawings of the houses or with photos taken from each view in the field. In cases where the original drawings do not exist and it is necessary to take photos, it is often difficult to obtain the photos of the four views, and as a result it is obtained a 3D polygon whose projection is incomplete. This projection is used as a template to look for a similar projection in a satellite image like the one in Fig. 3.1b. To carry out this search, a polygon matching technique is used in the proposed methodology. After the best match is found, the scale given by the satellite image is used to rectify the proportions of the 3D house model and to estimate the space available for a solar installation.

The rest of the chapter is organized as follows. Section 3.2 describes the proposed methodology for roof measurement in details. It includes the creation of rooftop models from elevation views and the description of the polygon matching (PM) problem. Meta-heuristic algorithms used to solve the PM problem are explained in Sect. 3.3.

3.2 Proposed Methodology for Roof Measurement

3.2.1 Related Works

For more than 20 years, rooftop has been measured by multiple aerial images [1–3], airborne laser scanning (ALS) data [4], and 2D ground plans. The literature is very broad and an exhaustive review of it goes beyond the scope of this work. Authors [1, 2, 4] recognized that the procedure for geometrical building modeling or city modeling encompasses three main steps, namely (1) recognition, (2) feature extraction, and (3) topology reconstruction with geometric modeling. Rather than automatic recognition, the most reliable and accurate results can normally be achieved by a building reconstruction system that integrates human-assisted visual interpretation capability. For a fully autonomous system, the integration of multiple data sources, such as multiple aerial images, airborne laser scanning (ALS) data, and 2D ground plans, might increase the reliability and degree of automation, but some constraints or limitations in certain aspects are unavoidable. The proposed methodology combines the creation of the roof model from the elevation views with a new framework for polygon matching based on geometric primitives extraction.

3.2.2 Creation of Rooftop Models from Elevation Views

The proposed methodology has been implemented in a system able to create 3D model using information from elevation views. Figure 3.2 shows an example of 2D drawings used for the creation of a rooftop model with the graphic user interface (GUI) of the developed system.

Four elevation views named North (N), South (S), East (E), and West (W) can be introduced independently or selected from a unique images by surround them inside a rectangular area. Two adjacent views (NE) are shown in Fig. 3.2a and a detail of the reconstructed rooftop is shown in Fig. 3.2b. The origin point (O) has to be introduced manually in at least two contiguous or adjacent views (e.g., NE, SE, NW, or SW). After the reference point in contiguous views has been set the user is able to create 3D polygons using two adjacent views. The user creates lines by left-clicking in any part of one of the views until a command is given to close the polygon (e.g., using right-click). The polygon will be closed between the last and

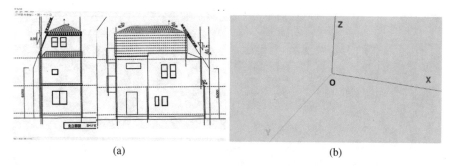

        (a)   (b)

Fig. 3.2 Creation of rooftop models from elevation views (**a**) North-East adjacent views (NE) provided by a user of the developed system and (**b**) 3D model of the rooftop

the first clicked points. The user should be careful about the input order given to the points in both polygons to allow correct correspondence. Given two images I_1 and I_2 corresponding to any of two adjacent views and two polygons represented by the set of points $P_1(u_1, v_1)$ and $P_2(u_2, v_2)$. The 3D coordinates of any polygon $P(x, y, z)$ will be represented as:

$$P(x, y, z) = (u_1, u_2, v_1) \tag{3.1}$$

where (u_n, v_n) are the polygonal point's coordinates in the image I_n. If the scale is available from the 2D drawings, the rooftop projection can be computed for the 3D model to compute its total area. However, there are cases where the scale is not available when taking pictures in the field. In these cases the scale must be obtained by matching the top view of the polygon with the ceilings in the satellite images.

3.2.3   Geometric Primitive Extraction for Polygon Matching

To solve above problem with the scale, we have posed the problem of polygon matching as a problem of geometric primitives extraction. An important problem in the field of model-based vision is the extraction of geometric primitives from predefined geometric data.

A geometric primitive is a curve or surface, which can be described by an equation with a number of free parameters. This equation has the implicit form $f(\mathbf{p}, \mathbf{a}) = 0$ where $\mathbf{p}$ is the datum point, and $\mathbf{a}$ defines the parameter vector for this particular primitive. For the polygon matching problem, the description of the polygon is a geometric primitive. Geometric data is an unordered list of points in two- or three-dimensional Cartesian space. In this work, such data are obtained by feature extraction operators. Figure 3.3a shows the original satellite image and the top view projection of the roof model over the results of the edge extraction with canny operator (see Fig. 3.3b).

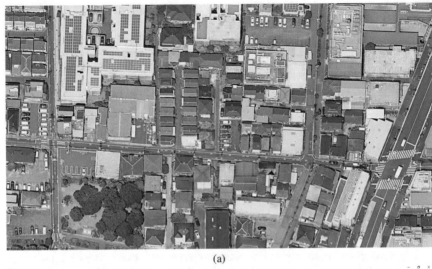

(a)

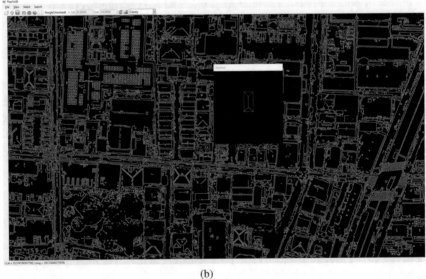

(b)

Fig. 3.3 Polygon matching as a geometric primitive extraction problem. (**a**) Original satellite image from Google maps. (**b**) Edge extraction and top view projection of the roof model

The inputs to the proposed extraction process are n geometric data points labeled $p_1, \ldots, p_n$, along with the definition of a geometric primitive in implicit form. The output consists of the parameter vector **a** of the best primitive, along with the subset of the geometric data that belongs to this primitive, which are called the inliers. The remaining points do not belong to the primitive and are called the outliers (or noise). The first step in proposed extraction algorithm is the computation of the closest distance of each point to the curves or surface. This distance is called the residual, and for point p_i of the geometric data it is labeled r_i. Ideally this residual should equal the closest distance of the point to the geometric primitive. For simple primitives this distance can be found in closed form. For more complex geometric primitives defined by an implicit equation $f(\mathbf{p}, \mathbf{a}) = 0$, the first-order approximation of the closest distance of a point $\mathbf{p_0}$ to the curve or surface is the absolute value of the function over the magnitude of the gradient vector ($\| f(\mathbf{p_0}, \mathbf{a})\|/\nabla f(\mathbf{p_0}, \mathbf{a})$). Given that residual $r_1, \ldots, r_n$ have been calculated, then extracting a single instance of a curve or surface is equivalent to finding the parameter vector $\mathbf{a}$ which minimizes/maximizes the value of a cost function $h(r_1, \ldots, r_n)$. The presence of multiple types of primitives (e.g., lines, circles, and polygons) makes the problem more complex. Since each primitives could be considered a local optima, the goal of proposed algorithm is to find the global optimum among several local optima.

Extracting geometric primitive is a prerequisite to solving other problems in model-based vision. It has been shown that extracting the best geometric primitive from a given set of geometric data is equivalent to finding the optimum value of a cost function [5]. Once it is understood that primitive extraction is such an optimization problem, the use of any technique for tackling optimization problem suggests itself. The key requirement of any global optimization method is that it must be able to avoid entrapment in local minima and continues the search to give a near-optimal final solution whatever the initial conditions. It is clear that this optimization model can describe various fitting and extraction algorithms by simply using different cost functions.

3.3   Meta-heuristic Algorithms

3.3.1   Genetic Algorithms

A major problem faced in geometric primitive extraction deals with the characteristics of the search space: multimodal, nonlinear, noisy, and highly dimensional; and with the possible variations of shapes of primitives. Genetic algorithms (GAs) [6] offer a domain-independent approach suitable for working with such complex problems. A genetic algorithm (GA) is an optimization approach based on evolutionary metaphor. GA starts from a set of N geometric primitives called population and combines them stochastically using selection, crossover, and mutation operators. Each geometric primitive is an individual in the population and compete with other individuals during the evolution of the algorithm. It consistently outperforms both gradient methods and random search in solving hard optimization problems [6, 7]. An optimization problem is hard if the cost function is noisy, multi-dimensional, and has many local optima. In our experiments, the images were passed through an edge detection process to extract the silhouette of the objects.

3.3.1.1 Minimal Subsets

All the points that belongs to the contour of the image were stored in an array as an ordered pair (x_i, y_i) $(i = 1 \ldots n)$ and GA codifies a primitive using the index of the points in the array. The parameters for each primitives are encoded by minimal subsets [7], i.e., the smallest number of points necessary to define a unique instance of a primitive. For example, for a line only two points are required to determine its equation and two integer numbers (n_1, n_2) are encoded. By randomly selecting two points from the geometric data $\{(x_i, y_i)\ (i = n_1, n_2)\}$, the parameter vector $\mathbf{a} = (a_2, a_1, a_0)$ is determined with the following equations:

$$
a_2 = \begin{bmatrix} y_1 & 1 \\ y_2 & 1 \end{bmatrix} \quad a_1 = \begin{bmatrix} x_1 & 1 \\ x_2 & 1 \end{bmatrix} \quad a_0 = \begin{bmatrix} x_1 & y_1 \\ x_2 & y_2 \end{bmatrix}
$$

Note that in a more general case, if roofs are represented by circles or ellipses the parameter vector $\mathbf{a} = (a_5, a_4, a_3, a_2, a_1, a_0)$ may include other types of conics which are removed from the population during the execution of GA by penalizing the primitives in the evaluation of the objective function.

3.3.1.2 Objective Function

The cost function returns a scalar which measures the "goodness" of the primitive with the given parameter vectors in terms of how well it matches the geometric data. The function proposed in this work counts the number of points from the geometric data that fall within a fixed-distance (d_{th}) of the geometric primitive:

$$
f_1(p_1, \ldots, p_n) = \sum_{i=1}^{n} s(r_{p_i}) \tag{3.2}
$$

where

$$
s(r_{p_i}) = \begin{cases} 1 \ if \ \ r_{p_i} \leq d_{th} \\ 0 \quad otherwise \end{cases}
$$

and r_{p_i} is the smallest distance from the point data p_i to the parameter vector of the randomly selected primitive. The larger the value of this cost function, the more significant the primitive. The advantage of evaluating the cost function only in the vector of parameters defined by minimal subsets is that the number of minimal subsets is limited and evaluating the cost function only in the minimal subsets guarantees to find the global optimum.

3.3.2   GA-Based Polygon Matching

Algorithm 1 describes the pseudocode of the proposed GA-based polygon matching. Given the polygons of the template **T**, the list of points **p**, the population size
(N), the crossover probability (p_c), and the mutation probability (p_m) as **Input**,
the **Output** is the best solution $S_{best} = (s, \theta, T_x, T_y)$. In step 1, as discussed in
Sect. 3.3.1.1, instead of the vector $\mathbf{a} = (s, \theta, T_x, T_y)$, two integer numbers (n_1, n_2)
are encoded in the individuals of the population. In the evaluation step 2, points
$A = \mathbf{p}_{n_1}(u_a, v_a)$ and $B = \mathbf{p}_{n_2}(u_b, v_b)$ are obtained from the data point **p**. Then s,
θ, T_x, and T_y are computed as $s = \|A - B\|$, $\theta = arctan\left(\frac{v_a - v_b}{u_a - u_b}\right)$, $T_x = u_a$, and
$T_y = v_a$. The rest of the points P in the template **T** are transformed accordingly and
the function f_1 in (3.2) is evaluated for all individuals in the population.

The improvements to the objective function are intended to reduce cost and
complexity in the calculation of the residuals. As rooftop polygons are closed,
only those points that are close to the bounding-box are considered. The bounding-
box is known once the projected polygon is transformed according to the vector
$\mathbf{a} = (s, \theta, T_x, T_y)$. To make a fairer distribution of the search space a normalization
was introduced in the objective function since the inclined lines, the circles, and
the ellipses of greater radius are favored by having a greater length and therefore
a greater number of points than others that could appear clearer within the image.
To solve the problem, a coefficient $C(\mathbf{a})$ is introduced into the objective function
to compute the number of points that the primitive described by the chromosome
$\mathbf{a} = (s, \theta, T_x, T_y)$ must have. This coefficient is introduced in f_2 as follows:

Algorithm 1: GA-based polygon matching

Input   : $\mathbf{T}, \mathbf{p}, N, p_c, p_m, d_{th}$
Output: S_{best}

1 *Population* $\leftarrow$ InitializePopulation$(N, \mathbf{p})$;
2 EvaluatePopulation$(Population, \mathbf{p}, \mathbf{T}, d_{th})$;
3 $S_{best} \leftarrow$ GetBestSolution$(Population)$;
4 **while** *not* StopCondition() **do**
5   *Parents* $\leftarrow$ SelectParents$(Population, N)$;
6   *Children* $\leftarrow \emptyset$;
7   **for** *Parent* $_1$, *Parent* $_2 \in$ *Parents* **do**
8   *Chield* $_1$, *Chield* $_2 \leftarrow$ Crossover$(Parent_1, Parent_2, p_c)$;
9   *Children* $\leftarrow$ Mutate$(Chield_1, p_m)$;
10   *Children* $\leftarrow$ Mutate$(Chield_2, p_m)$;
11   **end**
12   EvaluatePopulation$(Children, \mathbf{p}, \mathbf{T}, d_{th})$;
13     $S_{best} \leftarrow$ GetBestSolution$(Children)$;
14   *Population* $\leftarrow$ Replace$(Population, Children)$;
15 **end**

$$f_2 (p_1, \ldots, p_n) = \frac{1}{C(\mathbf{a})} \sum_{i=1}^{n} s\left(r_{p_i}\right) \qquad (3.3)$$

To determine the number of points that the primitive described by the vector of the parameters **a** has, it is necessary to consider the resolution of the image since the larger this is, the greater the number of points necessary to define it clearly in the image. Computer graphics algorithms can be used to estimate this number of points [8]. In this case, the cost of the objective function increases considerably since the polygon must be painted to estimate the number of pixels it has. The best solution to the problem will be the one that finds an analytical description for the coefficient $C(\mathbf{a})$, which can be reached as an extension of the calculation of the graphic length of a real function to the discrete case. The coefficient $C(\mathbf{a})$ is not more than an estimate of the discrete graphic length of the geometric primitive that gives us an idea of the number of holes that the primitive present in the picture. With the introduction of this coefficient, the amount of these holes is also minimized. At the moment of selecting the solutions, a threshold value is necessary to discriminate when there is a considerable primitive. If the solutions are normalized in this way, the selection of this threshold is facilitated.

3.3.3 Niching Methods

To ensure the extraction of all primitives at the same time there are two original alternatives. The first is to use operations defined for R-functions [9] and the second one is to use niching methods [10]. Applying the R-function method a single implicit equation could be obtained from the combination of multiple implicit equations and the GA can be run to look for the parameters of only one implicit function. On the other hand, niching methods extend the application of GAs to locate and maintain multiple implicit functions. This work follows the second approach because of its simpler implementation.

There are two ways for locating the best solution (evaluated by (3.2)) with niching methods: using *sequential location (SL)* or *parallel location (PL)*. Previous researches [10] have demonstrated that PL is more efficient than SL when the number of possible solutions is high. Within PL approach, the most widely used methods are *sharing (SH)* and *crowding (CW)*.

3.3.3.1 Sequential Location of Niches

Instead of slowing GA's convergence changing the population size (N), the crossover probability (p_c) or the mutation probability (p_m), SL encourages the

GA to converge quickly and then start another one. Goldberg suggests restarting GAs that have substantially converged, by reinitializing the population using both randomly generated individuals and the best individuals from converged population [6]. Reinitialization is somewhat similar to using high rate of mutation (p_m). Controlling the parameters of convergence of the population the selected primitive is removed from the image and GA continues searching with the residual data while it is significant [7]. The main disadvantage of this approach is that GA must be applied repeatedly to the geometric data for each extracted primitive but it offers good results when the number of local optima is not high [11].

3.3.3.2 Sharing (SH)

The objective function of an arbitrary individual can be altered in proportion with the vicinity average with other individuals of the population. This strategy is known as sharing. It causes the dispersion of highly similar individual in the search space and opposes to the early convergence. The behavior of the function that allows to share the space is defined as follows [10]:

$$f'(i) = \frac{f(i)}{\sum\limits_{j=1}^{n} sh\{d(i, j)\}} \tag{3.4}$$

where $sh(d) = 1 - (d/\sigma)^\alpha$ if $d \leq \sigma$ and 0 otherwise. The constant α (typically set to 1) is used to regulate the shape of the sharing function. The computation of $d(i, j)$ is carried out based on the Euclidean distance of phenotypes of the primitives (s, θ, T_x, T_y).

3.3.3.3 Crowding (CW)

Crowding methods attempt to replace population members in a way that maintain diversity. They insert new elements into a population by replacing similar elements. Two individuals are similar if the distance (either Euclidean or Hamming) between them is equal or less than some value σ taken as threshold. The removing of similar individuals from the population is known as *crowding with elimination (CE)* [10]. Alternatively some authors favor the mating of distant individuals instead of removing similar ones. This is known as *crowding with mating restrictions (CMR)* [10]. A crossover strategy where new individuals are created only if they are better than their parents is used to favor the formation of niches. This is known as *deterministic crowding (DC)* [10].

3.4   Preliminary Results

Figure 3.4 shows preliminary results obtained from the application of the proposed methodology. The photos obtained in the field in the South and East views (Fig. 3.4a) are combined with the zoom of the satellite image in Fig. 3.4b to obtain the 3D model of the roof in Fig. 3.4c. In Fig. 3.4b it is shown that there is no space to obtain photos in the North and West views because the house is surrounded by

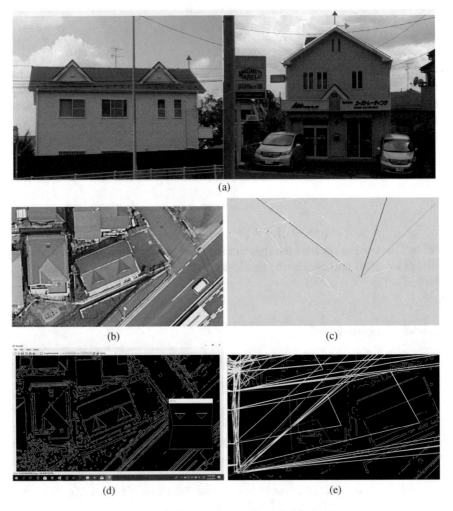

Fig. 3.4 Example of roof computed from photos taken in the field from (S)outh and (E)ast views. (**a**) Photos obtained in the field in the SE Contiguous Views with selected origin. (**b**) Zoom of Google map image, (**c**) 3D View of Reconstructed rooftop, (**d**) Edge extraction and top view projection, (**e**) GA Result using f_1

house areas that impede the passage to obtain good quality photos. Note also that when you get the photos from the ground, you cannot see the details of the roof and the peaks that protrude above the ceiling are only due to the perspective of the image. Figure 3.4d shows the result of the edge detection and the generated top view of the template used for matching with GA.

Table 3.1 shows a comparison of different niching methods defined by using function f_1 in Eq. (3.2), f_2 in Eq. (3.3), and f_{SL}, f_{SH}, and f_{CE} described in Sect. 3.3 as sequential location (SL), sharing (SH), and crowding (CE), respectively. The value $d_{th} = 1$ is used in all experiments for f_1 and f_2. The results are reported for the images in Figs. 3.3b and 3.4d. In Fig. 3.4d there is only one primitive and all methods can extract it except the function f_1 because it tries to favor the larger primitives and bad results are obtained (as reported in Fig. 3.4e). In Fig. 3.3d, there are eight primitives and Table 3.1 shows the average number of primitives found by the algorithm in five runs. For better use of the space we have shown this simple example that illustrates a clear superiority of CE over SH (see Table 3.1). CE found all primitives in ten generations(gen) with $pop = 100$, and SH found only 6.8 as average. The main goal of the experiments was to compare different niching algorithms under the same conditions (gen=10, pop=100, pc=0.9, pm=0.1, dth=1). Above conditions does not always guarantee the best solution (see Table 3.1 row 1, column 2). Increasing the number of generation the number of solution found increases (even using the classical GA with the function f_1 could find solution that failed in Table 3.1). However, increasing the number of generations to solve the problem in Fig. 3.3b also increases the computational time. In this way the number of solutions found increases but the results in the comparison of the algorithms are the same. The worst results are obtained with f_1, then f_2. The results using f_{SL} and f_{SH} were similar, with a little superiority using f_{SH}. The best results in this example were obtained with the function f_{CE}.

Table 3.1 Comparison of two objective functions and different niching methods to solve the polygon matching problem

Function	Fig. 3.4d	Fig. 3.3b
$f_1(gen = 10, pop = 100, p_c = 0.9, p_m = 0.1, d_{th} = 1)$	0	1.25
$f_2(gen = 10, pop = 100, p_c = 0.9, p_m = 0.1, d_{th} = 1)$	1	3.54
$f_{SL}(gen = 10, pop = 100, p_c = 0.9, p_m = 0.3, d_{th} = 1)$	1	6.2
$f_{SH}(gen = 10, pop = 100, p_c = 0.9, p_m = 0.1, \alpha = 1, \sigma = 2, d_{th} = 1)$	1	6.8
$f_{CE}(gen = 10, pop = 100, p_c = 0.9, p_m = 0.1, \sigma = 2, euclidean, d_{th} = 1)$	1	8

The parameters for each function are written in parenthesis

3.5   Conclusions and Future works

In this paper, a methodology able to create 3D models from 2D elevation views was proposed. In order to keep the real model proportions, a polygon matching algorithm was proposed. This polygon matching algorithm was based in image processing algorithms in combination with GA to search for the best matching. This GA uses the ceiling projection of the building as template to obtain the scale from a satellite image through the GA procedure. Later, this scale is used to rectify the 3D model dimensions and to estimate the space available for a solar installation.

More experiments should be done in order to obtain a deep understanding about the behavior of each objective function. However, the parameter of each function should be tuned in order to do a fair comparison in this case. The above tuning is currently under investigation.

References

1. Song, X., Huang, Y., Zhao, C., Liu, Y., Lu, Y., Chang, Y., Yang, J.: Energies **11**(11), 1 (2018). https://EconPapers.repec.org/RePEc:gam:jeners:v:11:y:2018:i:11:p:3172-:d:183168
2. Jaynes, C.O., Stolle, F., Collins, R.T.: Proceedings of 1994 IEEE Workshop on Applications of Computer Vision, pp. 152–159 (1994). https://doi.org/10.1109/ACV.1994.341303
3. Croitoru, A., Doytsher, Y.: Photogramm. Rec. **19**(108), 311 (2004). https://doi.org/10.1111/j.0031-868X.2004.00289.x. https://onlinelibrary.wiley.com/doi/abs/10.1111/j.0031-868X.2004.00289.x
4. Rau, J.Y.: ISPRS Ann. Photogramm. Remote Sens. Spatial Inf. Sci., 287–292 (2012). https://doi.org/10.5194/isprsannals-I-3-287-2012
5. Roth, G., Levine, M.D.: CVGIP Image Underst. **58**(1), 1 (1993). https://doi.org/10.1006/ciun.1993.1028
6. Goldberg, D.E.: Genetic Algorithms in Search, Optimization and Machine Learning. Addison-Wesley Longman Publishing, Boston (1989)
7. Roth, G., Levine, M.D.: IEEE Trans. Pattern Anal. Mach. Intell. **16**(9), 901 (1994). https://doi.org/10.1109/34.310686
8. Watt, A.: 3D Computer Graphics, 3rd edn. Person Education Limited, Harlow (2000)
9. Pasko, A., Savchenko, V., Adzhiev, V., Sourin, A.: Multidimensional geometric modeling and visualization based on the function representation of objects. Technical Report 92-1-008, The University of Aizu, Japan (1993)
10. Mahfoud, S.W.: Niching methods for genetic algorithms. Technical Report No.95001, IlliGAL Report (May 1995)
11. Diago, L.A., Ochoa, A.: Proceedings of III Ibero-American Workshop on Pattern Recognition TIARP-98, pp. 461–473 (1998)

Chapter 4
Solution of a Coupled Conduction–Radiation Inverse Heat Transfer Problem with the Topographical Global Optimization Method

Lucas Correia da Silva Jardim, Diego Campos Knupp,
Wagner Figueiredo Sacco, and Antônio José Silva Neto ⓘ

Abstract The method entitled topographical global optimization (TGO) is used to solve an inverse problem of heat transfer by coupled conduction and radiation. The TGO selects points from a random sample and uses them as initial solutions for a local optimization method. The selected points are considered as topographical minima. The parameters estimated with the solution of the inverse heat transfer problem were the thermal conductivity, the single scattering albedo, and the optical thickness. With simulated experimental data, four cases were tested and solved. The cases with lower sensitivity required data with lower experimental errors, and the sensitivity analysis provided a clear insight on the problem physical behavior. The results obtained confirm the good performance and effectiveness of TGO, associated with local optimization methods, in finding a good approximation to the global minimum in complex engineering problems.

Keywords Topographical global optimization · Coupled conduction and radiation heat transfer · Sensitivity analysis · Thermal parameters estimation

4.1 Introduction

Many authors describe the importance of radiative transfer analysis in different areas of interest, including research in optical tomography [2, 12] the design and control of

L. C. S. Jardim (✉) · D. C. Knupp · A. J. Silva Neto
Rio de Janeiro State University, Polytechnic Institute, Nova Friburgo, RJ, Brazil
e-mail: ljardim@iprj.uerj.br; diegoknupp@iprj.uerj.br; ajsneto@iprj.uerj.br

W. F. Sacco
Federal University of Western Pará, Institute of Engineering and Geosciences,
Santarém, PA, Brazil
e-mail: wagner.sacco@ufopa.edu.br

© Springer Nature Switzerland AG 2020
O. Llanes Santiago et al. (eds.), *Computational Intelligence in Emerging Technologies for Engineering Applications*, Studies in Computational Intelligence 872, https://doi.org/10.1007/978-3-030-34409-2_4

thermal systems with a predominance of radiation exchange [10], atmospheric and oceanic models [25] and oceanography [5]. The heat transfer modeling considering both conductive and radiative heat transfer is also used to estimate the thermal insulation performance of functionally graded thermal barrier coatings [8], as well as in the food industry, with the computational modeling of an electric furnace [6].

Most computational models dealing with parameter estimation do not contemplate heat transfer simultaneously on the forms of radiation and conduction. Such models have recently been used to study the heat loss in linear Fresnel reflector systems [11], the determination of the temperature dependent emissivity at the boundary of a two dimensional model [3], among others. In this work an inverse problem combining conduction and radiation is formulated [15, 21], and used for the determination of thermal and radiative properties, such as the thermal conductivity, single scattering albedo, and the optical thickness of a semi-transparent medium, i.e., a medium that emits, absorbs, and scatters radiation. That results in a global optimization problem, where it is desired to minimize residuals between predictions yielded by a mathematical/computational model and experimental measurements. For this purpose, it is necessary to use a computationally efficient optimization method that can overcome local minima.

The method used in this work is a clustering optimization technique based on the topographic information of the objective function, the so-called topographical global optimization (TGO), introduced by Törn and Viitanen [22]. TGO was recently used with success for solving problems dealing with the optimal design of nuclear reactors core [18], chemical equilibrium [9] and a constrained iterative version of its metaheuristics is proposed and used to optimize eight complex engineering design problems by Ferreira et al. [7]. It may be briefly described by three steps:

1. Sampling a closed search space with M uniformly distributed random points;
2. Selecting points with lower evaluations of objective function with respect to its closest K-neighbors;
3. Setting those selected points as initial candidate solutions for a local optimization method.

In order to obtain the results presented in this chapter, the well-known Nelder–Mead method [14] was used as the local optimization method.

To study the performance of TGO in the coupled conduction–radiation inverse heat transfer problem, four different cases were investigated. For two of them, TGO faced no trouble in estimating the parameters of interest, but for the other two, a sensitivity difficulty was found in regard to the single scattering albedo. For these cases we lowered the level of experimental error in the synthetic experimental data in order to better perform the search. To better understand this difficulty, a detailed sensitivity analysis is presented.

4.2 The Coupled Conduction–Radiation Heat Transfer Problem

Consider a one dimensional medium, subjected to steady-state combined heat transfer by conduction and radiation. Consider also that this medium absorbs, emits, and isotropically scatters radiation, having an optical thickness τ_0, with transparent boundary surfaces subjected to external incident radiation on the left boundary, i.e., $\tau = 0$, as shown in Fig. 4.1. The boundaries at $\tau = 0$ and $\tau = \tau_0$ are maintained at constant temperatures T_1 and T_2, respectively.

The mathematical formulation for the combined conduction–radiation heat transfer problem with the temperature in its dimensionless form, $\Theta(\tau)$, is given by the Poisson equation [13, 15]:

$$\frac{d^2\Theta}{d\tau^2} - \frac{(1-\omega)}{N}\left[\Theta^4 - G^*(\tau)\right] = 0, \text{ in } 0 < \tau < \tau_0 \tag{4.1a}$$

$$\Theta = \Theta_1 \text{ at } \tau = 0 \tag{4.1b}$$

$$\Theta = \Theta_2 \text{ at } \tau = \tau_0, \tag{4.1c}$$

where ω is the single scattering albedo, and the not yet defined dimensionless variables correspond to

$$G^*(\tau) = \frac{1}{2}\int_{-1}^{1} I(\tau\mu)d\mu \tag{4.2a}$$

$$N = \frac{k\beta}{4n^2\bar{\sigma}T_1^3} \tag{4.2b}$$

$$\Theta = \frac{T}{T_1} \tag{4.2c}$$

Fig. 4.1 Schematic representation of the physical system subjected to the coupled conduction–radiation heat transfer phenomena

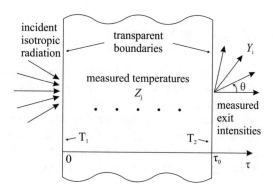

being G^* the dimensionless incident radiation, N the conduction–radiation parameter, and Θ the dimensionless temperature. In Eqs. (4.1) and (4.2), τ is the optical variable, μ is the cosine of the polar angle, i.e., the angle of the radiation beam with the positive τ axis, k is the thermal conductivity, I is the radiation intensity, β is the extinction coefficient, n is the refractive index, and $\bar{\sigma}$ is the Stefan–Boltzmann constant. For most test cases presented in this chapter, we consider Θ_2 being fixed as $\Theta_2 = 0.8$, but for two cases properly indicated, it was used $\Theta_2 = 0.2$. In all cases $\Theta_1 = 1.0$. The dimensionless radiation intensity, $I(\tau, \mu)$, is determined by the solution of the radiative transfer problem [13, 15]:

$$\mu \frac{\partial I(\tau, \mu)}{\partial \tau} + I(\tau, \mu) = R(\Theta) + \frac{\omega}{2} \int_{-1}^{1} I(\tau, \mu') d\mu' \tag{4.3a}$$

$$\text{in } 0 < \tau < \tau_0, \ -1 \leq \mu \leq 1$$

$$I(0, \mu) = 1 \text{ with } \mu > 0 \tag{4.3b}$$

$$I(0, -\mu) = 0 \text{ with } \mu > 0 \tag{4.3c}$$

with the dimensionless temperature dependent source term defined as

$$R(\Theta) = (1 - \omega)\Theta^4(\tau). \tag{4.4}$$

The problem described by Eqs. (4.3) and (4.4) is solved here using the built-in NDSolve routine of the software Wolfram Mathematica 11.0 [24], under automatic absolute and relative error control. For this solution to be obtained, the integral in Eq. (4.3a) is replaced by a Gauss–Legendre quadrature. The solution for the radiative intensities is considered converged when the maximum absolute error in two subsequent iterations is lower than a prescribed tolerance of $\epsilon = 10^{-4}$. For the heat conduction problem treated in Eq. (4.1), the solution is also obtained using NDSolve, with a pseudo-transient approach for the sake of stability.

4.3   Inverse Problem

Consider that a set of radiative and thermal properties of the medium are unknown, but experimental measurements of the intensity of the radiation leaving the medium at different polar angles θ_i are available, i.e., Y_i, with i= $1, 2, \ldots, N_I$. In addition, measurements of the temperature of the medium, Z_j with $j = 1, 2, \ldots, N_T$, are also available, where N_I and N_T represent the total amount of experimental data available for the radiation intensity and for the temperature, respectively. In the following sub-sections the inverse problem formulation and solution are presented.

4.3.1 Inverse Problem Formulation

The present work treat as unknowns the thermal conductivity k, which is included in the mathematical formulation of the heat transfer problem under analysis by means of the conduction–radiation parameter N, see Eq. (4.2a), the single scattering albedo ω, and the optical thickness τ_0. Therefore, the vector of unknowns $\mathbf{Q}$ to be estimated is

$$\mathbf{Q} = \{N, \omega, \tau_0\}^T. \tag{4.5}$$

Since real experimental data were not available, the inverse problem solution concerning the estimation of the parameters $\mathbf{Q}$ was obtained with a set of simulated experimental data, which, in this work, were generated by solving the direct problem, given by Eqs. (4.1)–(4.4), and adding noise drawn from a normal distribution, with known variance σ^2

$$Y_i = I_i(\mathbf{Q}) + e_i, \ e_i \sim N\left(0, \sigma^2\right), \ i = 1, 2, \ldots, N_I \tag{4.6a}$$

$$Z_j = \Theta_j(\mathbf{Q}) + r_j, \ r_j \sim N\left(0, \sigma^2\right), \ i = 1, 2, \ldots, N_T. \tag{4.6b}$$

Assuming uncorrelated measurement errors, and that their variance is constant, the solution of the inverse problem corresponds to the minimization of an objective function given by the summation of the squared residues between the radiation intensity and temperature measurements, i.e., $\mathbf{Y}$ and $\mathbf{Z}$, respectively, and the values calculated with the solution of the direct problem, i.e., I and Θ, respectively. That is, one must obtain the global minimum of the objective function [21]

$$S(\mathbf{Q}) = \sum_{i=1}^{N_I} [Y_i - I(\tau_0, \mu_i)]^2 + \sum_{j=1}^{N_T} [Z_j - \Theta(\tau_j)]^2. \tag{4.7}$$

From Eq. (4.7), one can observe that the exit radiation measurements considered in the formulation of the inverse problem are obtained at the exit boundary $\tau = \tau_0$.

4.3.2 Inverse Problem Solution via Topographical Global Optimization

Introduced by Törn and Viitanen [22], TGO uses a topographic heuristic to determine minima in a set of initial points. These found minima are then used as the initial estimates for a local optimization method [18]. To perform TGO, firstly a number of M points must be sampled, at random or uniformly, in the search space—they are represented by P_u, where $P_u = \{N_u, \omega_u, \tau_{0_u}\}$, with $u = 1, 2, \ldots, M$ [9, 22]. A poor distribution of these points throughout the search space may negatively influence the final outcome of the method, since all areas should be evenly covered [23]. So the array of sampled points **P** can be written as

$$\mathbf{P} = \begin{bmatrix} N_1 & N_2 & \dots & N_u & \dots & N_M \\ \omega_1 & \omega_2 & \dots & \omega_u & \dots & \omega_M \\ \tau_{0_1} & \tau_{0_2} & \dots & \tau_{0_u} & \dots & \tau_{0_M} \end{bmatrix}. \tag{4.8}$$

Now a distance matrix **D** is constructed where the element j of a row i is the distance between points P_i and P_j, with i and $j = 1, 2, \ldots, M$. The elements of **D** are ranked in ascending order and those ranks stored in another matrix **R**, but notice that the elements of the main diagonal of **D** must receive a sufficiently large number, e.g., 10^6, so they are included on the ranking as the last element.

The undirected graph **KNN** is obtained by taking the indices of the smallest K-ranks of **R**. For example, if one chooses to analyze the three closest neighbors, i.e., $K = 3$, and if the second line of **R** is $\{3, 5, 4, 2, 1\}$, the second line of **KNN** will be $\{5, 4, 1\}$, which means that the points P_5, P_4, and P_1 are the closest to point P_2.

The undirected graph becomes a directed graph by giving it a plus sign to elements with higher objective function evaluations than the associated row point, and a minus sign to elements with lower function evaluation [1]. Then the points associated with rows of **KNN** that only contain positive elements are considered the topographical minima. But on the other hand, if a row has only minus signs, the point associated with that row are topographical maxima. The algorithm for the steps above is presented in Algorithm 1 of Chap. 3.

Finally, all the points found as topographical minima (or maxima, depending on the problem) must be employed as the initial estimates for a local optimization method. In this last step, the optimum returned by the local search method is considered the global optimum of the objective function.

In the present work, the method of choice to perform this step is the Nelder–Mead (NM) [14]. Even though being considered a simple method, some authors have already pointed out the efficiency of its hybridization with other methods such as the particle collision algorithm, on the design problem of a nuclear reactor core [17], the differential evolution algorithm [19], and the ant colony algorithm in the field of finances [20], among others.

To minimize an objective function $f(\mathbf{Q})$, where **Q** is a vector of real variables, we define the set of points $\mathbf{P}^* = \{\mathbf{P}_1, \mathbf{P}_2, \ldots, \mathbf{P}_n, \mathbf{P}_{n+1}\}$ as initial candidate solutions for **Q**. In addition to the implementation of the Nelder–Mead method, four scalar parameters must be specified. They are the coefficient of reflection (a), expansion (b), contraction (c), and shrinkage (d). In this chapter we have also used a user-defined stopping criterion Δ. The NM algorithm is presented in Algorithm 2.

Algorithm 1: TGO for minimization

---

1 **choose:**
2   M: number of initial points
3   K: number of closest neighbors to analyze
4 **store** in **P**: M points randomly or uniformly distributed in a closed search space
5 **store** in **F**: the objective function evaluation at $\mathbf{P}_i$ with $i = 1, \ldots, M$
6 **assemble** the distance matrix **D** for the points of **P**
7 **rank** the elements of each line of **D** in ascending order and **store** in **R**
8 **for** *each line $i = 1, \ldots, M$* **do**
9 $\quad$ **for** *each rank $k = 1, \ldots, K$* **do**
10 $\quad\quad$ **for** *each element $j = 1, \ldots, M$* **do**
11 $\quad\quad\quad$ **if** $R_{i,j} = k$ **then**
12 $\quad\quad\quad\quad$ $KNN_{i,j} \leftarrow j$;
13 $\quad\quad\quad$ **end**
14 $\quad\quad$ **end**
15 $\quad$ **end**
16 **end**
17 **for** *each line $i = 1, \ldots, M$ of* **KNN** **do**
18 $\quad$ **for** *each element of the line $j = 1, 2, \ldots, K$* **do**
19 $\quad\quad$ **if** $F_{KNN_{i,j}} < F_i$ **then**
20 $\quad\quad\quad$ $KNN_{i,j} \leftarrow KNN_{i,j} \times (-1)$;
21 $\quad\quad$ **end**
22 $\quad$ **end**
23 **end**
24 **for** *each line $i = 1, \ldots, M$ of* **KNN** **do**
25 $\quad$ $j = 1$;
26 $\quad$ **while** $KNN_{i,j} > 0$ **do**
27 $\quad\quad$ $j \leftarrow j + 1$;
28 $\quad$ **end**
29 $\quad$ **if** $j = K$ **then**
30 $\quad\quad$ **append** j in **index**
31 $\quad$ **end**
32 **end**
33 P_{index_i} are the topographical minima, for all elements of **index**

4.3.3   Sensitivity Analysis

The sensitivity coefficients are defined as [4]

$$X_j^I = \frac{\partial I}{\partial Q_j}, \quad X_j^T = \frac{\partial \Theta}{\partial Q_j} \text{ with } j = 1, 2, \ldots, J, \tag{4.9a,b}$$

where J is the total number of parameters to be estimated.

This derivative is obtained for a chosen coordinate position, in $0 < \tau < \tau_0$, and polar angle, in $0 < \mu < 1$, with respect to the unknown parameter vector **Q**. It represents the sensitivity of the estimated potentials (observable variables) with respect to changes in the unknown parameters [16]. It can be shown that if the

---

Algorithm 2: Nelder-Mead simplex algorithm for minimization

---

1 choose: a, b, c, d and Δ
2 sort the points of $\mathbf{P^*}$ so that $f(\mathbf{P_1}) \leq f(\mathbf{P_2}) \leq \cdots \leq f(\mathbf{P_{n+1}})$
3 while $f(\mathbf{P_{n+1}}) - f(\mathbf{P_1}) > \Delta$ **do**
4 | **1. centroid:** compute the centroid $\mathbf{P_0}$ from all points except $\mathbf{P_{n+1}}$
5 | **2. reflection:** compute the reflected point $\mathbf{P_r}$
6 | $\mathbf{P_r} = \mathbf{P_0} + a\,(\mathbf{P_0} - \mathbf{P_{n+1}})$;
7 | **if** $f(\mathbf{P_1}) \leq f(\mathbf{P_r}) \leq f(\mathbf{P_{n+1}})$ **then**
8 | $\mathbf{P_{n+1}} \leftarrow \mathbf{P_r}$;
9 | **go to** step 6;
10 **else**
11 | **3. expansion:**
12 | **if** $f(\mathbf{P_r}) < f(\mathbf{P_1})$ **then**
13 | Compute the expanded point $\mathbf{P_e}$
14 | $\mathbf{P_e} = \mathbf{P_0} + b\,(\mathbf{P_r} - \mathbf{P_0})$;
15 | **if** $f(\mathbf{P_e}) < f(\mathbf{P_r})$ **then**
16 | $\mathbf{P_{n+1}} \leftarrow \mathbf{P_e}$;
17 | **go to** step 6
18 | **else**
19 | $\mathbf{P_{n+1}} \leftarrow \mathbf{P_r}$
20 | **go to** step 6
21 | **end**
22 | **end**
23 | **4. contraction**: Compute the contracted point P_c
24 | $\mathbf{P_c} = \mathbf{P_0} + c\,(\mathbf{P_{n+1}} - \mathbf{P_0})$;
25 | **if** $f(\mathbf{P_c}) < f(\mathbf{P_{n+1}})$ **then**
26 | $\mathbf{P_{n+1}} \leftarrow \mathbf{P_c}$
27 | **end**
28 | **5. shrinkage:**
29 | **for** *each point* $i = 2, 3, \ldots, n, n+1$ **do**
30 | $\mathbf{P_i} \leftarrow \mathbf{P_1} + d(\mathbf{P_i} - \mathbf{P_1})$;
31 | **end**
32 | **end**
33 | **6. sort** the points of $\mathbf{P^*}$ so that $f(\mathbf{P_1}) \leq f(\mathbf{P_2}) \leq \cdots \leq f(\mathbf{P_{n+1}})$
34 end
35 $\mathbf{P_1}$ is the solution

---

sensitivity coefficients vector of a given unknown parameter is linearly dependent of the sensitivity coefficients vector of others parameters, the inverse problem is ill-conditioned, thus not having a unique solution [4].

To calculate these derivatives, the central finite difference formula is used, as presented in Eq. (4.10), where the increment δ is the same for every unknown parameter, with $\delta Q_j = 10^{-4}$, $j = 1, 2, \ldots, J$. Temperature sensitivity coefficients are calculated for 31 points uniformly distributed in the spatial domain $0 < \tau < \tau_0$. But the coefficients for the radiation intensity are generated at the same points (discrete ordinates) of the Gauss–Legendre quadrature on the cosine of the polar angle domain, corresponding to the range $0 < \mu < 1$.

$$\frac{\partial I}{\partial Q_j}$$

$$= \frac{I\left(\{Q_1, Q_2, \ldots, Q_j + \delta Q_j, \ldots, Q_J\}\right) - I\left(\{Q_1, Q_2, \ldots, Q_j - \delta Q_j, \ldots, Q_J\}\right)}{2\delta Q_j}$$

(4.10a)

$$\frac{\partial \Theta}{\partial Q_j}$$

$$= \frac{\Theta\left(\{Q_1, Q_2, \ldots, Q_j + \delta Q_j, \ldots, Q_J\}\right) - \Theta\left(\{Q_1, Q_2, \ldots, Q_j - \delta Q_j, \ldots, Q_J\}\right)}{2\delta Q_j}$$

(4.10b)

with $j = 1, 2, \ldots, J$

4.4 Results and Discussion

In this section four different sets of values are considered for the unknown parameters $\mathbf{Q}$. These values are presented in Table 4.1. They were carefully chosen based on the difficulty encountered to obtain the solution of the inverse problem. Section 4.4.1 presents the sensitivity analysis and results obtained for all the four cases, and in the Sect. 4.4.2 the Case 3 is again solved with some changes on the sensors positioning and on the temperature at the boundary $\tau = \tau_0$. These changes led to the so-called Cases 5–7. For all cases, the experimental data are obtained with seven measured exit radiation intensities and two measured temperatures inside the medium. Therefore, $N_I = 7$ and $N_T = 2$ in Eq. (4.7).

To perform the first step of TGO, i.e., the generation and distribution of points in a closed search space, the built-in routine RandomReal of the software Wolfram Mathematica 11.0 was used. This routine is a pseudo-random point generator and tries to simulate true randomness. The number of initial points generated was $M = 200$ points for every optimization experiment performed. These points are distributed in a closed search space with its limits presented in Table 4.1.

Table 4.1 Summary of test cases with the exact values for the unknown parameters and respective space search

Cases	N	ω	τ_0	σ
1	0.05	0.90	1.00	0.010
2	0.10	0.90	1.00	0.010
3	0.05	0.30	1.00	0.005
4	0.10	0.30	1.00	0.005
Search limits	$0.005 \leq N \leq 0.5$	$0.00 \leq \omega \leq 1.0$	$0.25 \leq \tau_0 \leq 2.5$	–

σ is used to generate the synthetic experimental data (see Eq. (4.6))

When referring to the initial points generator, one can also use any other random or uniform point generator, as recommended by Törn and Viitanen [23], as long as it covers the whole search space. For example, Sacco et al. [18] used the Sobol sequence to generate the points when solving a nuclear reactor core design problem with TGO.

The amount of experimental data $N_I = 7$ and $N_T = 2$ are enough to solve Cases 1 and 2 with noise level of $\sigma = 0.01$. This level is lowered to perform Cases 3 and 4, where problems with sensitivity were encountered. For test Cases 5–7—later on explained—the level of noise used was the same $\sigma = 0.010$ higher one. The parameters used in NM were fixed as $a = 1, b = 2, c = -0.5$, and $d = 0.5$, as well as the stopping criterion $\Delta = 10^{-8}$.

4.4.1   Test Cases 1–4

For Cases 1–4 presented in Table 4.1, the exit intensities were measured from the polar angle $\theta = \pi/16$ up to $\theta = 7\pi/16$, with incremental steps of $\Delta\theta = \pi/16$. The temperatures were measured at $\tau = 0.4\tau_0$ and $\tau = 0.6\tau_0$.

In Fig. 4.2 the sensitivity coefficients for these four cases are presented. They are calculated with respect to the unknown parameters $\{N, \omega, \tau_0\}$ at $\tau = \tau_0$ ($\mu > 0$), for the radiation intensity, and $0 < \tau < \tau_0$, for the temperature. The vertical lines in each graph of Fig. 4.2 represent the detectors position, as described herein.

Although the sensitivity coefficients for the parameter N are smaller in Cases 1 and 2 when compared to Cases 3 and 4, notice how the behavior of these coefficients for N are very different from the ones with respect to ω in Cases 1 and 2—especially in Case 2. This explains why Cases 3 and 4 are harder to estimate, even with the small sensitivity for the parameter N in Cases 1 and 2.

In Cases 3 and 4 the sensitivity coefficients for the scattering albedo ω are much smaller, for both the radiation intensity and temperature, meaning that a change in that parameter yields a smaller variation on the outcome from the model calculation. This is not desired, to detect any parameter, i.e., the observable quantities must be sensible to changes in the parameters to be determined. This can become even worst when noise is added, in fact, all the tests performed for Cases 3 and 4 with noise level corresponding to $\sigma = 0.01$ led to values of ω approaching zero, which is a wrong solution, even though the NM reached the stopping criterion.

In Fig. 4.2 it is possible to see that τ_0 is the easiest parameter to estimate. On the other hand, N and ω can only be estimated with a poor level of confidence. The fact that the radiation intensity sensitivity coefficients present a correlated behavior may explain this difficulty. In order to better understand the correlation, the ratio between the sensitivities with respect to ω and N is presented for all the cases in Fig. 4.3.

Observe that the ratios for Cases 1 and 2 present a quasi-linear behavior in the range $0.5 < \mu < 1.0$ for the radiation intensity, thus yielding the difficulty described above on obtaining both parameters simultaneously. But that linear behavior gets even worst for Cases 3 and 4. Observe that for these two cases the sensitivity

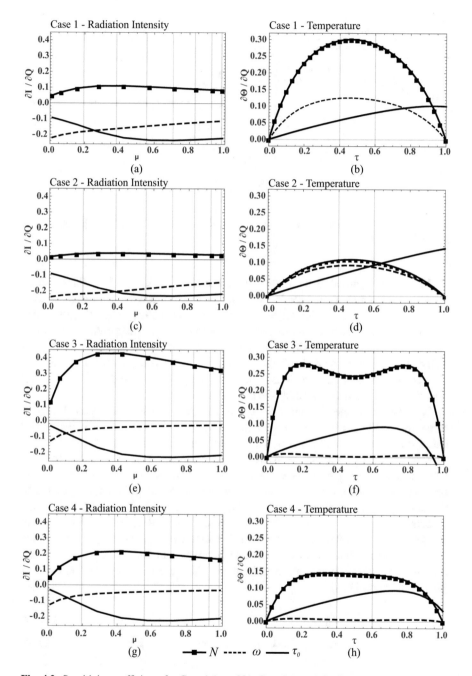

Fig. 4.2 Sensitivity coefficients for Case 1 (**a** and **b**), Case 2 (**c** and **d**), Case 3 (**e** and **f**) and Case 4 (**g** and **h**)

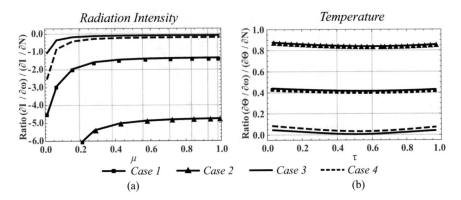

Fig. 4.3 Sensitivity coefficient ratios for (**a**) radiation intensity X_ω^I / X_N^I and (**b**) temperature X_ω^T / X_N^T

coefficient ratios are almost constant in the range $0.2 < \mu < 1.0$ for the radiation intensity. The detectors that renders the smaller linear dependency are positioned at $\theta = 7\pi/16$ and $\theta = 6\pi/16$ ($\mu = 0.1950$ and $\mu = 0.3826$, respectively), which are the ones farthest from the exit normal coordinate. The temperature sensitivity coefficient ratios are very similar in aspect and behavior for all the cases, so its analysis is not conclusive. As shown in Fig. 4.2, the sensitivity of the temperature with respect to the parameter N is higher for the test Cases 1 and 3. But in Fig. 4.3 it is confirmed the correlation for the parameters N and ω.

In Tables 4.2 and 4.3 are shown the results obtained for Cases 1 and 2, respectively. We performed five experiments with a different set of experimental data for each one, and the results shown are the average of ten executions for each experiment. In these tables σ_k is the standard deviation of these ten executions calculated for each parameter σ_N, σ_ω, and σ_{τ_0}, and the objective function value σ_S. The terms represented by $\bar{\sigma}_k/\bar{\mu}_k$ are the coefficients of variation, they represent the ratio between the average of the standard deviation of each parameter and the objective function value $\bar{\sigma}_k$ and their average $\bar{\mu}_k$. They illustrate how disperse that one parameter was relative to its own value. The terms I_{error} and Θ_{error} represent the maximum error obtained for the radiation intensity and temperature, respectively, for that particular experiment, in percentage. They are calculated as follows (see Eqs. (4.6a) and (4.6b)):

$$I_{\text{error}} = \max(e_i/I_i), \quad i = 1, 2, \ldots, N_I \tag{4.11a}$$

$$\Theta_{\text{error}} = \max(r_j/\Theta_j), \quad j = 1, 2, \ldots, N_T. \tag{4.11b}$$

Observe, in Tables 4.2 and 4.3, that indeed the optical thickness τ_0 was the easiest parameter to estimate—the coefficient of variation $\bar{\sigma}_{\tau_0}/\bar{\mu}_{\tau_0}$ was the smallest among the three parameters—with the conduction–radiation parameter N being the

Table 4.2 Average of ten executions obtained for Case 1 considering $\sigma = 0.01$

Exp.	$N = 0.05$ $(\sigma_N \times 10^{-4})$	$\omega = 0.90$ $(\sigma_\omega \times 10^{-4})$	$\tau_0 = 1.0$ $(\sigma_{\tau_0} \times 10^{-4})$	$S(\mathbf{Q}) \times 10^{-4}$ $(\sigma_S \times 10^{-9})$	$I_{error}(\%)$ $\Theta_{error}(\%)$
1	0.0691	0.9164	0.9893	5.8787	2.96
	(2.759)	(2.667)	(2.558)	(2.130)	0.91
2	0.1156	0.8161	1.0516	5.1734	4.14
	(3.221)	(4.724)	(3.687)	(8.337)	1.43
3	0.0823	0.7419	1.0961	2.9248	4.74
	(1.447)	(2.774)	(2.223)	(1.787)	0.86
4	0.0508	0.9682	0.9353	8.0087	2.66
	(3.606)	(2.128)	(2.349)	(1.363)	2.72
5	0.0267	0.9610	0.9533	4.6199	2.78
	(2.339)	(2.730)	(2.784)	(3.906)	0.88
$\bar{\mu}_k$	0.0689	0.8807	1.0051	5.3211	3.45
$\bar{\sigma}_k/\bar{\mu}_k$	0.388%	0.034%	0.027%	0.00065%	1.36

Table 4.3 Average of ten executions obtained for Case 2 considering $\sigma = 0.01$

Exp.	$N = 0.10$ $(\sigma_N \times 10^{-4})$	$\omega = 0.90$ $(\sigma_\omega \times 10^{-4})$	$\tau_0 = 1.0$ $(\sigma_{\tau_0} \times 10^{-4})$	$S(\mathbf{Q}) \times 10^{-4}$ $(\sigma_S \times 10^{-9})$	$I_{error}(\%)$ $\Theta_{error}(\%)$
1	0.1182	0.7997	1.0582	3.7814	2.79
	(2.919)	(3.035)	(2.786)	(1.884)	0.23
2	0.0713	0.9031	0.9759	3.8081	2.91
	(2.700)	(2.798)	(2.852)	(4.249)	0.20
3	0.0837	0.8546	1.0042	7.9526	3.27
	(2.149)	(3.296)	(2.709)	(2.445)	1.52
4	0.0471	0.9499	0.9468	6.7598	2.05
	(2.422)	(2.588)	(2.584)	(2.122)	0.63
5	0.0539	0.9229	0.9551	4.1944	3.58
	(3.190)	(2.873)	(2.617)	(2.196)	0.65
$\bar{\mu}_k$	0.0748	0.8860	0.9880	5.2993	2.92
$\bar{\sigma}_k/\bar{\mu}_k$	0.357%	0.033%	0.027%	0.00047%	0.64

hardest. The standard deviation values are in the order of 10^{-4} for the parameters and 10^{-9} for the objective function value, corroborating with the idea that the optimization method presents a strong feasibility in estimating the position of the global minimum. The maximum errors for the radiation I_{error} are considerably larger than those for the temperature Θ_{error}. This is acceptable because the exact radiation values are smaller than the ones for temperature, and the variance error is considered the same for both cases. The results obtained for Cases 3 and 4 are displayed in Tables 4.4 and 4.5. For these cases the experimental errors were generated with a smaller value for σ in Eq. (6). This can readily be seen in the errors columns.

Experiment 3 for Case 3, whose results are presented in Table 4.4, led to a scattering albedo value of $\omega = 0.1428$, which is not the solution, but that is expected

Table 4.4 Average of ten executions obtained for Case 3 considering $\sigma = 0.005$

Exp.	$N = 0.05$ $(\sigma_N \times 10^{-4})$	$\omega = 0.30$ $(\sigma_\omega \times 10^{-4})$	$\tau_0 = 1.0$ $(\sigma_{\tau_0} \times 10^{-4})$	$S(\mathbf{Q}) \times 10^{-4}$ $(\sigma_S \times 10^{-9})$	$I_{error}(\%)$ $\Theta_{error}(\%)$
1	0.0698	0.3439	1.0270	0.3914	0.52
	(1.359)	(9.879)	(2.817)	(3.828)	0.70
2	0.0415	0.2582	0.9775	0.1292	0.75
	(1.079)	(18.096)	(2.694)	(5.973)	0.39
3	0.0350	0.1428	0.9916	1.7767	0.97
	(0.925)	(14.764)	(2.095)	(0.7811)	1.15
4	0.0354	0.2997	0.9716	2.2397	1.58
	(0.822)	(8.785)	(2.078)	(2.896)	0.47
5	0.0533	0.3705	1.0025	1.2390	1.18
	(0.924)	(7.051)	(1.760)	(0.9867)	0.34
$\bar{\mu}_k$	0.0470	0.2830	0.9940	1.1612	1.00
$\bar{\sigma}_k/\bar{\mu}_k$	0.217%	0.414%	0.023%	0.00249%	0.61

Table 4.5 Average of ten executions obtained for Case 4 considering $\sigma = 0.005$

Exp.	$N = 0.10$ $(\sigma_N \times 10^{-4})$	$\omega = 0.30$ $(\sigma_\omega \times 10^{-4})$	$\tau_0 = 1.0$ $(\sigma_{\tau_0} \times 10^{-4})$	$S(\mathbf{Q}) \times 10^{-4}$ $(\sigma_S \times 10^{-9})$	$I_{error}(\%)$ $\Theta_{error}(\%)$
1	0.0993	0.2623	1.0099	1.6147	1.23
	(2.092)	(6.534)	(2.399)	(1.648)	0.17
2	0.0936	0.3550	0.9896	1.2816	1.12
	(1.891)	(8.251)	(1.549)	(2.288)	0.80
3	0.1315	0.4386	0.9955	0.1777	0.97
	(3.071)	(8.342)	(2.686)	(1.356)	1.15
4	0.1326	0.4198	0.9893	0.8612	0.62
	(2.850)	(6.637)	(2.621)	(2.039)	0.73
5	0.0638	0.2406	0.9725	1.4990	0.86
	(1.161)	(8.932)	(2.434)	(1.909)	1.47
$\bar{\mu}_k$	0.1041	0.3432	0.9913	1.0868	0.96
$\bar{\sigma}_k/\bar{\mu}_k$	0.212%	0.225%	0.023%	0.0017%	0.86

due to the low sensitivity with respect to that parameter—presented in Fig. 4.2e. The standard deviations have similar values as those obtained for Cases 1 and 2, on Tables 4.2 and 4.3, but observe how the coefficients of variation for Cases 3 and 4 are larger than those for Cases 1 and 2 with respect to ω. Therefore, it is safe to conclude that the estimates for ω in experiments from Cases 3 and 4 are more disperse than the ones in Cases 1 and 2. Again, the sensitivity analysis anticipated this outcome.

4.4.2   Test Cases 5–7

Now let us consider three extra cases, namely Cases 5–7. To overcome the problem of sensitivity previously described, we solved the same test Case 3, but with some changes. Therefore, the exact values for these extra cases are also $N = 0.05$, $\omega = 0.30$, and $\tau_0 = 1.0$.

For Case 5, the main change is related to the boundary conditions for the temperature, which were originally $\Theta(0) = 1.0$ and $\Theta(\tau_0) = 0.8$. For this new case, the boundary temperature at $\tau = \tau_0$, i.e. Θ_2 in Eq. (4.1c), was fixed as $\Theta_2 = 0.2$. The idea is to enhance the temperature sensitivity by increasing its variation along the physical domain.

In Fig. 4.4, the sensitivity coefficients for Case 5 are shown. Observe how the temperature sensitivity coefficient related to N is greater in magnitude than the one for Case 3—see Fig. 4.2e. This is also true for the single scattering albedo ω, but the main enhancement occurs for the conduction–radiation parameter N.

Now for Case 6, the change is made on the position of the radiation intensities detectors, but the temperature at the boundary $\tau = \tau_0$ remains the same as the original, i.e., $\Theta_@ = 0.8$. Since there is more sensitivity in the region $\mu < 0.4$, it is reasonable to reposition in such region the seven detectors used in the original experiment. The temperature detectors remain in the same original position. The exit intensities here are now obtained from the polar angle cosines $\mu = 0.1$ up to $\mu = 0.4$, by incremental steps of $\Delta\mu = 0.05$.

And finally, Case 7 is the combination of the changes made for Cases 5 and 6, i.e., it has the new temperature value at the boundary $\tau = \tau_0$, i.e., $\theta_2 = 0.2$, and the detectors are concentrated in the interval $0.1 < \mu < 0.4$. Combining these changes, it is expected that Case 7 yields the best results from all tests performed, even with noise level of $\sigma = 0.01$, which was impossible to estimate previously for Case 3.

The results obtained for Cases 5 and 6 are presented in Tables 4.6 and 4.7. They were all obtained with error levels corresponding to $\sigma = 0.01$. This particular level

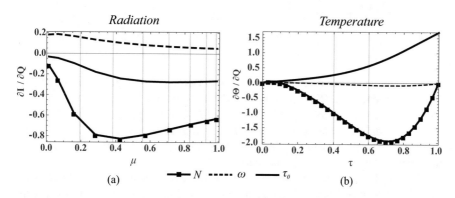

Fig. 4.4 Sensitivity coefficients for Case 5: Radiation intensity (**a**) and temperature (**b**)

Table 4.6 Average of ten executions obtained for Case 5 (changes on the temperature exit boundary condition) considering an experimental error corresponding to $\sigma = 0.01$

Exp.	$N = 0.05$ $(\sigma_N \times 10^{-4})$	$\omega = 0.30$ $(\sigma_\omega \times 10^{-4})$	$\tau_0 = 1.0$ $(\sigma_{\tau_0} \times 10^{-4})$	$S(\mathbf{Q}) \times 10^{-4}$ $(\sigma_S \times 10^{-9})$	$I_{error}(\%)$ $\Theta_{error}(\%)$
1	0.0572	0.3859	0.9940	6.9296	5.22
	(0.2207)	(2.123)	(1.233)	(3.943)	3.65
2	0.0485	0.1696	0.9379	9.7999	7.15
	(0.1297)	(2.061)	(0.5693)	(0.6652)	1.41
3	0.0556	0.1959	0.9566	6.4803	7.78
	(0.1703)	(2.864)	(0.9882)	(0.9552)	2.19
4	0.0480	0.3722	1.0504	6.2863	5.67
	(0.1731)	(1.149)	(0.7624)	(2.233)	0.94
5	0.0536	0.2547	0.9785	9.3412	7.94
	(0.1392)	(2.657)	(0.9754)	(1.184)	1.29
$\bar{\mu}_k$	0.0526	0.2756	0.9834	7.7675	6.75
$\bar{\sigma}_k/\bar{\mu}_k$	0.0316%	0.0787%	0.0092%	0.00023%	1.89

Table 4.7 Average of ten executions obtained for Case 6 (changes on the detectors positions) considering an experimental error corresponding to $\sigma = 0.01$

Exp.	$N = 0.05$ $(\sigma_N \times 10^{-4})$	$\omega = 0.30$ $(\sigma_\omega \times 10^{-4})$	$\tau_0 = 1.0$ $(\sigma_{\tau_0} \times 10^{-4})$	$S(\mathbf{Q}) \times 10^{-4}$ $(\sigma_S \times 10^{-9})$	$I_{error}(\%)$ $\Theta_{error}(\%)$
1	0.0188	0.0617	0.9534	7.9975	3.13
	(0.9274)	(16.418)	(3.345)	(3.391)	2.39
2	0.0366	0.2997	0.9460	1.8624	2.04
	(0.9813)	(6.517)	(2.055)	(2.686)	0.65
3	0.0325	0.2803	1.0030	8.5451	3.52
	(0.9568)	(6.186)	(2.628)	(1.332)	2.53
4	0.0460	0.5682	0.9028	1.2964	5.30
	(1.214)	(7.053)	(3.267)	(2.530)	0.95
5	0.0256	0.3223	0.8938	7.6333	4.44
	(0.7264)	(6.698)	(2.209)	(1.691)	0.99
$\bar{\mu}_k$	0.0319	0.3064	0.9398	7.7294	3.68
$\bar{\sigma}_k/\bar{\mu}_k$	0.301%	0.2798%	0.0287%	0.00030%	1.50

of noise was leading to totally wrong results in Case 3, but here the estimated values are properly obtained. Observe in Table 4.6 that the discrepancy for the parameter N is smaller when compared to the ones for Case 3. This is due to the fact that lowering the temperature value at the right boundary increased greatly the sensitivity for N. Even with more than 7% for radiation intensity error and 2% for temperature errors, the estimate for N is still very accurate.

The results obtained for Case 6 are poorer in accuracy than those obtained for Case 5. Observe in Table 4.7 that Experiment 4 did not lead to a good estimate for ω. Scenarios like these of Experiment 4 were normal when solving Case 3 with $\sigma = 0.01$. Although the repositioning of detectors is not the best approach to solve

Table 4.8 Average of ten executions obtained for Case 7 (changes on the temperature exit boundary condition and detectors positions) considering an experimental error corresponding to $\sigma = 0.01$

Exp.	$N = 0.05$ $(\sigma_N \times 10^{-4})$	$\omega = 0.30$ $(\sigma_\omega \times 10^{-4})$	$\tau_0 = 1.0$ $(\sigma_{\tau_0} \times 10^{-4})$	$S(\mathbf{Q}) \times 10^{-4}$ $(\sigma_S \times 10^{-9})$	$I_{error}(\%)$ $\Theta_{error}(\%)$
1	0.0561	0.3616	1.0676	2.3775	22.43
	(0.1805)	(2.462)	(1.963)	(3.163)	1.74
2	0.0528	0.3102	0.9585	1.3627	19.24
	(0.1477)	(1.280)	(0.804)	(1.278)	0.82
3	0.0620	0.3624	1.0330	14.174	16.99
	(0.1961)	(2.025)	(1.701)	(1.334)	1.06
4	0.0492	0.3554	0.9784	1.9897	12.34
	(0.1816)	(1.267)	(1.479)	(2.355)	1.04
5	0.0477	0.3096	0.9889	5.2890	13.99
	(0.1810)	(2.608)	(2.381)	(5.597)	0.93
$\bar{\mu}_k$	0.0535	0.3398	1.0052	5.0386	17.00
$\bar{\sigma}_k/\bar{\mu}_k$	0.033%	0.056%	0.016%	0.00054%	1.11

this problem, it is safe to say that it enhanced the confidence intervals for N. The coefficients of variation for Case 5 are generally smaller than the ones for Case 6, especially the ones for N and ω, this highlights the fact that changing the boundary temperature is the best approach to overcome the sensitivity problem. Comparing the coefficients variations of Cases 5 and 3, is yet another way to observe this: Case 3, with smaller level of noise, led to larger coefficients of variation.

Finally, in Table 4.8 the results for Case 7 are displayed. Observe firstly that the radiation intensity errors are larger than those for any of the cases previously treated. This is acceptable since in $\mu = 0.1$ the radiation intensity is smaller, and combining this with the boundary temperature $\Theta_2 = 0.2$, at $\tau = \tau_0$ Case 7 yielded the greater maximum radiative intensity errors. In spite of that, the estimations are very accurate when compared to Cases 3, 5, and 6, especially for the conduction–radiation parameter, which generated a good approximation, with an average of 0.0535. Observe also the standard deviations for this parameter that are smaller than those obtained for Cases 3, 5, and 6.

It is important to notice that for Cases 1–6, the TGO algorithm found, in average, seven points to use in the local optimization method. This is a good number for the NM when solving a problem with three dimensions. But for Case 7, several times TGO found four or less points, which cannot be used in the NM method for a problem of three dimensions. This can be readily solved by reducing the number of closest neighbors to analyze. For instance, in Cases 1–6, we used $K = 25$, but for Case 7 we used $K = 20$. This reduction allowed the TGO algorithm to find more than four points on average within the $H = 200$ points sampled. One possible way to explain this is that the objective function got smoother with the changes made, resulting in less local minima across the search space.

4.5 Conclusions

This chapter describes the problem of heat transfer by simultaneous radiation and conduction and, also, the formulation and solution of the inverse problem of estimating thermal and radiation properties via a technique known as topographical global optimization (TGO) combined with a local search method. The approach used here (TGO-NM) works well as long as the experimental noise is sufficiently small and the inverse problem has enough sensitivity so the parameters can be estimated. By further studying the sensitivity analysis presented in this chapter, it was possible to understand the difficulty encountered in Case 3, and tackle it with experimental improvements, leading to three new study cases, namely Cases 5–7. They represent better ways of designing the experiment and solving the inverse problem. It is important to note that TGO-NM found the global minimum for every experiment and execution. In other words, for the 350 optimization experiments performed in this chapter, the stopping criterion was reached in every single one. As one advantage of TGO it can be mentioned its simple control of parameters: the number of initial points and nearest neighbors must be in a reasonable match with the behavior of the objective function and the size of the search space. The experiments performed, for difficult test cases, show that TGO combined with a local search method performs well in global search problems involving the combined mode of heat transfer by conduction and radiation.

Acknowledgements The authors acknowledge the financial support provided by CAPES—Coordenaçã o de Aperfeiçoamento de Pessoal de Nível Superior (Finance Code 001), CNPq—Conselho Nacional de Desenvolvimento Científico e Tecnológico, and FAPERJ—Fundação Carlos Chagas Filho de Amparo à Pesquisa do Estado do Rio de Janeiro.

References

1. Ali, M.M.: Some modified stochastic global optimization algorithms with applications. Doctoral Dissertation, Loughborough University of Technology (1994)
2. Arridge, S.R.: Optical tomography in medical imaging. Inverse Prob. **15**, R41–R93 (1999)
3. Baghban, M., Hossein Mansouri, S., Shams, Z., Inverse radiation-conduction estimation of temperature-dependent emissivity using a combined method of genetic algorithm and conjugate gradient. J. Mech. Sci. Technol. **28**, 739–745 (2014)
4. Beck, J.V., Arnold, K.J., Parameter Estimation in Engineering and Science. Wiley, New York (1977)
5. Chalhoub, E.S., Campos Velho, H.F., Estimation of the optical properties of seawater from measurements of exit radiance. J. Quant. Spectrosc. Radiat. Transf. **72**, 551–565 (2002)
6. Chhanwal, N., Anishaparvin, A., Indrani, D., Raghavarao, K.S.M.S., Anandharamakrishnan, C., Computational fluid dynamics (CFD) modeling of an electrical heating oven for bread-baking process. J. Food Eng. **100**, 452–460 (2010)
7. Ferreira, M.P., Rocha, M.L., Neto, A.J.S., Sacco, W.F.: A constrained ITGO heuristic applied to engineering optimization. Expert Syst. Appl. **110**, 106–124 (2018)

8. Ge, W.A., Zhao, C.Y., Wang, B.X.: Thermal radiation and conduction in functionally graded thermal barrier coatings. Part II: Experimental thermal conductivities and heat transfer modeling. Int. J. Heat Mass Transf. **134**, 166–174 (2019)
9. Henderson, N., de Sá Rêgo, M., Sacco, W.F., Rodrigues, R.A.: A new look at the topographical global optimization method and its application to the phase stability analysis of mixtures. Chem. Eng. Sci. **127**, 151–174 (2015)
10. Howell, J.R., Daun, K.J., Erturk, H., Gamba, M., Sarvari, M., The use of inverse methods for the design and control of radiant sources. JSME Int. J. Ser. B Fluids Therm. Eng. **46**, 470–478 (2003)
11. Mohan, S., Saxena, A., Singh, S.: Heat loss analysis from a trapezoidal cavity receiver in LFR system using conduction-radiation model. Sol. Energy **159**, 37–43 (2018)
12. Montero, R.F.C., Roberty, N.C., Silva Neto, A.J.: Reconstruction of a combination of the absorption and scattering coefficients with a discrete ordinates method consistent with the source–detector system. Inverse Prob. Sci. Eng. **12**, 81–101 (2004)
13. Moura Neto, F.D., Silva Neto, A.J.: An Introduction to Inverse Problems with Applications. Springer Science & Business Media, New York (2013)
14. Nelder, J.A., Mead, R.: A simplex method for function minimization. Comput. J. **7**, 308–313 (1965)
15. Özisik, M.N.: Radiative Transfer and Interactions with Conduction and Convection. Wiley, New York (1973)
16. Özisik, M.N., Orlande, H.R.B.: Inverse Heat Transfer: Fundamentals and Applications. Taylor & Francis, New York (2000)
17. Sacco, W.F., Filho, H.A., Henderson, N., de Oliveira, C.R.E.: A metropolis algorithm combined with Nelder-Mead simplex applied to nuclear reactor core design. Ann. Nucl. Energy. **35**, 861–867 (2008)
18. Sacco, W.F., Henderson, N., Rios-Coelho, A.C.: Topographical global optimization applied to nuclear reactor core design: some preliminary results. Ann. Nucl. Energy **65**, 166–173 (2014)
19. Sacco, W.F., Rios-Coelho, A.C., Henderson, N.: Testing population initialisation schemes for differential evolution applied to a nuclear reactor core design. Int. J. Nucl. Energy Sci. Technol. **8**, 192–212 (2014)
20. Sharma, N., Arun, N., Ravi, V.: An ant colony optimisation and Nelder-Mead simplex hybrid algorithm for training neural networks: an application to bankruptcy prediction in banks. Int. J. Inf. Decis. Sci. **5**, 188–203 (2013)
21. Silva Neto, A.J., Özisik, M.N.: An inverse problem of estimating thermal conductivity, optical thickness, and single scattering albedo of a semi-transparent medium. In: Proceedings of the 1st Inverse Problems in Engineering Conference: Theory and Practice, Florida, USA, pp. 267–273 (1993)
22. Törn, A., Viitanen, S.: Topographical global optimization. In: Floudas, C.A., Pardalos, P.M. (eds.) Recent Advances in Global Optimization, pp. 384–398 (1992)
23. Törn, A., Viitanen, S.: Iterative topographical global optimization. In: Floudas, C.A., Pardalos, P.M. (eds.) State of the Art in Global Optimization, pp. 353–363 (1996)
24. Wolfram, S.: The Mathematica Book Version 5.2. Cambridge-Wolfram Media, Cambridge (2005)
25. Zhang, K., Li, W., Eide, H., Stamnes, K., A bio-optical model suitable for use in forward and inverse coupled atmosphere-ocean radiative transfer models. J. Quant. Spectrosc. Radiat. Transf. **103**, 411–423 (2007)

Chapter 5
Towards Intelligent Optimization of Design Strategies of Cyber-Physical Systems: Measuring Efficacy Through Evolutionary Computations

Soumya Banerjee, Valentina E. Balas, Abhishek Pandey, and Samia Bouzefrane

Abstract Designing of effective cyber-physical system (CPS) encompassing different vertical applications solicits different components of design. Most of the components are uncertain and dynamic in nature. They either could be in the form of hardware sensors, optimization process and their scheduling nature. In this chapter, we investigate various levels of CPS formulation driven by machine learning and evolutionary algorithms with their strategic similarities. We argue that how far intelligent optimization in the level designing a CPS should be viable? Thus, suitability of appropriate evolutionary and machine learning algorithms is discussed in the context of different design uncertainty of CPS. The efficacy of auto-adaptive or self-organization principle is also discussed.

Keywords Cyber-physical systems · Evolutionary computation · Intelligent optimization · Design uncertainty · Machine learning

S. Banerjee (✉)
CEDRIC Lab, Conservatoire National des Arts et Metiers, Paris Cedex 03, France
e-mail: soumyabanerjee@bitmesra.ac.in

V. E. Balas
Aurel Vlaicu University of Arad, Arad, Romania

A. Pandey
University of Petroleum and Energy Studies, Dehradun, India

S. Bouzefrane
CNAM-CEDRIC Lab, Conservatoire National des Arts et Metiers, Paris Cedex 03, France

© Springer Nature Switzerland AG 2020
O. Llanes Santiago et al. (eds.), *Computational Intelligence in Emerging Technologies for Engineering Applications*, Studies in Computational Intelligence 872, https://doi.org/10.1007/978-3-030-34409-2_5

5.1 Introduction

Cyber-physical systems (CPS) [1, 2] are evolving vertical systems orchestrated with suitable computation and overlapping physical procedures. Different flavors of applications of CPS are becoming pertinent, encompassing from smart grid to control of nuclear operations. Therefore, computation, communication, and control require more synchronized yet formal specifications. This requirement initiates to formulate intelligent CPS in design.

Routinely, CPS exhibits high-dimensional and scanty information, for example, content, video, and varieties of sensors information. We see that while structuring an intelligent CPS, the area information may impact the plan pattern of CPS. In this way, the procedure and message correspondences are evidently complicated for planning reason.

Intercommunication of various computing devices and physical objects through Internet is commonly known as the Internet of Things (IoT) [3], and is typical of machine-to-machine communication (M2M). Under this guideline, the adaptation process is qualified to implement [4] taking into consideration the basic CPS and IoT framework, the provision of similar service references and inferences.

To establish justified anticipation, the selection of an exact discrimination from high-dimensional data collection is obvious. In addition, the CPS context may be random, unspecific, and uncertain. Classification of CPS applications also involves substantial process categorization and scheduling classes of the components. This will ultimately help to infer to a particular group of decisions. For example, in terms of behavioral dynamics, design efficient and optimal multi-line transport networks (application-specific CPS) of passengers, relevant camera sensor data could be crucial.

Because of an unknown sample Z to be classified, we iteratively allocate it to each possible class j where $j = 1, 2 \ldots N$, and predict class membership [5] on the basis of predictive expression. The measuring expression could be a multi-tiered architecture. The archive of multi-objective optimization of different use cases can be built with different CPS applications and intelligent learning optimization can help measure appropriate classification of CPS's constraints to locate the solutions likely to be matched.

In this paper, we investigate various levels of CPS formulation driven by machine learning and evolutionary algorithms with their strategic similarities. Finally, the specific CPS style cluster could be instantly made available for plug-in to rapid prototype applications [6]. It is also observed that the principle of machine learning, intelligent optimization, and evolutionary algorithm can also be deployed while designing different CPS to address different levels of imprecise conditions, classify order of component scheduling positioning, and optimize the system's cost and performance maximization. There is no fixed deterministic formal model in most of the cyber-physical system design problem. However, formal optimization models are taken carefully with regard to constraints.

We are also studying various recent benchmarked applications of CPS design schemes, using either optimization principle or smart optimization. Recently, conceptual uncertainty model (known as the U-model) is one of CPS [7] subjective representation of uncertainty. U-model is subjective and reliant on the representation of knowledge and the presentation of rules. To solve embedded uncertainty in different stages of CPS, it cannot be a core CPS design model. Inspired by such necessity, this paper investigates bi-focally the scope and deployment of computational intelligence in CPS design, thereby deriving some core formal and practical issues for such smart and emerging CPS. From the point of view of application and implementation in real time, the key features of CPS are as follows (this will be an initial diagnosis that can demonstrate the persistence of CPS uncertainty and dynamic conditions):

- Large, frequently spatially distributed physical systems with complex dynamics
- Distributed control, monitoring and rendering
- Partial subsystem autonomy
- Dynamic reconfiguration on different overall system time scales
- Continuous overall system evolution during its operation
- Possibility of emerging behavior.

In order to obtain the desired optimization most of these functions will fully require adequate level of adaptive and computer intelligence. From all these features, the pioneering effort to optimize CPS was initiated. In addition, the paper elaborates the wide variety of heuristic algorithms for difficult optimization problems, which occurs occasionally in the design of CPS for different stages applications. The stages are in the form of strengthening learning and global optimization. The scale and measurement of contributions for both evolutionary optimization and machine learning is also coined with regard to CPS, interaction, and ongoing constraints in CPS design negotiations. This article discusses the scope of the auto-arrangement principle relating to machine learning and optimization before providing details on the objective of CPS design optimization. This study will also discuss several investigations into cyber-physical systems such as:

- What is the scope of uncertainty in CPS with respect to application, infrastructure, and integration?
- How the different optimization methods and machine learning algorithms can be tuned for effective CPS design?
- Does the self-organization principle assist for managing uncertainty?

The remaining parts of the paper are organized as follows: Sect. 5.2 describes the list of machine learning applications and evolutionary optimization in CPS design. Section 5.3 discusses hierarchy of measures in CPS uncertainty regarding mathematical proposals (classification and future selection). The emerging varieties of intelligent CPS design are proposed in Sect. 5.4. Finally conclusion is discussed in Sect. 5.5.

5.2 Machine Learning and Evolutionary Computations in Cyber-Physical Systems Design

As highlighted in Fig. 5.1, The cyber-physical system consists of access control capable of learning from the environmental attributes. Therefore, the optimization scope for computational tasks and processes is substantial. There are certain specific intelligent and heuristic components to achieve these goals, such as learning and optimization, anti-colony optimization, typical machine learning approaches, and fuzzy logic techniques [8]. Also to compensate for diversified uncertainty and impression, they are quiet persistent across large CPS variants categories.

Cyber-physical systems (CPS) have been defined as a holistic unit of design irrespective of different applications. Two important features of an efficient CPS design are self-adaptive and self-organization. Self-organization is the particular ability for processes, components, and structures to adjust their behavior in order to respond to their changes [9]. The changes are interpreted with regard to imprecise uncertainty in CPS (due to intercourse and differential sensor propagation) resulting in hardware failure, temporary termination of connectivity, differences in event coordination for exceptional run time metrics between different components [10].

Recently a new strategy for meta adoption has been introduced in CPS [11]. Meta-adaptation strategies will improve the entire CPS residence with considerable robustness and performance, despite various CPS applications and component behaviors. In CPS and IOT studies important causes of uncertainty have been identified [12, 13].

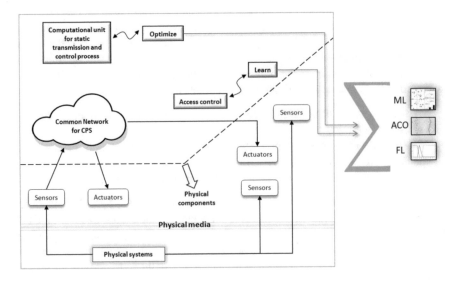

Fig. 5.1 Conceptual schematic of CPS and intelligent techniques

In CPS security, intelligence flavor is achieved including smart grid. The inclusion of game theory strengthens multiple strategies of attackers and defenders [14]. An appropriate mathematical analysis is conducted by formal security strategies and a number of components. In Aerospace Cyber-Physical Systems (ACPS), particle swarm optimization (PSO) [15] addressed a further direct involvement of computer intelligence into CPS design. This approach provides for an over actuated aircraft with a fault-tolerant approach in the event of actuator failures. CPS fault detection solutions are subsequently offered based on fuzzy logic as CPS is prone to faults. These systems also include process components, actuators, and sensors, and so the appearance of errors [16] is inevitable. Furthermore, cyber-physical systems may also be cyber attacked [17]. Attacks on cyber-physical systems therefore need to be detected. In most cyber-physical systems, differential and algebraic equations [18] may be modeled. These systems are known as systems with descriptors.

Takagi–Sugeno model is based upon fuzzy modeling. Gradually more complex CPS processes like complex production processes or (data-intensive) laser heat treatment are being developed with a machine learning approach [19]. Gaussian naive model is the direct application of this paradigm to control the auto-prediction of the process of laser thermal treatment. Another application with the cyber-physical system of the vehicle was planned in the time-varying prediction of the quality of cellular systems through the neighboring k-nearest algorithm or with the automated learning function of CPS [20].

A hybrid intelligent search process has been developed for the optimized K-means search space exploration clustering and the adapted binary search algorithm. In order to find a valid, optimum work piece position [8], the necessary number of collision assessments is therefore minimized. Experiments are carried out using an experimental design instead of a virtual machine tool.

The automated CPS [21] is also used for automatic classification of cutting process. Artificial neural networks were optimized using subsets of the recorded data to calculate the efficiency of the cutting process, and then applied to the independent validation data. Without using additional machine specific parameters, the established algorithm can identify the process efficiency. Also to improve system performance, we are examining an interesting combination of multi-target optimization using neural networks [22].

The principle of cost model optimization is seldom reflected in the machine learning approach. In particular, exponential function is used to implement unplanned cost modeling. Evolutionary computing has also recently encouraged distributed cyber-physical systems. In order to reach a nearly optimal solution for data centers, randomized search is used here and perspectives on energy and body sensor network cost management [23] have been discussed.

The optimization of the design demands a leading communication approach and infrastructure-based CPS for the exploration of modern technology. Timely networking, flow planning, and various stream optimization strategies also adopt the optimal smart approach [24]. One of the main objectives in the evolutionary computation (EC) is to find solutions to hard computational problems. The system to solve these problems are real decision support systems requiring continuous data flow, predictive components, and immediate recommendations to recover from sudden changes [25]. In general these types of problems deal with many variables, non-linear relations, varying constraints (e.g., real-world constraints often include "if-then" conditions), company rules, and many (usually conflicting) goals [10]. All of them are commissioned in a dynamic, noisy, and highly observed CPS environment. At the same time, it has not been studied the use of the domain knowledge to improve EC results, therefore a methodology is needed for injecting knowledge into evolutionary algorithms. It is also important to note that the CPS domain requires recommendations for optimization problems "the best" decision at the "moment." Co-evolutionary and genetic algorithm-based spatial and structural design has also evolved as a new CPS [26] vertical design. For an emerging need for trans-disciplinary applications [27, 28], cognitive aspects in the design of cyber-physical systems have been combined. In addition, an improved intelligent ant colony algorithm has been proposed to solve reliability optimization problem, concerned with manufacturing industry material cost distribution logistics [29]. Social network and sensing is a novel paradigm for CPS with large data applications. It is known as Scalable Uncertainty-Aware Truth Discovery (SUTD) in the social network [30]. Thus, real truth can be explored from CPS-based social networks.

5.2.1 Emerging Evolutionary and Machine Learning Algorithms

In this section, a set of formal description will be mentioned in tune with design of CPS. Broadly, we refer, principle of ant colony optimization, multi-objective evolutionary algorithm and the basic theme of machine learning algorithm for designing CPS.

5.2.2 Mathematical Analysis

The level of uncertainty in designing different cyber-physical systems includes a diversified domain such as Maximum Reward Collection Problem (MRCP), where agents collect time-dependent rewards in an uncertain environment with TARGETS (associated with data). There may be different topological features in the search space. The goal of such a problem could be to design CPS, which can process the agents as cooperative and collectively achieve the maximum possible reward with a given session time. Thereafter, uncertainty is also coined as a problem of data harvesting, where mobile agents collect data from stationary M targets in a 2D space. The aim here is to set up an effective CPS to minimize data collection and delivery delay movements over all targets within a fixed time interval [31]. This is analogous to the network of wireless sensors. In addition, the inherent uncertainty in CPS is presented conceptually in the form of content, time, occurrence, environmental context, and geographical locations [32, 33]. There are significant examples of uncertainty evolving from sensor networks and IoT devices in data transmission of big data [34]. In a rule-based framework [35], even data optimization at execution time was considered.

Hence, for intelligent optimization of CPS, there exists different mathematical propositions. Initially, the dynamic parameters are time, iteration counter, frequency of alterations of design with respect to different approaches of CPS applications.

Proposition 5.1 (On Dynamic Multi-Attribute Based CPS)

$$F(x, t) = (f_1(x, t), \ldots, f_{m(t)}(x, t))^T \tag{5.1}$$

subject to $x \in \Omega$, $t \in \Omega_t$, $m(t)$ *represents multi-attributes to be changed in time domain of t.*

Here t is a discrete time and Φ represents an iteration counter and frequency of changes in CPS components. The decision variable space Ω is expressed as:

$$\Omega = \prod_{i=1}^{s} [d_i] \subseteq \Re^s \tag{5.2}$$

Again s is the candidate solution of such CPS instances, d_i denotes all uncertain components that exist in the specific CPS design space. $\Re^s$ represents ordered s-tuple.

Proposition 5.2 (Under Uncertain Source, Sink and Transition Time Parameters for CPS Design) *The complete design system of application-specific CPS is represented as a source and sink with initial deterministic time domain depending on the different topology of CPS, the positioning of agents and their switching time. However, the switching time of the source and sink components cannot be measured if the design settings become uncertain: an optimization problem O is defined as: (t is the discrete time and S is one of the candidate solutions)*

$$\text{Max}_\Phi \sum_{k=1}^{T} R_\Phi(t_s, L_s, S) \tag{5.3}$$

where

$$R_\Phi(t_s, L_s, S) = \sum_{i=1}^{M} \sum_{j=1}^{N} \delta_i \psi_i(t_s) \natural\{d(L_{s_i}, S_i) \le S\} \tag{5.4}$$

In this phrase, Φ is the strategic policy, δ_i initial maximum changes in application, infrastructure and integration, M persists as bound value changes in the proposed CPS, ψ_i denotes a non-increasing bias function to compensate for sudden *changes in application, infrastructure and integration, t_s is the transitional time required for CPS components. The expression has a ♮. Indicator and control function that could map source and sink components of specific CPS design $\{L_{s_i}, S_i\}$. j was not instantiated in the expression, as j is embedded in the control space symbol.*

Proposition 5.3 *Typically, bi-level problems are non-convex, disconnected, and strongly NP-hard. We also investigate another version of the CPS design problem, where in the realm of constraints the design problem can be represented with a minimum requirement of reliability under uncertain decision variables. The problem with design can be either bi-level optimization [36] or tri-level optimization [37]. In power engineering in particular, CPS control faces severe uncertainties associated with load forecast error, unexpected generator, and transmission line failure. Here, optimization at three levels can support effective CPS design. However, bi-level optimization in conventional CPS design can express a generic balanced and reliable CPS while meeting uncertain constraints at both the upper and lower levels.*

For upper and lower level constraints, we define a high level CPS uncertainty problem, similar to:

Maximize $f(x, y) = x_2$
subject to $y = \text{argmax}\{f(x, y) = y_2\}$
and $\text{lc}_j(x, y) \geq 0$, where $j = 1, 2, \ldots J$
$U_j(x, y) \geq 0$
$-4 \leq x_1 \leq 10$, $-100 \leq x_2 \leq 200$, (these values are arbitrary)
Structurally, $U_J(x, y) = (\frac{y_1}{16} + \frac{18}{7})(x1 - 2)^2 - x_2$

Here, lc_j are lower level constraint function, whereas upper level constraints are expressed as U_j, x and y is the solution coefficient of CPS design. As mentioned uncertainties may arise upper level variable alone, or either in lower level or could be in both. Therefore, the normal distribution and variance of both these levels can be measured.

Proposition 5.3a *On the contrary to the bi-level optimization, tri-level introduces another additional level as middle level constraints problem and thus structurally it follows: $\sum\sum\sum$ (function denoting the parameter of design as cost, flow like of transmission).*

Proposition 5.4 *Resource scheduling for placing optimal network topology can address another vertical of CPS design optimization issue. Mathematically, it can be elaborated as:*

Given $G(E, V)$ as network topology, set of data flow from one component to other is D and the scheduled time to active the event as T, which will output as:

- *The optimal route for each data item flow as $d_i \in D \bigcup D^T$*
- *The optimal assignment of frame from source CPS to sink CPS*
- *Identifying the end value as NP-hard scheduling [38] with defined link, frame and flow transmission constraints.*

Thus, the strategy of CPS component scheduling demands an intelligent optimization for optimal distribution of resources. This signifies that evolutionary composition of different sub-components for CPS will be necessary.

Suppose, W_i denotes the ith modular element of CPS composition (e.g., sensors, actuators, controllers, timers, meta-accessories, they have marginal differences in terms of specification). Therefore, for the meta-components to be associated for effective scheduling, corresponds to P_i at the iteration of design I_D. It can be tested. if this meta-component is active in scheduling, then mean and standard deviation of individuals values in dimension H ($H \in I_D$) can be formulated:

$$m_{H,I_D} = \frac{1}{p} \sum_{t=1}^{p} q_{t,H,I_D} \tag{5.5}$$

$$\text{stddev}_{H,I_D} = \sqrt{\frac{1}{p} \sum_{t=1}^{p} (q_{t,H,I_D} - m_{H,I_D})^2} \tag{5.6}$$

where, p is the size of meta-components of CPS and q_{t,H,I_D} is the size of individual components. We refer to the standard theory [39] as if mean and standard deviation of meta-components of CPS is unaltered for several iterations of design, then the CPS design become stable for that specific application with optimal scheduling effort for placing the components.

5.2.3 CPS Design: Applications and Challenges

Modern CPSs vary in features, applications and operating levels. Several researchers have proposed different methods and design architecture to address these diverse requirements. However, several modeling techniques, semantics, programming tools have been proposed by researchers for effective CPS design. Recent CPS applications primarily cover medical information technology and safety paradigms [40, 41]. In CPS design, there are typically different levels of thematic design artifacts to ensure minimal uncertainty. Some of them are: model-driven [42], meta-architecture and programming [43], semantics [44], co-design [45]. Given CPS, broad design perspectives, classification on the basis of certain software engineering artifacts becomes mandatory:

Design Issues
- Architecture and Modeling
- Simulator/Simulation
- Tools/Programming Framework
- Verification

Aspects

- Security
- Resiliency
- Reliability
- QoS
- Real-time requirements

Applications

Extended CPS applications should also be referred to as: vehicle and transportation systems, medical and healthcare systems, smart homes and buildings, social and gaming networks, scheduling power/thermal management, cloud computing and data center power grids or power systems, networking systems, monitoring, industrial process control, aerospace and air traffic management, search engineering. A full survey can be found in a study [46]. The application domains indicate the relevance of reliable design with minimized uncertainty and meeting the maximum number of constraints depending on the design and aspect orientations.

5.3 CPS Design with Intelligence and Measures of Efficiency: Proposed Mathematical Proof and Algorithms

In this section, we propose different measures in tune with intelligent CPS design and present the following notations with corresponding semantics in Table 5.1.

The different perspectives of optimization strategy for CPS yield a scope of measures with either existing algorithms of evolutionary computing or from machine learning domain. All the prepositions mentioned in Sect. 5.2.2 are primarily optimization prone problems. The control variables are mostly the infrastructural vectors, as v, $v(t) = [v_1(t) \ldots v_N(t)]$ and $\theta(t) = [\theta_1 \ldots \theta_N(t)]$.

As the design process is random and expected outcome of design is optimized (either cost, resources, or scheduling time) under uncertainty $U \in [0, 1]$, hence

Table 5.1 List of notations

Notations	Semantics
v	Infrastructural vectors
θ	Control variables as vectors
U	Level of uncertainty
W	Expected result of design
z	Minimum time consumed for putting the components in CPS design space
dz_0^1	Population of components w.r.t minimum time required
$Cd_{z_1^1}$	Cumulative distribution function

$$\min_{v_{(t)}, \theta(t)} W(T) = \frac{1}{T} \int_0^T \left(\frac{U}{M_X} W_1(t) - \frac{(1-U)}{M_Y} W_2(t) \right.$$
$$\left. + \frac{1}{M_L} W_3(t) + \frac{1}{M_R} W_4(t) \right) dt + \frac{1}{M_P} W_f(t)$$

$$(5.7)$$

Here, U is the mapping of uncertainty in between 0 and 1, M_X, M_Y, M_L, M_R and M_P are the normalization biases, which will compensate the level of uncertain features in design of CPS. This will assist to maintain the same range of uncertainty as $U \in [0, 1]$. The uncertainty signifies that either cost, availability of resources and scheduling time could vary with respect to different applications and time domain.

In the second phase, the design strategy, should satisfy maximum numbers of design constraints, for different application based CPS, hence the expected measure can be given as:

$$m_1 = \int_0^\infty dz_0^1(z^1) C d_{z_1^1}(z^1) dz^1 \ldots \int_0^\infty dz_0^N(z^N) C d_{z_1^N}(z^N) dz^N \qquad (5.8)$$

Thus, on the basis of stochastic measure of design efficiency we can also write an asymptotic expression as:

$$Sm = \prod_{j=1}^N \int_0^\infty d_{z_0^1}(z^j) C d_{z_1^j}(z^j) dz^j - \sum_{j=1}^N (Cost^j \cdot Location^j) \qquad (5.9)$$

In Eq. (5.9) cumulative distribution function and population of components with minimum time for any CPS design can be directly related with cost and location of placing the components (as a product value), and thus efficacy of design in terms of success or failure could be asymptotic in nature.

For measuring performance, the objective is to minimize the data content and storage, whereas to maximize the resource deployment.

We redefine expected efficiency of the design (say λ) from source point to sink point, with a weight R, that represents comparative factor of design objective i (e.g., data content, cost and utilization aspect), where $X_i(t)$ and $Y_i(t)$ is source and sink parameters, respectively, for the components (to be placed):

$$\lambda_{isr} = \left[\sum_{i=1}^{Sink} R_i X_i(t) \right] \qquad (5.10)$$

Similarly, importance is provided towards sink also as:

$$\lambda_{isink} = \left[\sum_{i=1}^{Sink} R_i Y_i(t) \right] \qquad (5.11)$$

Finally, the problem appears to be as convex optimization problem of design, in turn as minimization problem (to be subtracted from λ source to λ sink of design):

$$\min_{v,\theta} \lambda = \frac{1}{T} \int_0^T U(\lambda_{is} - (1 - U)\lambda)dt \tag{5.12}$$

The proposed measure given in Eq. (5.12) is the measure of the data content from source to sink for any specific CPS. However, with the inclusion of collective intelligence (e.g., ant colony, termite, bees, fish and swarms), we can also indicate the measure about the movement and dynamics of these natural agents in CPS design space. It can only be possible if and only if, these agents can minimize their traversal time and inactive time while searching optimal points in design space with respect to given parameters. Referring the search of optimal points, where these agents could be inactive, can depend on Euclidean distance as $dist_{sr}$ (for source point) and $dist_{sink}$ (for sink point). To measure the inactive time $I_{inactive}$ we need to justify that for any of the social insects algorithm, both the distance from either source or from sink in design space should be strongly positive, Hence, we propose:

$$I_{inactive} = \log\left(1 + dist_{sink} \prod_{i=0}^{sink} dist_{sr}\right) \tag{5.13}$$

It is evident, that the value of inactive time of the agents. despite of different constraints, can only be 0, if product item inside the bracket evaluates to zero. Here we introduce a logarithmic function to combat the differences in efficiency value of λ for both sink and source points (depending on how the target CPS constraints are met either). The design exploration strategy is described in the next sections for multiple goals that change over time and the proposed design strategy is validated in the context of the efficiency of specific CPS application. A sufficient background of such changing goals is provided to assist readers.

5.3.1 Multi-Objective Instances and Associated Parameters

There is a large class of optimization problem that have multiple goals that change over time. This type of problem is usually referred to as dynamic multi-objective optimization problems (DMOPs). In the FDA test suite [47, 48], the Pareto Front or the PS change over time, while the number of decision variables, the number of goals and the search space boundaries remain fixed over the course of the run. The following instances are given for benchmark test functions as per search space from source components to sink components at physical level: (it will help formulate metrics for performance assessment): here, four dynamic multi-objective test problems are considered, e.g., Follow-up FDA, DMOP, F5, and F6. Please find

Table 5.2 Objective and design space

Category	Search space	Objective	Feature
FDA	$[0, 1] \times [-1, 1]^{n-1}$	$f_1(x) = x_1, f_2(x) = g \cdot h$ $g(x) = 1 + \sum_{i=2}^{n}(x_i - G(g(t)))^2,$ $h(x) - 1 - \sqrt{f_1/g}$	Fixed POF, POS varies
DMOP	$[0, 1] \times [-1, 1]^{n-1}$	$f_1(x) = x_1, f_2(x) = g \cdot h$ $g(x) = 1 + \sum_{i=2}^{n}(x_i - G(t))^2$ $G(t) = \sin(0.5\pi t)$	POF and POS varies
F5	$[0, 5]^n$	$f_1(x) = [x_1 - a]^{H(t)} + \sum_{i \in t} y_i^2$	POF varies
F6	$[0, 5]^n$	$f_1(x) = [x_1 - a]^{H(t)} + \sum_{i \in t} y_i^2$	POF varies

details in [47]. Proposed use cases are tested with different performance metrics in reference to these functions (Table 5.2).

5.3.2 Cyber-Physical System with Changing Number of Objectives: Use Cases-A

An approach to estimates of design, parameters, and the optimization principle could be the exploration of the design problem domain. However, it is also possible to define an appropriate design aim for the cyber-physical system in terms of the role of the control algorithm and bottom-up library and components evolution (Fig. 5.3). A design approach followed by an assessment of the efficiency of the design is the main aim of this article. There are also some specific applications in this context related to cyber-physical system transport and instances of security. The choice of such scenarios is essentially to examine the functionality and optimization effort in order to change the number of objectives within a given period of time.

5.3.2.1 Transportation

We are investigating a few CPS-related use cases with a number of goals. Dynamic properties and uncertainties of CPS are reflected in this context. For instance, the basic use case can be coined for a multimodal system of transport. Multimodal transportation is defined as at least two distinct modes of transportation; For the entire carriage, the carrier is (legally) liable, though it is carried out by several various modes. As the service model of traveling to the favorite area from different functional points, the primary transport objectives of CPS [10] shall be explained. This will be followed by the subsequent demand by the users for specific mode of transport and relevant assignments.

The physical parameters, however, are also becoming relevant such as vehicle density, vehicle flow rate, and speed. The parameters are used to know about CPS design dynamic attributes. Using this approach, density k traffic flow models, by the number of vehicles n, may at the time be formulated by $t0$, which may be occupying a given length, x, or a street interval in general, by Δx, at a particular moment in time of a road, as: $k = n/\delta x$. Of course it may also be possible for the unlimited driver speed during the traffic conditions, such as for the heavy traffic, even in the light of the attributes of the structural parameters of road friction, so the CPS control output can differ.

The driving situation is very different when road conditions are congested, since the speed for falling densities is no longer constant. Drivers can no longer choose free-flow speeds, since a safe distance to the front cars must be maintained. In this case, a congested velocity, vC linear function, can assume the safe distance (d). As the total vehicle density k is the opposite of the average vehicle space and this space is a ratio-weighted sum of type densities [4], the total density kT can be determined as a function of speed. These conditions, however, lead to different CPS control conditions. They are referred to as optimum solutions from Pareto. A solution (i.e., a specific set of decision variables values) for a set of goals is Pareto optimal, if there is no other solution for all the goals with one iteration in the decision area. The solution therefore becomes incoherent. This is because they are physical dynamic parameters of transportation CPS, where function $F(x, t) = (f_1(x, t) \ldots, f_{m(t)}(x, t))^T$ subject to $x \in \Omega$, $t \in \Omega_t$ here t is defined discrete time of the scheduling process of transportation based CPS. Therefore, the above parameters typically for the total density are modeled as decision space Ω, hence density as function of speed of particular vehicle:

$$k_T = \frac{1}{[\Omega]_{a_i,b_i} \sum p_T d_T} \tag{5.14}$$

$$= \frac{1}{\prod_{i=1}^{n}[a_i, b_i] \sum p_T d_T} \tag{5.15}$$

Including Ω indicating the time for the evaluation of traffic density under changing goals as a decision variable space: either the solution in nature is optimal or not optimal [10]. Coordination and control differs primarily (a) by increasing the target number and (b) by reducing the target number. First, if we increase the number of goals in this case, an optimization problem is requested. It is referred to by the status and speed of the CPS platform and all moving subsystems (for example, a payload). For each vehicle j in the team, a family of state-space vectors can describe the condition of the cooperative team. The Gx graph with edges Exi, j between pairs of cars I j) that communicate directly can be described as cooperative network connectivity planning and control [49]. Motion equations based on physics give dynamics to the vehicle, while connectivity to the network is usually a function of physical separation distance. Vehicles can carry sensors and/or install stationary sensors; vehicles and stationary sensor nodes [50, 51]. Therefore, when the problem of changing number of goals arises, it can be numerically described as:

- To minimize two objectives, each having one decision variable as shown in Eq. (5.15).

$$\min F(x) = [\text{objective}_1(x); \text{objective}_2(x)]x$$

where, $\text{objective}_1(x) = (x + 2)^2 - 10$, and $\text{objective}_2(x) = (x - 2)^2 + 20$.

For numerical solution, it is noted that at $x = -2$ and $x = +2$, respectively, the two objectives have their values at minimum. However, $x = -2$, $x = 2$, and any solution within the range of $-2 \leq x \leq 2$ is equally optimal in a multi-objective problem. There is no single solution to this multi-objective problem. The goal of the multi-objective algorithm is to find a set of solutions (ideally with a homogeneous distribution) within that range. The set of solutions, also known as the Pareto front. On the front of the Pareto, all solutions are optimal. In the case of transport-related CPS, there is also a multi-objective problem, as the road condition and congestion sometimes differ. Therefore, different design principles are included with changing goals in the actual design of CPS and the objective is to select optimum design pattern (ODP) focused on desirable CPS centric transport. We observe the variation in building effective CPS design while representing the design exploration problem as [52] convex optimization problem. The formal expression of such problem is given as follows: (referring Fig. 5.2), we consider the problem of minimizing a convex function f over a convex set C where C is given as the intersection of finitely many simple closed convex sets $C_1, \ldots, C_m (m \geq 2)$. Specifically, we focus on optimization problems of the following form: $mf* = \min f(x) + 1C_i(x)$, $(1)x \in X_i = 1$ where $1C_i$ is the indicator function for set C_i and $X(C \subset X)$ represents the domain of f. The convex sets are comprised of Euclidean projections concerning an oracle of that set.

In the following test scenario, the experimental section [10] is dealt with based on the congested conditions and driving contexts which also affects the working of the solution of CPS. Thus, a certain test problem is created by the above modification of CPS functions. The variable subset must converge on the new value

Fig. 5.2 Changing objectives

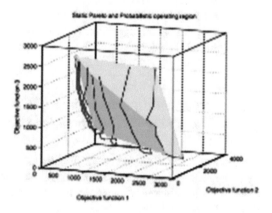

for each modification in order to track the dynamic and unknown solution. The above solution is built to correspond to a dominant solution for the true front of Pareto at any time interval value. Finally, the design solution becomes a further test scenario for transportation-driven CPS when it changes for an optimum solution value. Infringement of traffic in congested roads is also relevant. In the next section, the context is explained.

5.3.3 Use Cases: Case of Convex Optimization in Security-B

There are various instances of CPS representing it as a problem of convex optimization; it implies that it may follow a non-smooth objective function. The standard problem of convex optimization, however, can be represented as:

$$f_*^\lambda = inf_{(x \in \psi)} \equiv [f(x) + \lambda h_p(x)] \lambda > 0 \qquad (5.16)$$

where, h_p denotes the penalty function. It also returns as sub-gradient function for this optimization function. When all the parameters and data represents the potential damage and uncertainty as a priori, optimal attack resource allocation can also be expressed the mapping of attack resources committed to attack service s at time t.

A total of p sensors monitor the state of the physical system ($y_{(t)} \in R_p$) : $y_{(t)} = C_{x(t)} + \upsilon(t) + a(t)$. Few sensors are attacked: noise attack vector $a_i(t) = 0$ sensor i is attacked at time $t \in N$; If sensor i is attacked, $a_i(t)$ can be arbitrary (no boundedness assumption, no stochastic model, etc.). Set of attacked sensors is (unknown) and has cardinality s. The value of s is also (unknown), although we assume the knowledge of an upper bound s. Therefore the objective is to estimate the state of the physical system, which is given as: $x(t) \in$ Rn. While designing CPS with regard to this convex optimization in terms of security perspective, the objective function f should, if uncertain, minimize the value of f as a smooth saddle point problem (by definition, saddle a point on the graph surface of a function where the slopes (derivatives) of orthogonal function components that define the surface become zero). Impressive convergence can be achieved even through the static objective function [53, 54] for first order optimization algorithm. For design uncertainty in CPS, this will be more linear solution (Fig. 5.3).

5.3.4 Proposed Algorithm for Efficiency Measurement: Measuring Performance for Use Cases and Metrics

We propose a set of strategies for implementing the required paths in CPS design by adding a set of interconnection constraints to the original optimization problem. Constraints are obviously uncertain, but the proposed algorithm is not the algorithm design. Rather, it is the measure of efficiency when the uncertain constraints are

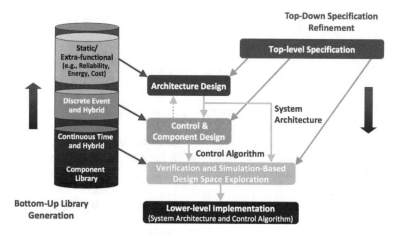

Fig. 5.3 Design levels of cyber-physical system

addressed by certain optimization principle, fuzzy logic or learning heuristics. CPS design proposals and strategies are subjected to the measurement of the convergence and distribution of the different parameters. In this regard, the method of inverted generational distance (IGD) [55] was proposed. Considering the scope of evolutionary computational intelligence, the computational complexity of different strategies is analyzed with regard to changing goals and problems with design optimization. The function's flow is as follows:

Initialize a Population of Evolutionary Agents
With a mix of multi-objective and dynamic problem scenario and referring to the inclusion of Pareto-optimal solutions for all CPS parameters, we envisage several possible uncertain variations of the pre-optimal front and the optimal setting of Pareto: for example, the POS changes over time, while the POF remains stationary, both the POF and the POS change over time or both the POF and the POS stationary. Suppose the design parameter changes in CPS, which can be expressed as a generic objective changeable function:

$$\min F(x, t) = (f_1(x, t) \dots f_M(x, t))^T \tag{5.17}$$

Subject to $h_i(x, t) = 0$ or

$$g_i(x, t) \geq 0 \text{ or } x \in \Omega_x \text{ where } t \in \Omega_t$$

where $\Omega_x \subseteq \mathrm{Rn}$ is the decision space and t is the discrete time instance defined as $t = \frac{1}{n_t} \frac{\tau}{\tau_t}$, where n_t represents the severity of change, τ is the iteration counter and τ_t represents the frequency of change and $\Omega_t \subseteq \mathrm{R}$ is the time space. $F(x, t)$: $\Omega_x \times \Omega_t \rightarrow R^M$ is the objective function vector that evaluates the solution x at time t. Therefore Pareto-optimal set is susceptible to variations in parameters. Inverted generational distance (IGD) metric is a set of uniformly distributed Pareto-optimal points of the PF in t time and let P_t be an approximation set of PF in t time. IGD metric is defined as follows:

$$\text{IGD}(\text{PF}_t, P_t) = \frac{\sum_{v \in \text{PF}_t} d(v, Pt)}{\text{PF}_t}$$

IGD is presented as performance efficiency indicator for CPS with changing numbers of objectives. The lower IGD values, the better the convergence and the distribution of the obtained solution. The statistical result including mean and standard deviation which can be shown in all instances of design for CPS. Total 150 changes are mapped according to:

$$1 \leq t \leq 40, \ 41 \leq t \leq 80, \ 81 \leq t \leq 150$$

The mean and STD values for $1 \leq t \leq 40$, $41 \leq t \leq 80$ and $81 \leq t \leq 150$.

Additionally, the IGD metric perspective could include different test function parameter settings for specific CPS design patterns. Like any CPS driven intelligent algorithm, environmental change frequency will follow. This is set to τT and varies from 25, 50, 100, 150 and 300 and so FPS parameters are also used as FPS' core working model. The cluster number is 5, the mean point series length is $M = 21$, and the prediction model probability is 0.9. However, if there may be more fluctuations in the frequency of changes in the design strategy, which are reflected in the IGD metric. Precision and performance are becoming more stable. Here, since the environmental changes of cyber-physical can be predictable, similar strategies are also adapted to predict POF and POS boundaries: mainly forward-looking prediction strategies (FPS) [56], population prediction strategies (PPS) [57], predictive gradient strategies (PGS) [58] and proposed dynamic evolution strategies (DES). It is evident that DES has better convergence over other three strategies. The important tracking point of all these strategies are to identify the boundary point of Pareto optimal front with respect to changing objectives and parameters and also to estimate the next immediate environmental change in CPS (as indicated by τ_t).

However, as scheduling and constraints, we focus on the decision variables for individual proposals and the environment has true multi-objective yet changing objective paradigms [59]. The values in each small cell denote average value and standard deviation. The inverted generational distance measurement is evaluated on the final distribution of the four proposals mentioned in Sect. 5.2.2. The plot shows that the time starts from $t = 0$ until it reaches $t = 90$. It is the fixed PS test problem in Fig. 5.5, six instances selected to observe results. CKPS has improved convergence and distribution in the early stages, which shows that CKPS can respond better to dynamic changes and inherent uncertainties.

This constraint learning function is therefore, instrumental and it follows the set of propositions mentioned in Sect. 5.2.2.

The algorithm is implemented by using standard C compiler and a synthetic data set is deployed.

Algorithm 3: Level of constraints in design

---

1 **Input:** Current uncertain constraints Cons, efficiency e, reliability requirement r∗, adjacency matrix m∗
2 **Output:** Final expected output of design with constraints
3 UncertainConstraints k ← PredictedDesignPath (e, r∗, m∗)
4 NewConstraints←[ ] S ←SinkDesignPoint(e∗)
5 (T1, . . . , Tn) ← AccquireTypes(e∗)
6   **for** all SourcePointsDesign S **do**
7   **if** k≥1 **then**
8   **for** all i ∈ (Tn-1, Tn-2,. . .,T1) **do**
9 NewConstraints ← GenerateDesign(v, i, k, NewCons, e∗)
10   **else**
11 i ← FindReducedDesignPath(v, e∗)
12 NewConstraints ←AddPath(v, i, 1, NewCons, e∗)
13   **if** derivedConstraints = [ ] **then return**
14 Constraints← Constraints ⋃ DerivedConstraints
15   **return** Constraints

5.3.5 Transport Centric Cyber-Physical System: Parameters and Data Set

Instances and traffic breach attributes include transportation-specific data characteristics:

- data and transportation schedule times
- Physical traffic breach locations
- traffic breaching agency
- Vehicle density for that particular location

Data Source[1]
This refers to the corresponding data set and generates the test function. The final distribution of the population from the data set and six subsequent instances with inverted generational distance as shown in Fig. 5.4. From the literature, with feed forward multi-objective prediction strategies, we referred to random initial strategy. All instances of time are shown. The 100 environmental changes are selected, these changes are subject to different transport goals and violations of traffic. Section 5.3.4 of various instances discusses different strategies, empirical convergence and their plots in detail for initialization and progressive values.

---

[1]https://data.montgomerycountymd.gov/api/views/4mse-ku6q/rows.csv?accessType=DOWNLOAD,

https://data.montgomerycountymd.gov/Public-Safety/Traffic-Violations/4mse-ku6q.

Algorithm 4: High level description for optimal design solution

---

1 Input: Objective functions of all CPS design for all design problem
2 T (number of ODP targeted)
3 Output: Final ODP
4 Generate $|R|(|R| \geq T)$
5 OPD=ϕ, ODP = Expected Optimal Solution
6   **while** $|ODP| < T$ **do**
7   **while** $|ODP| > 0$ **do**
8   Compute Expected utility function of each solution in ODP using R
9   Associate solutions to their closest reference points
10   Identify solution with highest EULF
11   **Select** a set of design solution (Ds, Ds$\subseteq ODP$) with highest
      EMUF
12   **if** $(|DS| < T) \vee (|DS| = |ODP|)$then
13   **break**
14   **else**
15   OPD=DS
16   **endif**
17   OPD=OPD$\cup OPD_{re}$
18   OPD$_{re}$=P/OPD

In dynamic environments it is necessary to discuss the performance of different strategies in different periods. The strategies are uncertain, initializing the strategies. Six different instances are evaluated on IGD metrics in the case of experimental evaluation (Fig. 5.4). Depending on the strategies and respective mean and standard deviation value of design exploration strategies, the density of scatter plots differs. Table 5.3 presents the elaborate comparison of performance and empirical validation. Assuming the CPS transport, another entropy relationship as shown in Fig. 5.5 With the density of the vehicle and the rate of transfer. The data set is synthesized from the earlier link, but the distribution of points is not sparse as the target (i.e., transfer rate) is also static with small changes and this single uniform objective values should be converted into the first level of design in CPS.

5.3.6 Discussion and Analysis

Referring to Figs. 5.6 and 5.7, analysis is presented on the level of uncertain constraints and minimization efficiency. Changing design goals in CPS has also been observed to be well correlated with the related or unlinked constraints. Figure 5.6 shows color indications in the CPS design paradigm for sparse and temporary learning of dynamic attributes. The learning epoch continues with error segments from 0.0 to 1.0 up to 20 iterations. This yields that in case of unlinked constraints, uncertain levels of intelligence constraints produce good convergence. It refers to time for cost and scheduling, use of resources and movement of data. The scenarios, as shown in Fig. 5.7, differ in terms of time and accuracy in design. Intelligent design

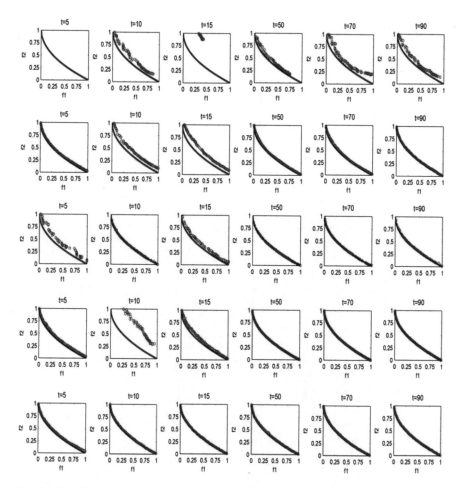

Fig. 5.4   Time driven optimal design solution

Table 5.3   Comparison of design strategies

Strategy	$0 \leq t \leq 150$	τ_t
FPS	0.0292 (0.00060)	25
PGS	0.0143 (0.0017)	50
PPS	0.0686 (0.0295)	100
DES	0.0069 (0.00004)	150–200

optimization denoted in black color, which is a linear relationship. In addition to task versus accuracy and task versus time, the design exploration strategy's efficiency is considered.

We introduce some generic mathematical proposals for efficient CPS design in the previous section followed by their efficiency measures for smart design perspectives. Under uncertainty, we envisage CPS design problem as multiple

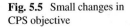

Fig. 5.5 Small changes in CPS objective

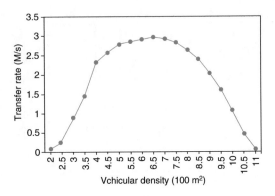

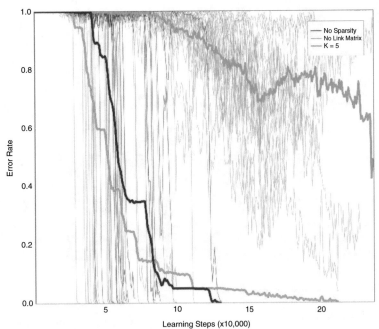

Fig. 5.6 Learning versus error rate in design

objective optimization problem (which means different goals in CPS design such as infrastructural variable, scheduling time from source to sink problem, design cost and ultimately effective use of resources). Space exploration strategies are incorporated to measure such important changes. Contribution is made through various proximity measurements where measurement can identify relative affinity from the ideal optimum theoretical point without knowing the exact location of the optimum point [60, 61]. The measure is referred to as the proximity measure Karush–Kuhn–Tucker (KKTPM). Such proximity measures can also under uncertainty locate suitable yet optimal selection of design parameters for smart

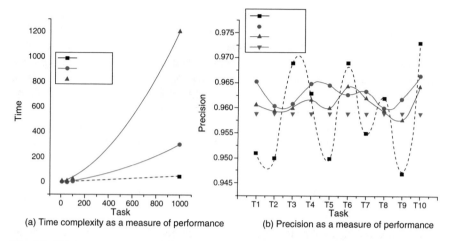

(a) Time complexity as a measure of performance   (b) Precision as a measure of performance

Fig. 5.7 Efficiency measures of performance for CPS with intelligence. (**a**) Time complexity as a measure of performance. (**b**) Precision as a measure of performance. Best optimal: Blue; Medium: Red; Linear: Black; Random points: Pink

CPS. Even in recent times For multi-objective design problems (in the case of CPS as well) [60], more computationally faster measurement is suggested. We found that these measures could be effective as many of the constraints for uncertain and dynamic combinations of parameters can be either positive or negative. To satisfy vector v and θ. Negative or positive values, we can set them to zero and can again differentiate those objective functions for evaluation.

5.4 Resilience and Self-organizing Pattern in CPS Design: Role of Intelligence

We discuss the principle of autonomy and resilience, as mentioned in the introduction, which can significantly reduce the degree of uncertainty in CPS. The classification framework for adaptive system uncertainty, which is quite similar to CPS design issues: location, level, time and sources [62]. In general the repetitive design patterns of the linked parameters can even predict the closest design attribute (which still has to be functions in the actual design area of the CPS). We are investigating more elements in CPS-related design problems. Sparsity [10] is this particular phenomenon. Sparse network is a network with fewer connections to the same network than the maximum number of connections. Evolutionary and other contemporary algorithms control task allocation, scheduling and dynamic reconfiguration [63–65], which is represented as self-organization and control. When exceptions occur, uncertainty can be removed by the self-organized and re-configurable components. This is also one resilience mechanism mode and thus the parameters such as cost, time execution, unreliability if any, can even be predicted throughout the development of CPS and service time [66]. Significantly, the data points are involved in all CPS parameters when the links are identified. This is possible if exact Pareto-optimal solution is known due to any multi-objective optimization convergence of solution. In such issues, the proximity measure of Karush–Kuhn–Tucker persists [67]. Due to CPS design uncertainty, the optimal objective value ϵk with all combinations of CPS design parameters could be direct optimal or adjusted value. According to the KKTPM principle, an expected value of the optimum objective is followed as an average (referring randomly to the objective value changes):

$$\epsilon^{\text{antiptaed}} = \frac{1}{3}(\epsilon^{Dir} + \epsilon^P + \epsilon^{\text{adjoining}}) \tag{5.18}$$

For a n variable, M-objective optimization problem with J inequality constraints:

$$\text{Minimize}_{(x)}\{f_1(x), f_2(x), \ldots f_M(x)\} \tag{5.19}$$

Subject to $g_j(x) \leq 0$, $j = 1, 2, \ldots J$ and hence final condition for KKTPM becomes:

$$\sum_{m=1}^{M} \text{multip(first function)}(x^k) \tag{5.20}$$

$$+ \sum_{j=1}^{J} \text{multip(2nd function)}(x^k) = 0 \tag{5.21}$$

where, $g_j(x^k) \leq 0$.

All multipliers are non-negative and can provide optimal KKT solution for x^k. Considering that this KKTM convergence contributes to the learning rate according to the value of sparsity. Due to random changes in objective functions, the learning rate fluctuates. The number of services commonly invoked by users is usually quite small due to the sparsity and presence of the large number of services and limited invocation experiences. Service Quality (QoS) can be affected easily [68]. While plotting the condition and rate of learning associated with KKT, Fig. 5.6 shows learning steps as opposed to error rates. The plot, however, also shows no sparsity, no connected constraints and finally the value of $k = 0$.

5.5 Conclusion

This paper investigates certain specific causes of cyber-physical system design uncertainty. Due to many environmental changes and CPS design strategies, the uncertainty is present to ensure applicability in the respective applications. While referring to the design aspects of precise CPS, it was also observed that there is a significant trend of multi-objective optimization in terms of resource utilization, scheduling, and even learning the dynamic attributes of design. The paper calls for certain popular computationally intelligent techniques to support the design under uncertainty with intelligent optimization and the principle of machine learning. However, no specific application has been presented, rather the proposal focuses on the analytical aspects of the design paradigm of CPS. Since the chapter is the amalgamation of the existing as well novel algorithm and design aspects, therefore all the necessary pros and cons of optimization algorithms are navigated.

We present certain mathematical proposals that coincide with the above-mentioned properties and then present appropriate mathematical processes for measuring the efficiency of such proposals referring broadly to different application-centered CPSs. The different strategies of design where multi-objectives and may change randomly. Pareto-optimization and design exploration methods are also compared in this study. The proposed multi-and changing objectives have been validated with CPS design for transportation and security, and it can be found from the perspective of convergence of objective function, proposed dynamic evolutionary strategies (DES) can foster the generic design principle for CPS. However, cyber-physical system categories of applications are also becoming influential in tracking the frequency of changing goals. Five basic modules, i.e., sensor module, actuator module, network module and interference module, are available in more comprehensive CPS design. Except for the sensor and controller module (both clock driven modules), the actuator module is event driven from these functional modules. This means that the actuator will only trigger in CPS control when the controller receives the control signal.

Uncertainty may therefore be inherent in the controller itself. Consequently, CPS reliability and efficiency is most frequently regulated by the precise objective function that will meet CPS operational requirements. First order transfer function with specified gain and time constant in the generic concepts of CPS design. Therefore, evolutionary strategies of prediction Can improve CPS efficiency. Other network module parameters such as packet drop-out and delay in transmission can also degrade CPS performance [69]. Thus the mechanism of resilience can be an appropriate approach. In order to analyze and improve such efficiency measures, the levels of proximity measures are also emphasized. There are a variety of research directions with learning and intelligence in CPS design. Recent advances in learning criteria and data volume [70], deep learning is capable of learning hierarchical attributes from large data sets and producing effective improvement in learning performance. Product Restricted Boltzmann Machine (RBM), which is the basis for deep learning, can train with uncertain dynamic design attributes and can predict suitable design principles. Learning in high-dimensional data, with an optimal selection framework for features, can also be developed to select the most discriminatory classification features. Many CPS require distributed high-performance computing to support spatial and temporal decision-making. In such cases, the Hidden Markov model can be used in a probabilistic manner to model each event's spatial-temporal pattern. Representing a dynamic process characterized by a stochastic Markov chain with non-observed or hidden states is a powerful probabilistic tool. From Markov chain theory, the hidden variable moves with a certain probability at each discrete interval of time from one state to another (or to the same state). This will also eradicate significant uncertainty in the design of CPS.

References

1. Ning, H.S., Liu, H., Ma, J.H., Yang, L.T., Huang, R.H.: Cyber-physical-social-thinking hyperspace based science and technology. Futur. Gener. Comput. Syst. **56**, 504–522 (2016)
2. Botta, A., De Donato, W., Persico, V., Pescap, A.: Integration of cloud computing and Internet of things: a survey. Futur. Gener. Comput. Syst. **56**, 684–700 (2016)
3. Jiang, L., Da Xu, L., Cai, H., Jiang, Z., Bu, F., Xu, B.: An IoT-oriented data storage framework in cloud computing platform. IEEE Trans. Ind. Inf. **10**(2), 1443–1451 (2014)
4. Tang, B., He, H., Ding, Q., Kay, S.: A parametric classification rule based on the exponentially embedded family. IEEE Trans. Neural Netw. Learn. Syst. **26**(2), 367–377 (2015)
5. Hastie, T., Tibshirani, R., Friedman, J.: The Elements of Statistical Learning: Data Mining, Inference, and Prediction. Springer, New York (2013)
6. Valero-Mas, J.J., Calvo-Zaragoza, J., Rico-Juan, J.R.: On the suitability of prototype selection methods for kNN classification with distributed data. Neurocomputing **203**, 150–160 (2016)
7. Simula Research Laboratory: Understanding uncertainty in cyber-physical systems: a conceptual model. Technical Report 2015-3 Feb (2016)
8. Wang, G.G., Cai, X., Cui, Z., Min, G., Chen, J.: High performance computing for cyber physical social systems by using evolutionary multi-objective optimization algorithm. IEEE Trans. Emerg. Top. Comput. (2017) in press. https://doi.org/10.1109/TETC.2017.2703784
9. Andersson, J., de Lemos, R., Malek, S., Weyns, D.: Modeling dimensions of self-adaptive software systems. In: Cheng, B.H.C., de Lemos, R., Giese, H., Inverardi, P., Magee, J. (eds.) Software Engineering for Self-adaptive Systems. Schloss Dagstuhl - Leibniz-Zentrum fuer Informatik, Germany, Dagstuhl, Germany, pp. 27–47 (2009). https://doi.org/10.1007/978-3-642-02161-9_2
10. Banerjee, S., Qaheri, H., Bhatt, C.: Handling uncertainty in IoT design: an approach of statistical machine learning with distributed second-order optimization. In: Healthcare Data Analytics and Management Advances in Ubiquitous Sensing Applications for Healthcare, pp. 227–243. Elsevier BV, Amsterdam (2019)
11. Gerostathopoulos, I., Bures,T., Hnetynka, P., Hujecek, A., Plasil, F., Skoda, D.: Strengthening adaptation in cyber-physical systems via meta-adaptation strategies. ACM Trans. Cyber-Phys. Syst. **1**(3), 13 (2017)
12. Ciccozzi, F., Spalazzese, R.: MDE4IoT: Supporting the Internet of things with model-driven engineering. In: International Symposium on Intelligent and Distributed Computing, pp. 67–76 (2016)
13. Guerriero, M., Tajfar, S., Tamburri, D.A., Di Nitto, E.: Towards a model-driven design tool for big data architectures. In: ACM Proceedings of the 2nd International Workshop on BIG Data Software Engineering, pp. 37–43 (2016)
14. Pal, R., Prasanna, V.: The STREAM mechanism for CPS security the case for the smart grid. IEEE Trans. Comput. Aided Des. Integr. Circuits Syst. **36**(4), 537–550 (2017)
15. Khan, A.H., Khan, Z.H., Khan, S.H.: Optimized reconfigurable autopilot design for an aerospace CPS. In: Khan, Z.H., Shawkat Ali Zahid Riaz, A.B.M. (eds.) Computational Intelligence for Decision Support in Cyber-Physical Systems, vol. 540. Springer, Heidelberg (2014)

16. Olteanu, S.C., et al.: Fuel cell diagnosis using Takagi-Sugeno observer approach. In: International Conference on Renewable Energy for Developing Countries (REDEC), pp. 1–7 (2012)
17. Neuman, C.: Challenges in security for cyber-physical systems. In: DHS Workshop on Future Directions in Cyber-Physical Systems Security (2009)
18. Pasqualetti, F., Dorfler, F., Bullo, F.: Attack detection and identification in cyber-physical systems –part I: models and fundamental limitations (2012). arXiv preprint arXiv:1202.6144
19. Diaz, J., Bielza, C., Ocana, J.L., Larranaga, P.: Development of a cyber-physical system based on selective Gaussian naïve Bayes model for a self-predict laser surface heat treatment process control. In: Niggemann, O., Beyerer, J. (eds.) Machine Learning for Cyber Physical Systems. Selected Papers from the International Conference ML4CPS (2015)
20. Bockenkamp, A., Weichert, F., Stenzel, J., Lunsch, D.: Towards autonomously navigating and cooperating vehicles in cyber-physical production systems. In: Niggemann, O., Beyerer, J. (eds.) Machine Learning for Cyber Physical Systems. Selected Papers from the International Conference ML4CPS (2015)
21. Walther, C., Beneke, F., Merbach, L., Siebald, H., Hensel, O., Huster, J.: Machine-specific approach for automatic classification of cutting process efficiency. In: Niggemann, O., Beyerer, J. (eds.) Machine Learning for Cyber Physical Systems. Selected Papers from the International Conference ML4CPS (2015)
22. Walther, C.: Multikriteriell evolutionär optimierte Anpassung von unscharfen Modellen zur Klassifikation und Vorhersage auf der Basis hirnelektrischer Narkose-Potentiale. Shaker Verlag, Aachen (2012)
23. Abbasi, Z., Jonas, M., Banerjee, A., Gupta, S., Varsamopoulos, G.: Evolutionary green computing solutions for distributed cyber physical systems. In: Evolutionary Based Solutions for Green Computing. Studies in Computational Intelligence, vol. 432, pp. 1–28. Springer, Berlin (2013)
24. Pop, P., Raagaard, M.L., Craciunas, S.S., Steiner, W.: Design optimisation of cyber-physical distributed systems using IEEE time-sensitive networks. IET Cyber-Phys. Syst. Theory Appl. $\mathbf{1}$(1), 86–94 (2016)
25. Michalewicz, Z.: Quo vadis, evolutionary computation? In: IEEE World Congress on Computational Intelligence, pp. 98–121. Springer, Berlin/Heidelberg (2012)
26. Hofmeyer, H., Davila Delgado, J.M.: Coevolutionary and genetic algorithm based building spatial and structural design. Artif. Intell. Eng. Des. Anal. Manuf. $\mathbf{29}$, 351–370 (2015)
27. Van der Vegte, W.F., Vroom, R.W.: Considering cognitive aspects in designing cyber-physical systems: an emerging need for transdisciplinarity. In: Proceedings of the International Workshop on the Future of Transdisciplinary Design TFTD, vol. 13, pp. 41–52 (2013)
28. Ray, A.: Autonomous perception and decision-making in cyber-physical systems. In: 2013 8th International Conference on Computer Science & Education, pp. 1–10. IEEE, Piscataway (2013)
29. Luo, S.: An improved intelligent ant colony algorithm for the reliability optimization problem in cyber-physical systems. J. Softw. $\mathbf{9}$(1), 20–25 (2014)
30. Huang, C.,Wang, D., Chawla, N.: Scalable uncertainty-aware truth discovery in big data social sensing applications for cyber-physical systems. IEEE Trans. Big Data PP(99), 1–1 (2017). https://doi.org/10.1109/TBDATA.2017.2669308
31. Khazaeni, Y., Cassandras, C.G.: Event-driven Trajectory optimization for data harvesting in multiagent systems. IEEE Trans. Control Netw. Syst. $\mathbf{5}$(3), 1335–1348 (2017)
32. Taylor, B.N.: Guidelines for Evaluating and Expressing the Uncertainty of NIST Measurement Results. rev. Diane Publishing, Darby (2009)
33. Cimatti, A., Micheli, A., Roveri, M.: Timelines with temporal uncertainty. In: AAAI (2013)
34. Chattopadhyay, S., Banerjee, A., Banerjee, N.: A data distribution model for large-scale context aware systems. In: International Conference on Mobile and Ubiquitous Systems: Computing, Networking, and Services, pp. 615–627 (2013)
35. Chattopadhyay, S., Banerjee, A., Yu, B.: A utility-driven data transmission optimization strategy in large scale cyber-physical systems. In: 2017 Design, Automation & Test in Europe Conference & Exhibition (DATE) (May 2017)

36. Sinha, A., Malo, P., Deb, K.: A review on bilevel optimization: from classical to evolutionary approaches and applications (2017), arXiv:1705.06270v1
37. Wu, X., Conejo, A.J.: An efficient tri-level optimization model for electric grid defense planning. IEEE Trans. Power Syst. **32**(4), 2984–2994 (2017)
38. Yang, Z., Cai, L., Lu, W.S.: Practical scheduling algorithms for concurrent transmissions in rate-adaptive wireless networks. In: INFOCOM, IEEE 2010 Proceedings (2010)
39. Yang, M., Li, C., Cai, Z., Guan, J.: Differential evolution with auto-enhanced population diversity. IEEE Trans Cybern. **45**(2), 302–315 (2015). https://doi.org/10.1109/TCYB.2014.2339495
40. Bogdan, P., Jain, S., Goyal, K., Marculescu, R.: Implantable pace-makers control and optimization via fractional calculus approaches: a cyber-physical systems perspective. In: ICCPS, pp. 23–32 (2012)
41. Hong, J., et al.: Integrated anomaly detection for cyber security of the substations. IEEE Trans. Smart Grid **5**(4), 1643–1653 (2014)
42. Yue, K., et al.: An adaptive discrete event model for cyber-physical system. In: Analytic Virtual Integration of Cyber-Physical Systems Workshop, pp. 9–15 (2010)
43. Hang, C., Manolios, P., Papavasileiou, V.: Synthesizing cyber-physical architectural models with real-time constraints. In: Computer Aided Verification, pp. 441–456. Springer, Berlin (2011)
44. Bujorianu, M.C., Bujorianu, M.L., Barringer, H.: A formal framework for user centric control of probabilistic multi-agent cyber-physical systems. In: International Workshop on Computational Logic in Multi-Agent Systems, pp. 97–116. Springer, Berlin/Heidelberg (2008)
45. Goswami, D., Schneider, R., Chakraborty, S.: Co-design of cyber-physical systems via controllers with flexible delay constraints. In: ASPDAC, pp. 225–230 (2011)
46. Khaitan, S.K., Mccalley, J.: Design techniques and applications of cyberphysical systems: a survey. IEEE Syst. J. **9**(2), 350–365 (2014)
47. Farina, M., Deb, K., Amato, P.: Dynamic multi-objective optimization problems: test cases, approximations, and applications. IEEE Trans. Evol. Comput. **8**(5), 425–442 (2004)
48. Goh, C.K., Tan, K.C.: A competitive-cooperative coevolutionary paradigm for dynamic multi-objective optimization. IEEE Trans. Evol. Comput. **13**(1), 103–127 (2009)
49. Ren, W., Beard, R.W., Atkins, E.M.: Information consensus in multivehicle cooperative control. IEEE Control Syst. **27**, 71–82 (2007)
50. Cao, X., Cheng, P., Chen, J., Sun, Y.: An online optimization approach for control and communication codesign in networked cyber-physical systems. IEEE Trans. Ind. Inf. **9**, 439–450 (2013)
51. Song, Z., Chen, Y., Sastry, C.R., Tas, N.C.: Optimal Observation for Cyber-Physical Systems: A Fisher-Information-Matrix-Based Approach. Springer Science & Business Media, Berlin (2009)
52. Shoukry, Y., Nuzzo, P., Vincentelli, S., Seshia, S.A., Pappas, G.J., Tabuada, P.: SMC: satisfiability modulo convex optimization. In: HSCC'17, Pittsburgh, PA, USA, April 18–20 (2017)
53. Juditsky, A., Nemirovski, A.: First order methods for nonsmooth convex large-scale optimization, II: utilizing problems structure. Optim. Mach. Learn. **30**, 149–183 (2011)
54. Guigues, V., Juditsky, A., Nemirovski, A.: Non-asymptotic confidence bounds for the optimal value of a stochastic program. Optim. Methods Softw. **32**(5), 1033–1058 (2017)
55. Yuan, Y., Xu, H., Wang, B., et al.: Balancing convergence and diversity in decomposition-based many-objective optimizers. IEEE Trans. Evol. Comput. **20**(2), 180–198 (2016)
56. Hatzakis, I., Wallace, D.: Dynamic multi-objective optimization with evolutionary algorithms: a forward-looking approach. In: Proceedings of the 8th Annual Conference on Genetic and Evolutionary Computation, pp. 1201–1208. ACM, New York (2006)
57. Zhou, A., Jin, Y., Zhang, Q.: A population prediction strategy for evolutionary dynamic multi-objective optimization. IEEE Trans. Cybern. **44**(1), 40–53 (2014)
58. Koo, W.T., Goh, C.K.: A predictive gradient strategy for multi-objective evolutionary algorithms in a fast changing environment. Memetic Comput. **2**(2), 87–110 (2010)

59. Peng, Z., Zheng, J., Zou, J.: A population diversity maintaining strategy based on dynamic environment evolutionary model for dynamic multiobjective optimization. In: 2014 IEEE Congress on Evolutionary Computation (CEC), pp. 274–281. IEEE, Piscataway (2014)
60. Deb, K., Abouhawwash, M.: A optimality theory based proximity measure for set based multi-objective optimization. IEEE Trans. Evol. Comput. 20(4), 515–528 (2016)
61. Birbil, S.I., Frenk, J.B.G., Still, G.J.: An elementary proof of the Fritz-John and Karush-Kuhn-Tucker conditions in nonlinear programming. Eur. J. Oper. Res. 180(1), 479–484 (2007)
62. Mahdavi-Hezavehi, S., Avgeriou, P., Weyns D.: A classification framework of uncertainty in architecture-based self-adaptive systems with multiple quality requirements. In: Mistrik, I., Ali, N., Kazman, R., Grundy, J., Schmerl, B. (eds.) Managing Trade-offs in Adaptable Software Architectures, pp. 45–78. Morgan Kaufmann, San Francisco (2016)
63. Bussmann, S., Schild, K.: Self-organizing manufacturing control: an industrial application of agent technology. In: Proceedings Fourth International Conference on MultiAgent Systems, Boston, MA, pp. 87–94 (2000)
64. Leitao, P., Barbosa, J.: Adaptive scheduling based on self-organized holonic swarm of schedulers. In: Proceedings of the 23rd IEEE International Symposium on Industrial Electronics, Istanbul, pp. 1706–1711 (2014)
65. Vrba, P., Marik, V.: Capabilities of dynamic reconfiguration of multiagent-based industrial control systems. IEEE Trans. Syst. Man Cybern. A 40(2), 213–223 (2010)
66. Zhang, Y., Qian, C., Lv, J., Liu, Y.: Agent and cyber-physical system based self-organizing and self-adaptive intelligent shopfloor. IEEE Trans. Ind. Inf. 13(2), 737–747 (2017)
67. Deb, K., Abouhawwash, M., Seada, H.: A computationally fast convergence measure and implementation foe single, multiple- and many-objective optimization. IEEE Trans. Emerg. Top. Compt. Intell. 1(4), 280–293 (2017)
68. Yin, Y., Yu, F., Xu, Y., Yu, L., Mu, J.: Network location-aware service recommendation with random walk in cyber-physical systems. Sensors 17, 2059 (2017)
69. Fang, Z., Mo, H., Wang, Y., Xie, M.: Performance and reliability improvement of cyber-physical systems subject to degraded communication networks through robust optimization. Comput. Ind. Eng. 114, 166–174 (2017)
70. Akkaya, I.: Data-driven cyber-physical systems via real-time stream analytics and machine learning, Technical Report No. UCB/EECS-2016-159, Electrical Engineering and Computer Sciences University of California at Berkeley, October 25 (2016)

Chapter 6
A Novel Approach for Leak Localization in Water Distribution Networks Using Computational Intelligence

Maibeth Sánchez-Rivero, Marcos Quiñones-Grueiro, Alejandro Rosete Suárez, and Orestes Llanes Santiago (iD)

Abstract This article provides a new approach to localize leaks in water distribution networks (WDN). The model-based leak localization is formulated as an inverse problem solved by using optimization tools. In most approaches to leak localization currently described in the scientific literature, the performance depends on the labeling method for nodes in the parameter space of objective functions. The main contribution of this article lies in a new model-based leak localization approach for water distribution networks whose performance does not depends on the node labeling method. As a result, the accuracy and efficiency of the optimization tools used for leak location task are improved. In addition, the proposal does not depend on the sensitivity matrix used in other methods to localize leaks. In order to achieve all the above, the method hereby proposed builds upon a localization strategy based on the network topological configuration. The Hanoi WDN is used to validate the proposed methodology where the metaheuristic algorithms differential evolution, particle swarm optimization, and simulated annealing are used as optimization tools.

Keywords Leak detection and localization · Water distribution networks · Optimization algorithm · Computational intelligence · Inverse problems

M. Sánchez-Rivero (✉) · M. Quiñones-Grueiro · O. Llanes Santiago
Automation and Computing Department, Universidad Tecnológica de La Habana José Antonio Echeverría, CUJAE, La Habana, Cuba
e-mail: maibeth@automatica.cujae.edu.cu; orestes@tesla.cujae.edu.cu

A. Rosete Suárez
Artificial Intelligence Department, Universidad Tecnológica de La Habana José Antonio Echeverría, CUJAE, La Habana, Cuba
e-mail: rosete@ceis.cujae.edu.cu

© Springer Nature Switzerland AG 2020
O. Llanes Santiago et al. (eds.), *Computational Intelligence in Emerging Technologies for Engineering Applications*, Studies in Computational Intelligence 872, https://doi.org/10.1007/978-3-030-34409-2_6

6.1   Introduction

According to the United Nations Population Fund, the percent of people expected
to live in urban areas will reach 70% by 2050 [1]. Therefore, the density increase of
urban areas demands efficient management of the water resources. The goal of water
distribution networks (WDNs) is to distribute drinking water to customers. Leakage
detection and location is one of the fundamental tasks that must be performed to
guarantee the efficient operation of WDNs.

Water losses detection and location strategies are usually classified into reactive
and proactive [2]. The reactive approach relies on customer contacts and additional
information such as network regular examination to detect water losses and a
full network survey to locate them. Proactive strategies encompass measurement
analysis methods together with the principle of district metering proposed by
the International Water Association (IWA). According to the IWA proposal, the
distribution system is subdivided into discrete zones or district meter areas (DMA),
by the permanent closure of valves, with the idea of allowing the characterization
of the normal flow and pressure variations [3]. A DMA generally comprises an area
containing between 200 and 3500 properties and a maximum pipe length of 30 km
[3, 4].

Proactive strategies are divided into either direct observation methods or inference methods [5]. Direct methods use highly specialized hardware equipment such
as those based on leak noise correlations [6], gas injection [7], and pig-mounted
acoustic sensing [8] for water losses location. Most of this specialized equipment
requires trained personnel and is expensive. Moreover, the application over a large-
scale DMA is labor-intensive, time-consuming and may require the partial shutdown
of DMA operations.

Inference methods build models to represent the DMA behavior with measurement data from permanently installed sensors. Traditional water balance method
together under minimum night flow regime allows the estimation of the water losses
in the network. This approach allows leakage detection in real-life WDNs [9]. The
water balance method requires the installation of flow meters in some pipes of the
network for collecting flow data. However, water companies usually measure flow
only in the input and output of the DMA. Moreover, the installation and maintenance
of these sensors are expensive and time-consuming. As an alternative solution,
pressure head sensors are considered in other works for the leakage monitoring task
because their price is lower than flow meters as well as they are easier to install and
maintain.

Residual-based analysis has been proposed in previous works for leakage
location [10, 11]. A residual signal represents the difference between the outputs
of a hydraulic model and the measurements obtained from the pressure head
sensors installed in the network. The deviation of these residuals from pre-defined
thresholds indicates the presence of a leak. Afterwards, the location of the leak is
determined by comparing the residual vector with a set of leak signatures that form
a sensitivity matrix. The results of the leak location task depend on the similarity

metric used for comparing the residual with each vector of the sensitivity matrix. Thus, different metrics have been tested such as the Pearson correlation, the angle between vectors and the Euclidean distance.

Overall, the performance in the leakage location task for the above-mentioned approaches depends on the leakage size used to create the sensitivity matrix. Hence, if the magnitude of the leak is close to the one used for creating the matrix, the results are satisfactory; otherwise, the performance worsens.

The leak localization task can also be formulated as the solution of an inverse problem [12]. Recent works have followed this approach and optimization tools have been used for solving the inverse problem [13]. The performance of this approach depends mainly on: (1) the distance metric used for computing the objective function and (2) the labeling method for the leakage positions in the network (how the nodes are labeled for solving the inverse problem).

The above-mentioned issues motivate this chapter where the main contribution is a leak localization approach that does not depend on the sensitivity matrix neither the labeling method for the nodes. The proposed approach solves the inverse problem by using metaheuristic optimization algorithms. The search strategy for these metaheuristics is adapted for taking into account the topological configuration of the network. Specifically, three optimization methods are used: differential evolution (DE) [14], particle swarm optimization (PSO) [15], and simulated annealing (SA) [16, 17]. The Hanoi WDN is used as case study to develop the experiments. The simulated scenarios consider a varying leakage size and measurement noise. The results confirm that by considering the topological information of the network the search strategy is independent of the labeling method. The best results are achieved with the DE method.

6.2   Leak Localization as the Solution of an Inverse Problem Using Optimization Algorithms

Figure 6.1 shows the leak localization model proposed in this article. It is based on the use of pressure differences among the pressure measurements from a WDN and their estimates obtained from a hydraulic simulator. The optimization algorithm determines the most probable leak produced by such differences.

The size of the r residual vector depends on the number of pressure sensors in the hydraulic network. The number of potential leaks, l, is considered to be equal to the number of nodes n_n in the network. Such criterion was set forth in the work submitted in [12] and it is the one presumed in.

The simulated hydraulic model considers a known network structure in regard to pipes, nodes, and valves. Likewise, the network parameters relative to the pipe loss coefficients, its dimension and nominal demands (using the historical consumption records), are assumed to be known in the nodes $(nd_1, nd_2, \ldots, nd_{n_n})$ since, in general, the node demands are not measured by flow sensors.

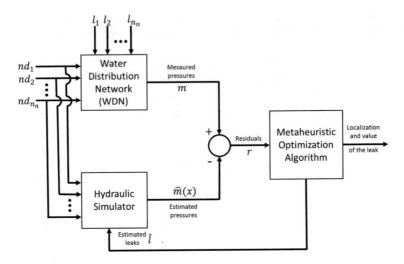

Fig. 6.1 Proposed methodology for localization and size determination of leaks in a WDN

Model-based leak localization is based on minimizing residuals obtained from the difference between m measurements and the corresponding $\hat{m}(x)$ estimated values found due to potential leaks obtained by using the network hydraulic model.

The objective is to find the leak position (L_P) and the leak value (L_L). From a mathematical point of view, an arbitrary metric can be used to describe the difference $d(m, \hat{m}(x))$. The problem can be formulated as follow:

$$f(x) = d(m, \hat{m}(x)), \tag{6.1}$$

where $f(x)$ is a one-dimensional function called the fitness or objective function, and $x \in R^2$ is defined as:

$$x = \begin{pmatrix} L_L \\ L_p \end{pmatrix}. \tag{6.2}$$

The leak position and size can be estimated by looking for a vector which minimizes the objective function in the $L_L - L_P$ space

$$\min_x f(x) \tag{6.3}$$

$$s.a. \quad 0 < L_L < L_{Lmax}$$

$$0 < L_P < n_n$$

with $L_L \in \Re$ where L_{LMax} is the maximum outflow of leak to localize, and $L_P \in Z^+$, where n_n is the number of nodes of the network. The objective function may have many local minimums resulting in the failure of a descending gradient based algorithm to find the global minimums. This problem can be satisfactorily solved by resorting to optimization methods based on the use of metaheuristic tools.

L_L is a real value and L_P a discrete value. Therefore, the localization strategy should take into account both types of variables. If L_P is introduced as an optimizing parameter and the nominal value of the node label (where the leak takes place) is taken, an in-depth search is required in order to find the global optimum. The results obtained from an in-depth search depend on the amount of information collected of the network configuration introduced in the node labeling. Besides, the labeling strategies to solve the optimization problem do not take into account the specific relationship between the network topological configuration and the node label thus causing the localization method performance to be sensitive to the label method which constitutes a disadvantage [13].

6.3   Application of the Proposed Leak Localization Strategy with Three Optimization Algorithms

The above-mentioned inconvenience led to the formulation of the new localization strategy submitted in this article. Figure 6.2 shows the general scheme of an algorithm of optimization using a metaheuristic tool, where the reproduction step debit must be adapted.

Fig. 6.2 Pseudo-code for optimization algorithm

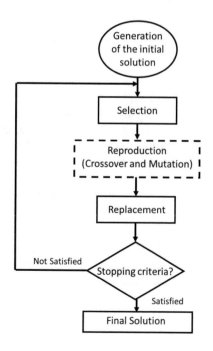

The proposal states that the leak magnitude coding and the leak position should be dealt with in different ways for every localization algorithm. The leak magnitude should be developed in a continuous space while the leak position should take into account the network topological configuration instead of the node label.

The strategy proposed can be considered for adapting metaheuristic algorithms applied for optimization purposes.

6.3.1 Topological-Differential Evolution (T-DE)

The differential evolution (DE) algorithm [14, 18] is an evolutionary method based on populations, thus providing a friendly implementation structure. It is based on three operators: mutation, crossing, and selection of vectorial operations. The general idea behind DE is to provide a new solution by varying the solutions relative to the population up to that moment.

The solution model can be expressed by using the following notation $DE/x/y/z$, where x denotes the type of the mutated vector (also called the base vector), y indicates the number of pairs of solutions to be used in order to vary the current solution, and z represents the distribution function which will be used during the crossing. In this chapter, the configuration $DE/rand/1/bin$ was used where a random vector ($x = rand$) and the difference of a vector pair ($y = 1$) were applied in order to generate a mutation. This can be obtained through a binomial crossing model ($z = bin$). The mutation operator is expressed as:

$$\bar{X}_i^{(j,t)} = X_i^{(\alpha,t)} + F_S(X_i^{(\beta,t)} - X_i^{(\gamma,t)}) \qquad j = 1, \ldots, N; i = 1, \ldots, D, \qquad (6.4)$$

where $X_i^{(\alpha,t)}, X_i^{(\beta,t)}, X_i^{(\gamma,t)} \in \Re^D$ are elements of the population generated in the previous iteration, and α, β, and γ are chosen randomly in the interval $[0, N]$ where N is the population size and D is the localization space dimension.

F_S is a real and constant value known as scaling factor which determines the influence of the vector pair used in the mutation. The crossing operator for each component of the solution vector, is defined as:

$$\bar{\bar{X}}_i^{(j,t)} = \begin{cases} \bar{X}_i^{(j,t)} & if \quad q_{rand} < C_R \quad i = 1, \ldots, D \\ X_i^{(j,t)} & otherwise \end{cases} \qquad (6.5)$$

where $0 < C_R < 1$ is the crossing constant and q_{rand} is a random number generated when using the z distribution function; in this case the binomial distribution function. Finally, the selection operator is:

$$X^{(j,t+1)} = \begin{cases} \bar{\bar{X}}^{(j,t)} & if \quad F(\bar{\bar{X}}^{(j,t)}) < F(X^{j,t}) \\ X^{(j,t)} & otherwise \end{cases} \qquad (6.6)$$

where $F(\bar{\bar{X}}^{(j,t)})$ and $F(X^{(j,t)})$ are the objective values for the $\bar{\bar{X}}^{(j,t)}$ test vector and the $X^{(j,t)}$ current vector, respectively.

In this problem, the search vector is made up by a real value (L_L), whose optimal value can be found by applying the traditional DE method previously analyzed and a discrete value (L_P) is determined by the node label where the leak takes place [13].

Even when the L_P optimal value is localized by applying the strategy proposed by [19]—which is a modification of the traditional DE using both real values and discreet values—the localization algorithm also depends on the labeling method of nodes in the localization space.

In this article [19], DE is reformulated, called topological-differential evolution (T-DE), such that the mutation operator for L_P will select one of the $X^{(j,t)}$, neighboring nodes (connected to it through a pipe) in the WDN defined as

$$\bar{X}_i^{(j,t)} = X_i^{(\theta)}, \tag{6.7}$$

where $X_i^{(\theta)} \in P = v_1, v_2, \ldots, v_l$ such that v_l is $X_i^{(j,t)}$ neighbor, θ is randomly selected in the $[0, L]$ interval, and L is, in general, the number of neighbors to the node $X_i^{(j,t)}$. Figure 6.3 shows the pseudo-code for the T-DE metaheuristic algorithm.

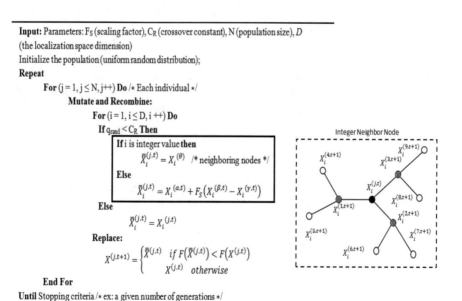

Input: Parameters: F_S (scaling factor), C_R (crossover constant), N (population size), D (the localization space dimension)
Initialize the population (uniform random distribution);
Repeat
    For $(j = 1, j \leq N, j{+}{+})$ **Do** /* Each individual */
        Mutate and Recombine:
        For $(i = 1, i \leq D, i{+}{+})$ **Do**
          If $q_{rand} < C_R$ **Then**
            If i is integer value **then**
              $\bar{X}_i^{(j,t)} = X_i^{(\theta)}$ /* neighboring nodes */
            Else
              $\bar{X}_i^{(j,t)} = X_i^{(\alpha,t)} + F_S\left(X_i^{(\beta,t)} - X_i^{(\gamma,t)}\right)$
          Else
            $\bar{X}_i^{(j,t)} = X_i^{(j,t)}$
        Replace:
        $X^{(j,t+1)} = \begin{cases} \bar{X}^{(j,t)} & \text{if } F(\bar{X}^{(j,t)}) < F(X^{(j,t)}) \\ X^{(j,t)} & \text{otherwise} \end{cases}$
    End For
Until Stopping criteria /* ex: a given number of generations */
Output: Best population or solution found.

Fig. 6.3 Pseudo-code for T-DE algorithm

6.3.2 Topological-Particle Swarm Optimization (T-PSO)

Particle swarm optimization (PSO) is a stochastic population-based metaheuristic inspired from swarm intelligence. It mimics the social behavior of natural organisms such as bird flocking and fish schooling to find a place with enough food. Indeed, in those swarms, a coordinated behavior using local movements emerges without any central control [15].

In PSO, the searching of solutions for every particle is defined by updating its velocity and position inside the swarm. The velocity and position are defined as follows:

$$V_i^{(j,t)} = \omega V_i^{(j,t-1)} + C_1 r_1 (Pb_i^j - X_i^{(j,t-1)}) + C_2 r_2 (Gb_i - X_i^{(j,t-1)})$$
$$j = 1, \ldots, N \qquad i = 1, \ldots, D \tag{6.8}$$
$$X_i^{(j,t)} = X_i^{(j,t-1)} + V_i^{(j,t)},$$

where i, j, and t are the number of variables to optimize by the algorithm, particles in the swarm, and iterations of the algorithm, respectively, V and X are the velocity and position of the particle (solution of the problem), C_1 and C_2 are the acceleration coefficients (individual and social), Pb_i is the best position found in the previous iteration, Gb is the best global position found until the i iteration, r_1 and r_2 are random numbers in the interval [0, 1] used for maintaining the diversity of the population, and ω is the inertial weight.

A large value of ω implies more exploration of the searching space and longer convergence time, while a small value reduce the convergence time but increases the risk of falling in a local optimal. The determination of the value for ω is made considering a contiguity index (CI). It provides evidence of how close each particle is to the optimal solution and it is calculated as follows:

$$\omega^{(j,t-1)} = \frac{1}{1 + e^{-(\alpha CI^{(j,t-1)})-1}}, \tag{6.9}$$

where α is a positive constant in the range (0, 1]. This parameter controls the speed at which ω decreases through the iterations and CI can be calculated as follows:

$$CI^{(j,t-1)} = \frac{f(Pb_i^{(j,t=1)})}{f(Pb_i^{(j,t-1)})} - 1. \tag{6.10}$$

This metaheuristic, just like the classic differential evolution can be used for searching the leak magnitude, but it will be reformulated for estimating the position of the leak the topological configuration of the network. Therefore, the original method is reformulated as follows, and it will be called topological-particle swarm optimization.

$$V_i^{(j,t)} = \omega V_i^{(j,t-1)} + C_1 r_1 \| Pb_i^j - X_i^{(j,t-1)} \| + C_2 r_2 \| Gb_i - X_i^{(j,t-1)} \|$$

$$j = 1, \ldots, N \qquad i = 1, \ldots, D \qquad\qquad (6.11)$$

$$X_i^{(j,t)} = X_i^{(\theta,t-1)},$$

where $\| Pb_i^j - X_i^{(j,t-1)} \|$ is the topological distance between the node of the best position of particle and the actual position (minimum pipe distance between these elements), in the same way $\| Gb_i - X_i^{(j,t-1)} \|$ is the topological distance between the node of the best position of swarm and the actual position of particle. The next position of the particle is a random neighbor where θ is randomly select from the $[0, L]$ interval and L is the number of neighbors to the node $X_i^{(j,t-1)}$ at the neighborhood of $\| X_i^{(j,t-1)} + V_i^{(j,t)} \|$. Figure 6.4 shows the strategy for the T-PSO metaheuristic algorithm.

Input: Parameters: C_1 (acceleration individual coefficients), C_2 (acceleration social coefficients), ω (inertial weight), N (population size), D (the localization space dimension)
Initialize the population (uniform random distribution);
Repeat
        For (j = 1, j ≤ N, j++) **Do** /∗ Each individual ∗/
        Update velocities:
        For (i = 1, i ≤ D, i ++) **Do**

> **If** i is integer value **then**
>
> $$V_i^{(j,t)} = \omega V_i^{(j,t-1)} + C_1 r_1 \| Pb_i^{(j)} - X_i^{(j,t-1)} \|$$
>
> $$+ C_2 r_2 \| Gb_i - X_i^{(j,t-1)} \| \quad /*\text{Topological distance}*/$$
>
> Move to the new position: $X_i^{(j,t)} = X_i^{(\theta,t-1)}$ /∗ neighboring nodes∗/
>
> **Else**
>
> $$V_i^{(j,t)} == \omega V_i^{(j,t-1)} + C_1 r_1 \left(Pb_i^{(j)} - X_i^{(j,t-1)} \right) + C_2 r_2 \left(Gb - X_i^{(j,t-1)} \right)$$
> Move to the new position: $X_i^{(j,t)} = X_i^{(j,t-1)} + V_i^{(j,t)}$;

        End For
        If $f(X^j) < f(Pbest^{(j)})$ **Then** $Pbest^{(j)} = X^j$
        If $f(X^j) < f(Gbest)$ **Then** $Gbest = X^j$
        Update(X^j, V^j) ;
    End For
Until Stopping criteria /∗ ex: a given number of generations ∗/
Output: Best population or solution found.

Fig. 6.4 Pseudo-code for T-PSO algorithm

6.3.3   Topological-Simulated Annealing (T-SA)

Simulated annealing (SA) applied to optimization problems emerges from the work of Cerny and Kirkpatrick et al. [16, 17]. The SA algorithm simulates the energy changes in a system subjected to a cooling process until it converges to an equilibrium state (steady frozen state).

SA is a stochastic algorithm that enables, under some conditions, the degradation of a solution. The objective is to escape from local optima and so to delay the convergence. SA is memoryless in the sense that it does not use any information gathered during the search. At each iteration, a random neighbor is generated. Moves that improve the cost function are always accepted with a given probability.

Otherwise, the neighbor is selected with a probability that depends on the parameter called temperature and the amount of degradation ΔE of the objective function. ΔE represents the difference in the objective value (energy) between the current solution and the generated neighboring solution. As the algorithm progresses, the probability that such moves are accepted decreases. This probability follows, in general, the Boltzmann distribution.

In this case the proposed methodology for the leak position, topological simulated annealing (T-SA), the random neighbor, generated is a neighboring nodes easily connected to it through a pipe or in the WDN. Figure 6.5 shows the strategy for the T-SA metaheuristic algorithm.

Input: Parameters: T_{min} (minimal temperature), T_{max} (maximum temperature), D (the localization space dimension)

$X = X_0$; /* Generation of the initial solution */

$T = Tmax$; /* Starting temperature */

Repeat /* At a fixed temperature */

     For $(i = 1, i \leq D, i ++)$ **Do**

         If i is discrete value **Then**

               $X_i^{(j,t)} = X_i^{(\theta,)}$ /* neighboring nodes*/

         Else

             Generate a random neighbor X' :

     End for

     $\Delta E = f(X') - f(X)$;

If $\Delta E < 0$ **Then** $X = X'$ /* Accept the neighbor solution */

Else Accept X' with a probability $e^{\frac{-\Delta E}{T}}$;

Until Equilibrium condition

/* e.g. a given number of iterations executed at each temperature T */

$T = g(T)$; /* Temperature update */

Until Stopping criteria satisfied /* e.g. T < Tmin */

Output: Best solution found.

Fig. 6.5 Pseudo-code for T-SA algorithm

6.3.4 Objective Functions

The fitness landscape depends on the metric used to calculate the objective function. Different metrics can be applied to model-based leakage localization. Thus, the most popular metrics described below are used in this chapter for assessing the performance achieved by the proposal [13, 20].

One family of metrics is the Minkowski distance metrics. Its general form is

$$d(m, \hat{m}(x)) = \left(\sum_{i=1}^{n} |m_i - \hat{m}_i(x)|^p \right)^{1/p}. \tag{6.12}$$

Special forms of the Minkowski metric are the Euclidean ($p = 2$), which is equivalent to sum of the square of the deviations between the observed and calculated values (SSQ). Other types of metrics are the angles between vectors like Pearson's correlation coefficient:

$$d(m, \hat{m}(x)) = 1 - \frac{(m - \bar{m}) * (\hat{m}(x) - \bar{m}(x))}{\| m - \bar{m} \|_2 * \| \hat{m}(x) - \bar{m}(x) \|_2}. \tag{6.13}$$

The Sorensen metric is also considered, which is defined as follows:

$$d(m, \hat{m}(x)) = \frac{\sum_{i=1}^{n}(m_i - \hat{m}_i(x))}{\sum_{i=1}^{n}(m_i + \hat{m}_i(x))}. \tag{6.14}$$

All these metric and node's labeling methods were tested on a Hanoi network, in order to evaluate the performance of the proposal.

6.3.5 Case Study. Hanoi Network

The water distribution trunk network in Hanoi, Vietnam, introduced by Fujiwara and Khang in 1990 is shown in Fig. 6.6 [21]. The network consists of 34 pipes, 32 nodes, and 3 loops. It is fed by gravity from a single fixed head source and it is designed to satisfy the customers demand at specific minimum pressures. The network configuration is the one proposed in [22] for a total demand of 2770 lps.

6.3.6 Labeling the Search Space

Different node's labeling methods, tested previously by Steffelbauer et al. [13], are used for demonstrating the dependency between the performance of the leak localization methods and the labeling strategy. The first strategy is an alphabetical

Fig. 6.6 Hanoi network

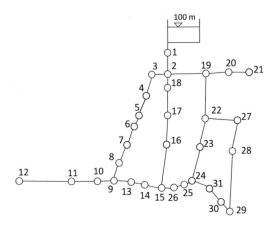

sorting of the name of the junctions (Alphabetical). Second, the nodes are sorted completely randomly (Random). Third, the sorting is defined through the Cuthill–McKee algorithm which is a nodal numbering scheme for reducing the bandwidth of the adjacency matrix. The connections between nodes and pipes of the WDN are described by the adjacency matrix such that rows and columns correspond to the node sets, and the position (i, j) has a value of one if there is an edge between nodes i and j, and zero otherwise. Finally, a depth-first search (DFS) algorithm is tested.

6.4 Experiments and Results

In this section, experiments conducted had the following objectives:

- Demonstrate that the performance of the method proposed in this article for leak localization does not depend on the node labeling method considering each distance metric.
- Demonstrate that the strategy used in the leak localization model proposed in this chapter improves the performance of traditional metaheuristics while decreasing the computational cost. As examples are used T-DE, T-PSO, and T-SA.
- Demonstrate the improvement achieved by the method proposed when compared with some of the most popular leak localization methods based on models.

Single leakages are assumed to occur at the nodes of the network. The leak nominal value used to build the sensitivity matrix is 50 lps. The magnitude of the simulated leaks is $L_L = [10 \quad 20 \quad 30 \quad 40 \quad 50 \quad 60 \quad 70]$ lps. This set of leak levels is taken into consideration when analyzing the behavior of algorithms used in the experiments when leak magnitudes are above or below the leak nominal value. In the case of hydraulic simulations, the EPANET software package is used [23].

The parameters used in the T-DE algorithm are the $C_R = 0.9$ to ensure a low influence of the population solution obtained up to that moment and $F_s = 0.6$, which determines the influence of the mutation operator. The population is made up by 20 individuals ($N = 20$). The stop criteria constitute an error below 10^{-4} and the highest number of generations is 100.

The parameters used in the T-PSO algorithm are $\alpha = 0.5$ to achieve a decrease of ω during the iterations. The population is formed by 20 individuals ($N = 20$). The stopping criteria are an error below 10^{-4} and, in this case, the highest number of generations is 100. The individual coefficient, C_1, takes values between 2.5 and 0.5, while the social coefficient C_2 is between 0.5 and 2.5 [18, 24].

In case of T-SA, it has an initial temperature $T = 10$ and final temperature $t_f = 0.001$ with a maximum number of 150 iterations and the stop criteria constitutes an error below 10^{-4} [24].

The confusion matrix is used to evaluate the performance of leak localization methods. The confusion matrix is a square matrix with a number of rows and columns equal to the number of network nodes (potential leak positions) where each coefficient (i, j) indicates how many times a leak in the node i is recognized as a leak in the j node. In case of perfect classification, the confusion matrix should be diagonal. In practice, values different from zero will appear outside the main diagonal. The global performance (accuracy) of the leak localization method is defined as the percentage of leaks accurately localized in relation to the total number of simulated leaks.

In this article, the experiments were conducted by measuring pressures in nodes 12 and 21 [25].

Three different metrics were used to formulate the objective function and four different forms of node labeling were tested in the Hanoi network test to evaluate the performance of the method proposed. Each optimization algorithm was run ten times for each combination of the distance, node labeling, and leak magnitude metric.

For the algorithm T-DE the Friedman non-parametric test is applied for determining if there are significant differences in the performance of the localization method when four node labeling methods are employed. The p-value obtained in the Friedman test (with an $\alpha = 0.05$ significance level) shows no significant differences in the method performance for every leak magnitude (Table 6.1). This demonstrates empirically that the proposed method does not depend on the node labeling.

Selecting the sorting defined through the Cuthill–McKee algorithm as node labeling method, in order to analyze the performance of the distance metrics, the Friedman test was applied to the results and significant differences were found (Table 6.2). The Wilcoxon test was subsequently applied and it showed that the performance of Pearson's correlation coefficient is a statistical significant difference with distance metric Euclidean and Sorensen. In Fig. 6.7 is showed the global performance (accuracy) in percent (%) achieved in each metric for each analyzed level of leak.

Selecting the Euclidean distance and Sorensen metrics for T-DE algorithm, no significant differences were found when comparing the obtained results.

Table 6.1 Results of the Friedman tests for T-DE ($\alpha = 0.05$)

Level of leak (lps)	Euclidean		Pearson's correlation		Sorensen	
	Value in χ^2	p-value	Value in χ^2	p-value	Value in χ^2	p-value
10	1.44	0.70	2.86	0.41	0.94	0.82
20	5.25	0.16	1.95	0.58	0.27	0.97
30	5.21	0.16	1.42	0.70	1.58	0.67
40	2.74	0.43	5.53	0.14	4.34	0.23
50	4.73	0.19	5.12	0.16	7.73	0.06
60	1.47	0.69	2.93	0.40	1.77	0.62
70	1.76	0.62	7.06	0.07	0.77	0.86
All	2.89	0.41	1.36	0.71	0.03	0.99

Table 6.2 Results of the Friedman tests for T-DE for three different distance metric (Euclidean, Pearson's correlation and Sorensen)

Level of leak (lps)	Cuthill-McKee	
	Value in χ^2	p-value
10	15.2	0.0005
20	15.8	0.0004
30	16.9	0.0002
40	16.3	0.0003
50	17.6	0.0002
60	15.4	0.0004
70	16.9	0.0002
All	112.2	$4.23 * 10^{-25}$

Labeling method: Cuthill-McKee ($\alpha = 0.05$)

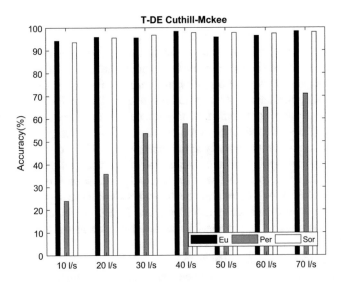

Fig. 6.7 Performance comparison of T-DE for three different distance metric (Euclidean, Pearson's correlation, and Sorensen). Labeling method: Cuthill-McKee

Table 6.3 Results of the Friedman tests for T-PSO ($\alpha = 0.05$)

Level of leak (lps)	Euclidean		Pearson's correlation		Sorensen	
	Value in χ^2	p-value	Value in χ^2	p-value	Value in χ^2	p-value
10	1.67	0.64	1.52	0.68	1.82	0.61
20	4.02	0.26	1.33	0.72	1.59	0.66
30	1.22	0.75	3.16	0.37	2.50	0.48
40	2.38	0.50	1.47	0.69	1.57	0.67
50	5.78	0.12	2.48	0.32	6.93	0.08
60	6.60	0.09	3.64	0.31	1.30	0.73
70	0.46	0.93	0.28	0.96	0.49	0.92
All	3.27	0.35	2.73	0.43	1.52	0.68

Table 6.4 Results of the Friedman tests for T-PSO for three different distance metric (Euclidean, Pearson's correlation, and Sorensen)

Level of leak (lps)	Cuthill-McKee	
	Value in χ^2	p-value
10	15.44	0.0004
20	15.80	0.0004
30	16.89	0.0002
40	16.27	0.0003
50	15.44	0.0004
60	15.44	0.0004
70	15.44	0.0004
All	109	$1.8 * 10^{-24}$

Labeling method: Cuthill-McKee ($\alpha = 0.05$)

A similar analysis is performed for the T-PSO algorithm (Table 6.3) and it is corroborated that regardless of the distance metric used, the performance of the algorithm is independent of the node labeling method.

In T-PSO, the distance metric Pearson's correlation coefficient presents a very low performance, which causes that it to be statistically different from the others (Table 6.4).

For using T-PSO algorithm in future comparisons, the Euclidean and Sorensen distance can be used, since their results do not present a significant statistical difference.

In case of T-SA algorithm, it is also proved that the performance of the method is independent of the node labeling method (Table 6.5) but it is sensitive to the distance metric used (Table 6.6).

Again, the Pearson's correlation coefficient metric shows a very bad performance, and between the other two there is no statistical difference and they can be used in future comparisons.

Although, for each level, of leak the performance of the algorithm is independent of the node labeling method, and it can be appreciated that exist differences among the performance of every optimization algorithms. Table 6.7 summarizes the performance of every algorithm for the Cuthill–McKee node labeling method,

Table 6.5 Results of the Friedman tests for T-SA ($\alpha = 0.05$)

Level of leak (lps)	Euclidean		Pearson's correlation		Sorensen	
	Value in χ^2	p-value	Value in χ^2	p-value	Value in χ^2	p-value
10	0.41	0.94	1.92	0.59	0.67	0.88
20	5.94	0.11	1.38	0.71	2.74	0.43
30	4.15	0.24	1.81	0.61	3.10	0.11
40	0.88	0.83	1.08	0.78	2.23	0.52
50	2.61	0.46	0.66	0.88	3.90	0.09
60	1.47	0.69	5.60	0.13	0.98	0.81
70	5.91	0.12	3.29	0.35	0.54	0.91
All	0.03	0.99	2.43	0.49	0.21	0.97

Table 6.6 Results of the Friedman tests for T-SA for three different distance metric (Euclidean, Pearson's correlation, and Sorensen)

Level of leak (lps)	Cuthill-McKee	
	Value in χ^2	p-value
10	15.2	0.0005
20	15.4	0.0004
30	17.9	0.0001
40	16.0	0.0003
50	15.9	0.0004
60	17.6	0.0002
70	15.9	0.0004
All	108.7	$2.5 * 10^{-24}$

Labeling method: Cuthill-McKee ($\alpha = 0.05$)

Table 6.7 Performance comparison of T-DE, T-PSO, and T-SA (Euclidean distance-Cuthill McKee)

Level of leak (lps)	T-DE	T-PSO	T-SA
10	94.19	73.23	40.00
20	95.81	79.68	52.26
30	95.48	81.94	51.61
40	98.39	87.74	54.62
50	95.81	87.42	51.61
60	96.45	86.45	54.19
70	98.39	86.13	52.58

using one of the metrics with the best performance for each of the algorithms, which in this case coincides with the Euclidean distance. From the previous analysis it is established that the algorithm with better performance is T-DE.

The second experiment demonstrates that the strategy used in the leak localization model proposed in this chapter improves the performance of traditional metaheuristics while decreasing the computational cost.

The metric used to formulate the objective function was the Euclidean distance, because the best localization results have been obtained by using this distance, and the node labeling was Cuthill–McKee. They were tested in the Hanoi network in order to evaluate the performance of the proposed strategy.

Figure 6.8 shows the performance for each metaheuristic with and without the proposed search strategy.

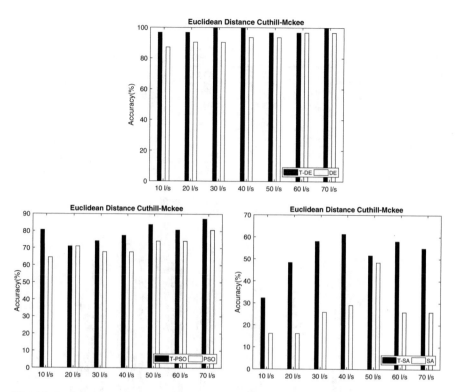

Fig. 6.8 Performance comparison of DE, PSO, and SA with and without proposed strategy

It is necessary to remark that the proposed strategy modifies only the way of generating new individuals for the leak position. Thus, it does not increase the computational complexity of the algorithm (considered here as the times that the objective function is evaluated). Moreover, the average times of execution decrease with our proposal since the optimization methods converge faster to the solution. Table 6.8 shows the average iterations for each metaheuristic method for the Hanoi network. These results show that less objective function evaluations are required for each method, thereby decreasing their total execution time.

The third experiment was conducted to make a comparison between the performance of proposal using T-DE and other leak localization methods which have achieved satisfactory performances according to the literature. The method proposed by Steffelbauer et al. is also formulated as the solution to an inverse problem independent from the sensitivity matrix which does not require any knowledge about the nominal leak magnitude [13].

The method proposed in this chapter inherits previous advantages and eliminates its main disadvantage due to the reformulation of the leak localization strategy which allows the independence of the optimization algorithm from the labeling method used in the nodes, with the objective to improve the performance and the efficiency of the used optimization algorithms.

Table 6.8 Comparison of the average number of evaluations for the objective function in the Hanoi network

Level of Leak (lps)	T-DE	DE	T-PSO	PSO	T-SA	SA
10	1644	1769	1439	1457	155	167
20	1674	1791	1396	1489	156	167
30	1671	1839	1199	1383	156	168
40	1627	1865	1609	1670	157	168
50	1632	1854	1453	1572	157	168
60	1623	1795	1158	1385	155	167
70	1625	1851	1454	1599	157	168

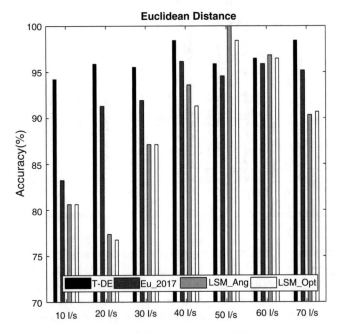

Fig. 6.9 Performance comparison of Euclidean distance-Cuthill McKee proposed method (T-DE), Euclidean distance-Cuthill McKee method by Stefellbauer (Eu-2017), LSM with angle (LSM-Ang), and LSM using optimization (LSM-Opt)

In the article written by Steffelbauer et al., the best results were obtained for the Euclidean distance and the labeling method based on the Cuthill–McKee algorithm. Therefore, this combination was selected to conduct the experiments and to compare them with results obtained with the proposal submitted in this work.

Likewise, the leak sensitivity matrix (LSM) methods were selected for comparison purposes due to their good findings. It is worthwhile to remark that these methods depend on the sensitivity matrix and the leak nominal value [10, 11, 20].

Figure 6.9 shows the performance of the method proposed as well as the Steffelbauer et al. method [13] which considers the Euclidean distance and the

Cuthill–McKee method as the distance metric and node labeling, respectively. These methods were jointly assessed against two methods based on the sensitivity matrix, namely one using the angle between vectors as the metric and another one using an optimization strategy with genetic algorithms.

Figure 6.9 illustrates that, in general terms, the performance of the proposed method is higher than others except when the leak magnitude is 50 lps, that is, the value used to build the sensitivity matrix.

The above-mentioned fact indicates that the proposal submitted in this article shows a satisfactory performance which is independent from the leak size unlike the LSM-based methods whose performance is gradually degraded to the extent in which the real leak magnitude is different from the value used to build the LSM.

In addition, the method proposed show a performance with over 95% accurate classification for every leak magnitude thus improving over 10% the remaining methods in case of low magnitude leaks (10 lps).

6.5   Conclusions

A new leak localization approach based on the solution of an inverse problem by using metaheuristic tools has been presented in this chapter. This approach attempts to adapt optimization tools so that the topological configuration of the network can be used in leak localization instead of the nominal value of node labels. In our case, this was formulated by changing the reproduction step for the differential evolution, particle swarm optimization, and simulated annealing.

The proposal was applied to the Hanoi benchmark which demonstrated that the performance is independent from the strategy used to label the nodes of the network. Besides, the method exposed improves the efficiency of the optimization tools by decreasing of the average number of evaluations of the objective function. Likewise, the results obtained show that the performance of the localization method proposed does not require the nominal value of the leak size and is independent from the sensitivity matrix. The findings confirm the viability of this proposal. Differential evolution is the algorithm with better performance among the analyzed metaheuristics.

As future works, it is interesting to apply this approach in the analysis of other network types as for example gas distribution networks.

References

1. UN: World urbanization prospects: the 2007 revision population database (2008). http://esa.un. org/unup/
2. Puust, R., Kapelan, Z.S., Savic, D., Koopel, T.: J. Urban Water **7**(1), 25 (2010). https://doi.org/10.1080/15730621003610878
3. Loveday, M., Dixon, J.: Proceedings IWA Leakage Conference, Halifax (2005)

4. MacDonald, G., Yates, C.: Proceedings IWA Leakage Conference, Halifax (2005)
5. Covas, D., Ramos, H., de Almeida, A.B.: J. Hydraul. Eng. **131**(12), 1106 (2005)
6. Muggleton, J.M., Brennan, M.J.: J. Sound Vib. **278**, 527 (2004)
7. Farley, M., Trow, S.: Losses in Water Distribution Networks A Practitioner's Guide to Assessment, Monitoring and Control. IWA Publishing, London (2003)
8. Mergelas, B., Henrich, G.: Proceedings of the Leakage 2005 Conference, Halifax (2005)
9. Farah, E., Shahrour, I.: Water Res. Manag. **31**(15), 4821 (2017)
10. Pérez, R., Puig, V., Pascual, J., Quevedo, J., Landeros, E., Peralta, A.: Control Eng. Pract. **19**(10), 1157 (2011). https://doi.org/10.1016/j.conengprac.2011.06.004
11. Pérez, R., Puig, V., Peralta, A., Landeros, E., Jordanas, L.: Water Sci. Technol. Water Supply **9**(6), 715 (2009). https://doi.org/10.2166/ws.2009.372
12. Pudar, R.S., Liggett, J.A.: J. Hydraul. Eng. **118**(7), 1031 (1992)
13. Steffelbauer, D.B., Günther, M., Fuchs-Hanusch, D.: Proc. Eng. **186**, 444 (2017). https://doi.org/10.1016/j.proeng.2017.03.251
14. Storn, R., Price, K.: Technical Report TR-95-012, International Computer Science Institute, University of California (1995)
15. Kennedy, J., Eberhart, R.: IEEE International Conference on Neural Networks, Perth, pp. 1942–1948 (1995)
16. Cerny, V.: J. Optim. Theory Appl. **45**(l), 41 (1985)
17. Kirkpatrick, S., Gelatt, C.D., Vecchi, M.P.: Science **220**(4598), 671 (1983)
18. Camps Echevarría, L., Llanes-Santiago, O., da Silva Neto, A.J.: In: Proceedings of the IEEE Congress on Evolutionary Computation, CEC 2010 (2010), September 2015. https://doi.org/10.1109/CEC.2010.5586357
19. Datta, D., Rui, J.: Appl. Soft Comput. J. **13**(9), 3884 (2013). https://doi.org/10.1016/j.asoc.2013.05.001
20. Casillas Ponce, M.V., Castañón, L.E.G., Puig, V.: J. Hydroinf. **16**(3), 649 (2014). https://doi.org/10.2166/hydro.2013.019
21. Fujiwara, O., Khang, D.B.: Water Resour. Res. **26**(4), 539 (1990)
22. Tospornsampan, J., Kita, I., Ishii, M., Kitamura, Y.: Int. J. Comput. Inf. Syst. Sci. **1**(4), 28 (2007)
23. Rossman, L.A.: Water supply and water resources division. National Risk Management Research Laboratory. EPANET 2 User's Manual. Technical Report. September (2000). http://www.epa.gov/nrmrl/wswrd/dw/epanet.html
24. Talbi, E.G.: Metaheuristics from Design to Implementation (Wiley, London, 2009)
25. Soldevila, A., Blesa, J., Tornil-Sin, S., Duviella, E., Fernandez-canti, R.M., Puig, V.: Control Eng. Pract. **55**, 162 (2016). https://doi.org/10.1016/j.conengprac.2016.07.006

Chapter 7
Determination of Nano-aerosol Size Distribution Using Differential Evolution

Lucas Camargos Borges, Eduarda Cristina de Matos Camargo, João Jorge Ribeiro Damasceno, Fabio de Oliveira Arouca, and Fran Sérgio Lobato

Abstract Nanotechnology characterizes an important area in engineering due to various applications that can be found, such as electronics and pharmaceutical industries, development of air filters, among others. From the environmental point of view, because nanometric particles provide special characteristics to the products, emission of these particles into the air must be limited. Among the approaches proposed in the literature, the electrical mobility technique is an emerging strategy used to ensure an aerosol stream with monodispersed particles. This technique is based on the ability of a charged particle to cross an electrical field. Thus, depending on the size of the particles, the bigger ones will arrive later in the central electrode than the smaller ones, and only a narrow band of sizes will be collected in a slit located at the bottom of the equipment. In order to characterize the relation between the monodispersed and polydispersed aerosol stream, an inverse problem is formulated and solved by using differential evolution. The objective function consists of determining transfer functions that minimize the sum of difference between predicted and experimental concentrations of NaCl obtained by a differential mobility analyzer. The results demonstrated that the proposed methodology was able to obtain a good approximation for two classical transfer functions.

Keywords Monosized nanoparticles · Differential mobility analyzer · Inverse problem · Differential evolution

L. C. Borges · E. C. M. Camargo · J. J. R. Damasceno · F. O. Arouca
NUCAPS - Laboratory of Separation Processes, School of Chemical Engineering, Federal University of Uberlândia, Uberlândia, Brazil
e-mail: damasceno@ufu.br; arouca@ufu.br

F. S. Lobato (✉)
NUCOP - Laboratory of Modeling, Simulation, Control and Optimization, School of Chemical Engineering, Federal University of Uberlândia, Uberlândia, Brazil
e-mail: fslobato@ufu.br

© Springer Nature Switzerland AG 2020
O. Llanes Santiago et al. (eds.), *Computational Intelligence in Emerging Technologies for Engineering Applications*, Studies in Computational Intelligence 872, https://doi.org/10.1007/978-3-030-34409-2_7

123

7.1 Introduction

The study of nanotechnology characterizes an important emerging research area due to various applications that can be found. Among them, we can cite synthetic materials, biotechnology, semiconductor manufacturing, pharmaceuticals, nanocomposites, and ceramics [1–3]. As mentioned by Pui and Chen [4], nanotechnology consists of using technologies to create and/or manipulate materials to develop nano-sized products with very particular characteristics.

From the environmental point of view, the effects of nanoparticles on human health are gaining wide recognition and research is required in areas such as biology, medicine, and engineering [5]. In addition, monodisperse nanoparticles have received considerable attention because of their potential applications in catalysis, energy storage, encapsulation of active material, among others [6, 7].

To characterize the physical behavior of nanoparticles, the particle size distribution is used [8, 9]. The particle size can be determined in an online mode of operation through a number of techniques, such as optical, electrical, and combined physical techniques [10]. Widely used equipment is the electrostatic precipitator (ESP) whose function consists of charging particles of a gas passing through an electric field. As a result, the particles are attracted to the electrode [11]. Among the instruments available to measure particles by electric mobility, we can mention the electrostatically enhanced fibrous filter (EEFF) and the scanning mobility particle sizer (SMPS)—which consists of a condensation particle counter (CPC) and a differential mobility analyzer (DMA).

From the mathematical point of view, the particle trajectory can be predicted considering phenomenological and empirical models. In the phenomenological context, models based on transfer functions are the most used. Thus, Hagwood [12] evaluated the particle trajectories for both non-diffusing and diffusing particles by using the Stolzenburg's transfer function in a DMA. In this case, the effect of particle diffusion is assessed by using a Monte-Carlo method considering particles of sizes 1, 3, 10, 30, and 100 nm. This author concluded that the effect of particle diffusion on the DMA transfer function can be conveniently assessed by using Monte-Carlo simulations without having to neglect axial diffusion or wall losses.

Martinsson et al. [13] developed a method to estimate the non-ideal features of triangular transfer function of individual DMA. This approach considers three DMAs of unknown characteristics, which are used in three rounds of experiments with two DMAs according to a fixed schedule. The method was tested experimentally considering four DMAs of four different designs and manufacturers. The authors concluded that the width of the transfer function could be unambiguously apportioned.

Seol et al. [14] proposed a new DMA to increase the particle-measurement size range considering a variable column length (between 0 and 300 mm). The obtained results considering the Stolzenburg's transfer function demonstrated that the proposed equipment was able to measure particle size from 1 nm to hundreds of nanometers by adjusting the column length.

Karlsson and Martinsson [15] carried out an investigation on the size dependence of the DMA transfer function considering Stolzenburg's model. For this purpose, the woven wire mesh used in DMAs to distribute the sheath-air flow over the cross section of the DMA was analyzed. The authors concluded that the DMA performance is influenced by properties of the wire mesh and that the magnitude of this problem was strongly dependent on the sheath-air flow-rate.

Song et al. [16] studied the experimental and numerical performance of a long differential mobility analyzer (LDMA) in measurements of silver particles in a size range of 5–30 nm. These authors determined the particle trajectory considering Stolzenburg's transfer function. In addition, by using the computational fluid dynamics (CFD), the flow around the aerosol inlet slit was analyzed, and the resulting particle mobility distribution was compared with one for an ideal flow. The experimental measurements of mobility distributions were in good agreement with the theoretical prediction of particle size ranges over 10 nm.

Song and Dhaniyala [17] developed a nanoparticle cross-flow differential mobility analyzer (NCDMA) to characterize the aerosols' size distribution. This new equipment presents a cross-flow configuration for aerosol and sheath flows. To evaluate the NCDMA flow and electrical setup, Stolzenburg's transfer function was considered. These authors concluded that the numerical results obtained by using the NCDMA transfer function in the non-diffusing regime are similar to that of conventional DMA designs.

Ramechecandane et al. [18] studied the dispersion of ultrafine/nano particles considering a medium DMA (M-DMA) and a long DMA (L-DMA). For this purpose, the Langevin equation was proposed to represent the movement of a nanoparticle subjected to an electric field considering the effect of Brownian force on dispersion of ultrafine/nano particles. In addition, the experimental points were modeled by using Knutson & Whitby and Stolzenburg's transfer functions.

Cai et al. [19] studied the effective particle sizing in aerosols with diameters less than 3 nm. These authors developed a miniature cylindrical differential mobility analyzer (mini-cyDMA) for tetra-alkyl ammonium cations. The concentration profiles were represented by using Stolzenburg's transfer function considering particles with electrical mobility diameters of 1.16 nm, 1.47 nm, and 1.78 nm, respectively.

In order to determine the DMA transfer function, i.e., the relation between monodispersed and polydispersed aerosol stream, an inverse problem is formulated and solved by using the differential evolution (DE) algorithm. The objective function consists of determining the transfer functions that minimize the sum of difference between predicted and experimental concentrations of sodium chloride (NaCl). This chapter is structured as follows. Section 7.2 presents a brief description of the nano-DNA and experimental procedure, respectively. In Sect. 7.3, the classical triangular transfer function is described. Section 7.4 presents the proposed methodology and a brief description of the DE strategy, respectively. The obtained results are shown in Sect. 7.5. Finally, the conclusions of this work are drawn in the last section.

Table 7.1 presents the nomenclature used in this chapter.

Table 7.1 Nomenclature

Acronyms	
CFD	Computational fluid dynamics
CPC	Condensation particle counter
DE	Differential evolution
DMA	Differential mobility analyzer
EEFF	Electrostatically enhanced fibrous filter
ESP	Electrostatic precipitator
L-DMA	Long DMA
M-DMA	Medium DMA
Mini-cyDMA	Miniature cylindrical differential mobility analyzer
NCDMA	Nanoparticle cross-flow differential mobility analyzer
SMPS	Scanning mobility particle sizer
Roman symbols	
A	Area
C	Concentration
CR	Crossover probability
F	Perturbation rate
$FWHM$	Full width at half maximum
h	Height
L	Column length
M	Number of experimental data
N	Order for each approximation
NP	Population size
N_0	Particle concentration at the outlet
N_i	Particle concentration at the inlet
N_1, N_2	Number of unknown parameters
OF	Objective function
Q_a	Incoming flows of poly-disperse aerosol
Q_{sh}	Particle free sheath air
Q_m	Outgoing flows of nearly monodisperse aerosol
Q_{ex}	Excess flow
R_1	Inner radius
R_2	Outer radius
V	Voltage
$\mathbf{x}$	Vector of design variables
$\mathbf{X}$	Resulting vector (candidate)
Z_0	Midpoint electrical mobility
$Z_p, \tilde{Z}_p$	Particle mobility
Z_p^*	Particle electrical mobility (reference point)
Greek symbols	
θ	Relative full width at half maximum
λ	Penetration factor
μ	Broadening parameter
Ω	Transfer function

7.2 Nano-differential Mobility Analyzer and Experimental Procedure

A DMA is commonly used to measure the particles' distribution size. In general, DMA particle classifiers pass charged particles through concentric electrodes, as well as those in which an electric field is applied. The particle motion in this electric field is governed by particle mobility, charge, and diffusion. The measurement range of DMAs is limited by two phenomena: particle diffusion and electrical discharge between the electrodes [14].

7.2.1 Nano-DMA

In this contribution, the nano-DMA used to obtain the experimental points was built at the Chemical Engineering School at the Federal University of Uberlñdia, Brazil. The experimental unit is shown in Fig. 7.1.

In this figure, we can observe the following inputs/outputs: A—air coming from a compressor; B—monodisperse flow; C—excess flow; D—sheath flow; E—polydisperse flow. The experimental unit comprised the following elements: 1—high-efficiency air filters (TSI model 3074B); 2—a commercial atomizer (TSI

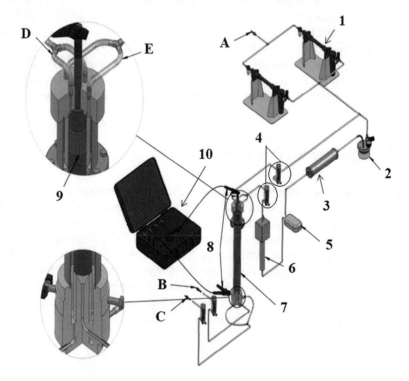

Fig. 7.1 Experimental unit for nanoparticle classification [20]

Model 9302-responsible for generating a polydisperse flow); 3—a diffusion dryer (consisting of two concentric cylinders, with the gap between them filled with silica gel, designed to remove excess moisture of the aerosol flow from the atomizer); 4—rotameters; 5—a mini air compressor (Regent 8500); 6—an aerosol neutralizer (TSI Model 3087); 7—a nano-differential mobility analyzer (nano-DMA); 8—a cable that connects a high-voltage source to the nano-DMA; 9—a perforated acrylic plate to homogenize the dilution air; and 10—a high-voltage source [20]. In addition to the devices mentioned, a condensation particle counter (CPC-TSI Model 3007) identified particle concentration in both the input and output of the nano-DMA.

The nano-DMA consisted of four parts: central electrode, head, nano-DMA body, and base of the equipment, and all parts were concentric stainless steel cylinders. There were four airflow inlets at the top of the DMA, two of them dilution air (enveloped) inlets near the walls of the central electrode, and the other two polydisperse aerosol inlets touching the wall of the DMA body. The internal cylinder, also called central electrode, was solid and had a diameter of 29.9 mm. The external cylinder, or nano-DMA body, had a diameter of 83.9 mm and the head (top cylinder) had an external diameter of 107.5 mm. Both the external cylinder and the top cylinder had 9.2 mm thick walls. The nano-DMA used in this study had a total height of 28 cm. The equipment's top and base were made of technyl because this material has desirable resistance, isolation, and machining characteristics.

The internal electrode was attached to the nano-DMA's head in such a way that it was suspended and separated from the technyl base by only approximately 0.71 mm. This spacing was called a particle classification slit, and the base of the nano-DMA had a central orifice through which the monodisperse aerosol came out. The excess air came out by passing through eight equally arranged orifices of circular shape. A plate was used for distributing the entering enveloped air to avoid the mixture of dilution air and polydisperse aerosol, thereby ensuring a laminar regime. This plate was made of acrylic and was 3 cm thick with small, evenly distributed holes.

7.2.2 Experimental Procedure

To produce particles within the nanometer diameter range, we used ultra-pure water as solvent and aqueous solutions of NaCl prepared at different concentrations (the concentrations were not higher than $3.0 \, gL^{-1}$ due to caking and possible corrosion with salt at high concentrations, which could result in equipment and piping damage). In the nano-DMA, voltages applied in the central cylinder ranged from 0 to 4000 V. Thus, particle diameter in the monodisperse flow ranged from 10 nm to 120 nm depending on the voltage applied. To evaluate the equipment performance, we collected particle concentrations in the monodisperse and excess flows using the NaCl concentration of $0.01 \, gL^{-1}$, $0.1 \, gL^{-1}$, and $0.2 \, gL^{-1}$.

7.3   Transfer Function

As mentioned earlier, the particle trajectory can be predicted considering different kinds of models. In general, these can be written using the concept of transfer function $\Omega(Z_p)$. This is defined as the relation between the particle concentration at the outlet (N_0) and the particle concentration at the inlet (N_i) [16]:

$$N_0 = \Omega(Z_p)N_i, \tag{7.1}$$

where Z_p is the particle mobility.

In the literature, various kinds of equations to represent the $\Omega(Z_p)$ can be found. In this contribution, the classical triangular transfer function will be presented.

7.3.1   Triangular Transfer Function

A DMA transfer function is ideal when it is function of flow rates (Q_a—the incoming flows of polydisperse aerosol; Q_{sh}—particle free sheath air; Q_m—outgoing flows of nearly monodisperse aerosol; and Q_{ex}—excess flow). The shape of the ideal, symmetrical transfer function is triangular (symmetrical flows are used, i.e., $Q_a = Q_m$ and $Q_{sh} = Q_{ex}$) [15]. The (ideal) triangular transfer function may be characterized by its height (h), full width at half maximum ($FWHM$) and area (A) [13]:

$$h = 1 \tag{7.2}$$

$$FWHM = \theta Z_0 = \frac{Q_a}{Q_{sh}} Z_0 \tag{7.3}$$

$$A = h \times FWHM = \frac{Q_a}{Q_{sh}} Z_0, \tag{7.4}$$

where θ is the relative full width at half maximum and Z_0 is the midpoint electrical mobility.

In order to describe the degradation of the transfer function in relation to the ideal case, Martinsson et al. [13] introduced two parameters, i.e., the broadening parameter (μ) and the penetration (λ). Mathematically, this model is given by

$$h = \mu\lambda \tag{7.5}$$

$$FWHM = \theta Z_0 = \frac{Q_a}{\mu Q_{sh}} Z_0 \tag{7.6}$$

$$A = \lambda \frac{Q_a}{Q_{sh}} Z_0, \tag{7.7}$$

where h is obtained from the triangular distribution relation between the area and the half-width, $h = A(FWHM)^{-1}$. The transfer function, which includes the parameters describing the non-ideal behavior, is given by

$$\Omega\left(Z_p\right) = \begin{cases} \lambda\mu\left(1 + \mu\left(Q_{sh}/Q_a\right)\left(\tilde{Z}_p - 1\right)\right), & \text{if } 1 - \mu\left(Q_{sh}/Q_a\right) \leq \tilde{Z}_p \leq 1 \\ \lambda\mu\left(1 + \mu\left(Q_{sh}/Q_a\right)\left(1 - \tilde{Z}_p\right)\right), & \text{if } 1 \leq \tilde{Z}_p \leq 1 + \mu\left(Q_{sh}/Q_a\right) \\ 0, & \text{otherwise} \end{cases},$$
(7.8)

where $\tilde{Z}_p$ is the particle mobility in flow coordinates:

$$\tilde{Z}_p = \frac{Z_p}{Z_p^*} = \frac{4\pi L V Z_p}{Q_{sh} + Q_{ex}}\left(\frac{1}{\ln\left(R_2/R_1\right)}\right),$$
(7.9)

where L is the column length, V is the voltage, Z_p is the particle electrical mobility, Z_p^* is the particle electrical mobility defined in a particular point (reference), and R_1 and R_2 are the inner and outer radius, respectively.

7.4 Methodology

As mentioned earlier, this study aims to estimate a mathematical model to represent the particle distribution size considering experimental data obtained by using the nano-DMA. For this purpose, it is necessary to formulate and solve an inverse problem considering the differential evolution (DE) strategy as the optimization tool. Initially, the user chooses the mathematical model considered, the design variables, and the parameters of the DE algorithm. Then, the optimization strategy generates potential candidates. Each candidate is evaluated according to the mathematical model and the objective function is calculated. This procedure is repeated until a stop criterion is satisfied. The following subsections present the definition of the inverse problem of interest and a brief description of DE, respectively.

7.4.1 Formulation of the Inverse Problem

The proposed inverse problem consists of determining the design variable vector (unknown variables) that minimizes the functional OF (objective function), i.e., the sum of difference between the prediction model and the experimental data:

$$OF \equiv \sum_{i=1}^{M}\left(\Omega_i^{exp}(\tilde{Z}_p) - \Omega_i^{pred}(\tilde{Z}_p, \mathbf{x})\right)^2,$$
(7.10)

where Ω_i^{exp} and Ω_i^{pred} represent the value of experimental transfer function (dependent variable) and predicted transfer function by the model, respectively, $\tilde{Z}_p$

is the electrical mobility, M is the number of experimental data, and $\mathbf{x}$ is the vector of design variables (that depends on each mathematical model considered).

For the triangular transfer function, the predicted transfer function is given by

$$\Omega^{pred}\left(\tilde{Z}_p\right) = \begin{cases} x_1\left(1 + 1\big/x_2\left(\tilde{Z}_p - 1\right)\right), & \text{if } 1 - \beta \leq \tilde{Z}_p \leq 1 \\ x_1\left(1 + 1\big/x_3\left(1 - \tilde{Z}_p\right)\right), & \text{if } 1 \leq \tilde{Z}_p \leq 1 + \beta \\ 0, & \text{otherwise} \end{cases} \tag{7.11}$$

The design variables are x_1 (height) and x_2 and x_3 (broadening parameters). In this case, at $\tilde{Z}_p$ equal to 1, the continuity of this equation is guaranteed considering the same height of the triangle. Thus, for this kind of model, three parameters should be computed by using DE.

The triangular transfer function is a linear polynomial defined by parts. In order to increase the quality of this approximation, in this contribution, a N-order polynomial function defined by parts is proposed. Mathematically, this model is given by

$$\Omega^{pred}\left(\tilde{Z}_p\right) = \begin{cases} \sum\limits_{i=1}^{N_1} x_i\left(\tilde{Z}_p - 1\right)^{N_1-i}, & \text{if } \tilde{Z}_p \leq 1 \\ \sum\limits_{i=1}^{N_2} x_{i+N_1}\left(\tilde{Z}_p - 1\right)^{N_2-i}, & \text{if } \tilde{Z}_p \geq 1 \end{cases} \tag{7.12}$$

where N_1 and N_2 are the number of unknown parameters (x_i, $i=1, \ldots, N_1+N_2$) in the model considered. Therefore, the order for each polynomial is N_1-1 and N_2-1, respectively. For this model, both continuity and differentiability should be guaranteed at $\tilde{Z}_p$ equal to 1.

7.4.2 Differential Evolution

In order to estimate the models that represent the particle size distribution in nano-DMA considered in this contribution, the DE algorithm proposed by Storn and Price [21], is considered. This algorithm consists of the following steps:

- To begin with, an initial population is randomly generated with NP feasible solutions, i.e., the design variable vector satisfies the limits established by the user;
- In general, an individual ($\mathbf{X}_1$) is randomly selected in the population to be replaced. Two other individuals ($\mathbf{X}_2$ and $\mathbf{X}_3$) are randomly selected in the population to perform the vector subtraction;
- The result of the subtraction operation between $\mathbf{X}_2$ and $\mathbf{X}_3$ is weighed by a parameter, namely the perturbation rate (F). This result ($F \times (\mathbf{X}_2 - \mathbf{X}_3)$) is added to individual ($\mathbf{X}_1$). Therefore, the new (potential) candidate ($\mathbf{X}$) is given by: $\mathbf{X} = \mathbf{X}_1 + F \times (\mathbf{X}_2 - \mathbf{X}_3)$. It is important to emphasize that other schemes to generate potential candidates can be used [21];

- If the resulting vector (**X**) has a better value in terms of the objective function, it can replace the previously chosen candidate. This operation happens if a random number generated is less than the crossover probability (CR), also defined by the user. Otherwise, the previously chosen candidate survives in the next generation. This procedure is repeated until NP completes the candidates (formed by new and current individuals);
- To finalize the algorithm, the stopping criteria is defined by the user (generally the maximum number of generations).

7.5   Results and Discussion

To evaluate the proposed methodology in this study, some points should be highlighted.

- To formulate both the inverse problems (triangular and N-order polynomial), the experimental data obtained by Camargo [22] were considered. For this purpose, three NaCl concentrations were considered ($C = [0.01\ 0.1\ 0.2]\,\mathrm{gL}^{-1}$);
- Nano-DMA parameters ($L = 23.5\,\mathrm{cm}$, $R_1 = 1.496\,\mathrm{cm}$, $R_2 = 4.914\,\mathrm{cm}$, $Q_a = 0.5\,\mathrm{L/min}$, $Q_{sh} = 5\,\mathrm{L/min}$, $Q_m = 0.5\,\mathrm{L/min}$, and $Q_{ex} = 5\,\mathrm{L/min}$ [22];
- For the triangular transfer function (Eq. (7.11)), the following design space was considered: $[10^{-9}\ 10^{-9}\ 10^{-9}] \leq [x_1\ x_2\ x_3] \leq [10\ 10\ 10]$;
- For the N-order polynomial transfer function (Eq. (7.12)), $N_1 = 3$ ($\tilde{Z}_p \leq 1$) and $N_2 = 4$ ($\tilde{Z}_p \geq 1$). In this case, the following design space was considered: $[-10\ -10\ -10\ -10\ -10\ -10\ -10] \leq [x_1\ x_2\ x_3\ x_4\ x_5\ x_6\ x_7] \leq [10\ 10\ 10\ 10\ 10\ 10\ 10]$;
- DE parameters in all test cases [21]: population size equal to 50 individuals, number of generations equal to 2500, crossover probability and perturbation rate equal to 0.8, respectively. For these parameters, the $50 + 50 \times 2500$ objective function evaluations are required in each algorithm execution;
- Each case study was simulated ten times considering different seeds for the initialization of the random number generator;
- The stopping criterion adopted in this work was the maximum number of generations.

Table 7.2 presents the results obtained by using the DE strategy to estimate the parameters in both models (Eqs. (7.11) and (7.12)). In general, the results obtained in all the simulations were considered satisfactory, as observed in values obtained for the objective function (average value). In addition, as expected, the increase in the N value implies in a reduction in the objective function value.

The experimental curves showed a characteristic behavior of the asymmetric distribution of nanoparticle sizes in aerosols. For the voltage range worked, particles with diameters between 10 and 120 nm were classified, and the particles with smaller diameters (between 10 and approximately 50 nm) showed an increasing concentration until a maximum concentration at the point critical. For particles

Table 7.2 Parameters estimated by using DE strategy for each concentration considering the polynomial models

$C = 0.01\,\mathrm{gL}^{-1}$					
Triangular	x_1	x_2	x_3		FO
	0.0680	1.2794	2.4253		5.3011×10^{-4}
N-Order	x_1	x_2	x_3		
	−0.1596	−0.0293	0.0680		
	x_4	x_5	x_6	x_7	FO
	−0.0100	−0.0287	0.0118	0.0680	9.3718×10^{-5}
$C = 0.1\,\mathrm{gL}^{-1}$					
Triangular	x_1	x_2	x_3		FO
	0.0786	1.2330	2.4397		6.7887×10^{-5}
N-Order	x_1	x_2	x_3		
	−0.2593	−0.0711	0.0783		
	x_4	x_5	x_6	x_7	FO
	−0.0091	−0.0253	0.0184	0.0783	3.7559×10^{-5}
$C = 0.2\,\mathrm{gL}^{-1}$					
Triangular	x_1	x_2	x_3		FO
	0.0672	1.4238	1.8020		4.554×10^{-4}
N-Order	x_1	x_2	x_3		
	−0.1055	−0.0168	0.0657		
	x_4	x_5	x_6	x_7	FO
	−0.0220	−0.0460	0.0168	0.0657	2.6655×10^{-4}

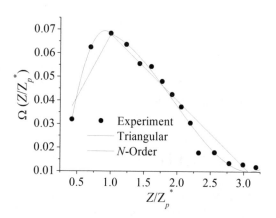

Fig. 7.2 Triangular and N-order transfer functions $(C = 0.01\,\mathrm{gL}^{-1})$

with larger diameters, a gradual reduction of the concentration forming a longer *tail* was observed to the right of the ordinate, producing an asymmetry. Increasing the salt concentration of the NaCl solution causes the particle size distribution peaks to be higher, resulting in a higher concentration of nanoparticles in the aerosol produced. This good agreement between the experimental and predicted values can be observed in Figs. 7.2, 7.3, and 7.4 for each concentration. Finally, it is important

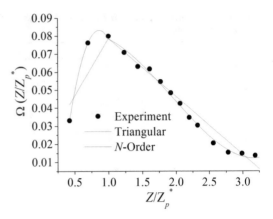

Fig. 7.3 Triangular and N-order transfer functions ($C = 0.1\,\mathrm{gL^{-1}}$)

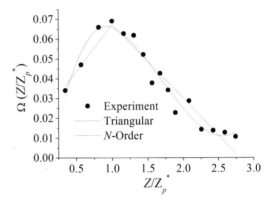

Fig. 7.4 Triangular and N-order transfer functions ($C = 0.2\,\mathrm{gL^{-1}}$)

to emphasize that in all runs, the standard deviation obtained was, approximately, equal to 10^{-9}. This implies that the DE strategy always converged to the solution presented in this table.

7.6 Conclusions

This work proposed and solved inverse problems used to estimate transfer functions in nano-DMA considering a separation of a NaCl polydispersed aerosol stream and the DE strategy. For this purpose, the empirical and phenomenological model parameters were identified to represent the relation between monodispersed and polydispersed concentrations. For both models, the proposed methodology was able to obtain consistent results, as observed by the average and standard deviation values found. From the physical point of view, the nanoparticles presented an asymmetrical size distribution in terms of their concentration in the aerosol produced. Such behavior is in accordance with the standards observed in the literature. Furthermore, the maximum particle concentration is a function of the salt concentration in the solution prepared for atomization. A maximum point is observed in the size distribution curve.

As suggestions for future work, we can mention the formulation and resolution of an inverse problem related to identifying other models to represent this separation process and the solution of an inverse problem considering the insertion of uncertainties (robustness and reliability techniques).

Acknowledgements The authors would like to acknowledge the financial support from FAPEMIG and CNPq agencies.

References

1. Kauffeldt, T., Kleinwechter, H., Schmidt-Ott, A.: Absolute on-line measurement of the magnetic moment of aerosol particles. Chem. Eng. Commun. **151**(1), 169–185 (1995)
2. Gonzalez, D., Nasibulin, A.G., Jiang, H., Queipo, P.: Electrospraying of ferritin solutions for the production of monodisperse iron oxide nanoparticles. Chem. Eng. Commun. **194**, 901–912 (2007)
3. Lee, H., You, S., Pikhitsa, P.V., Kim, J., Kwon, S., Woo, C.G., Choi, M.: Three-dimensional assembly of nanoparticles from charged aerosols. Nano Lett. **11**(1), 119–124 (2011)
4. Pui, D.Y.H, Chen, D.R.: Nanometer particles: a new frontier for multidisciplinary research. J. Aerosol Sci. **28**, 539–554 (1997)
5. Shi, J., Votruba, A.R., Farokhzad, O.C., Langer, R.: Nanotechnology in drug delivery and tissue engineering: from discovery to applications. Nano Lett. **10**, 3223–3230 (2010)
6. Seto, T., Kawakami, Y., Suzuki, N., Hirasawa, M., Aya, N.: Laser synthesis of uniform silicon single nanodots. Nano Lett. **6**, 315–318 (2001)
7. Rosati, J.A., Leith, D., Kim, C.S.: Aerosol Sci. Technol. **37**, 528–535 (2003)
8. Intra, P., Tippayawong, N.: Brownian diffusion effect on nanometer aerosol classification in electrical mobility spectrometer. Korean J. Chem. Eng. **26**(1), 269–276 (2009)
9. Luond, F., Schlatter, J.: Improved monodispersity of size selected aerosol particles with a new charging and selection scheme for tandem DMA setup. J. Aerosol Sci. **62**, 40–55 (2013)
10. Kievit, O., Weiss, M., Verheijen, P.J.T., Marijnissen, J.C.M., Scarlett, B.: The online chemical analysis of single particles using aerosol beams and time of flight mass spectrometry. Chem. Eng. Commun. **151**(1), 79–100 (1995)
11. Zhao, Z.M., Pfeffer, R.: A semi-empirical approach to predict the total collection efficiency of electrostatic precipitators. Chem. Eng. Commun. **148–150**(1), 315–331 (1995)
12. Hagwood, C.: The DMA transfer function with Brownian motion a trajectory/Monte-Carlo approach. Aerosol Sci. Technol. **30**(1), 40–61 (1999)
13. Martinsson, B.G., Karlsson, M.N.A., Frank, G.: Methodology to estimate the transfer function of individual differential mobility analyzers. Aerosol Sci. Technol. **35**(4), 815–823 (2001)
14. Seol, K.S., Yabumoto, J., Takeuchi, K.: A differential mobility analyzer with adjustable column length for wide particle-size-range measurements. Aerosol Sci. **33**, 1481–1492 (2002)
15. Karlsson, M.N.A., Martinsson, B.G.: Methods to measure and predict the transfer function size dependence of individual DMAs. Aerosol Sci. **34**, 603–625 (2003)
16. Song, D.K., Lee, H.M., Chang, H., Kim, S.S., Shimada, M., Okuyama, K.: Performance evaluation of long differential mobility analyzer (LDMA) in measurements of nanoparticles. Aerosol Sci. **37**, 598–615 (2006)
17. Song, D.K., Dhaniyala, S.: Nanoparticle cross-flow differential mobility analyzer (NCDMA): theory and design. Aerosol Sci. **38**, 964–979 (2007)

18. Ramechecandane, S., Beghein, C., Allard, F., Bombardier, P.: Modelling ultrafine/nano particle dispersion in two differential mobility analyzers (M-DMA and L-DMA). Build. Environ. **46**, 2255–2266 (2011)
19. Cai, R., Chen, D-R., Hao, J., Jiang, J.: A miniature cylindrical differential mobility analyzer for sub-3 nm particle sizing. J. Aerosol Sci. **106**, 111–119 (2017)
20. Dalcin, M.G., Nunes, D.M., Damasceno, J.J.R., Arouca, F.O.: Project and construction of a differential mobility analyzer to produce monosized nanoparticles materials. Sci. Forum **802**, 197–202 (2014)
21. Storn, R., Price, K.: Differential evolution: a simple and efficient adaptive scheme for global optimization over continuous spaces. International Computer Science Institute, 12, pp. 1–16 (1995)
22. Camargo, E.C.M.: Performance evaluation of a low cost electrostatic analyzer for classification of nanoparticles sizes. Federal University de Uberlândia (2019, In Portuguese)

Chapter 8
Estimation of Timewise Varying Boundary Heat Flux via Bayesian Filters and Markov Chain Monte Carlo Method

Wellington B. da Silva, Julio C. S. Dutra, Diego C. Knupp, Luiz A. S. Abreu, and Antônio José Silva Neto ⓘ

Abstract This chapter addresses the reconstruction of timewise varying functions within a Bayesian framework. The Markov Chain Monte Carlo method implemented with the Metropolis-Hastings sampler is implemented with a total variation prior model and compared against the results obtained with the Bayesian filter known as SIR (Sampling Importance Resampling). Besides, it is proposed a combination of the Bayesian filter solution (supposed to be obtained online) with the Markov Chain Monte Carlo method solution, consisting of employing the SIR filter solution as the initial state for the Markov Chain Monte Carlo method, allowing for an offline solution refinement with reduced CPU time. An application is presented considering the reconstruction of a boundary heat flux applied to a thermally thin plate. The good results obtained for this application indicate the feasibility of the proposed methodology.

Keywords Inverse problems · Inverse heat conduction · Boundary heat flux estimation · Bayesian filters · Markov chain Monte Carlo methods

8.1 Introduction

The inverse heat conduction problems of estimating timewise and/or spacewise varying functions, like boundary heat fluxes, heat transfer coefficients, or internal sources, have been investigated for a long time with different methods [4, 19, 24, 25, 27, 33, 34, 36–38], motivated by several applications in engineering [2, 9, 10, 12, 17, 20, 28] and diagnosis and therapy in medicine [21]. This subject has gained

W. B. da Silva (✉) · J. C. S. Dutra
Chemical Engineering Program, CCAE-UFES, Alegre, ES, Brazil

D. C. Knupp · L. A. S. Abreu · A. J. Silva Neto
Mechanical Engineering and Energy Department, IPRJ-UERJ, Nova Friburgo, RJ, Brazil
e-mail: diegoknupp@iprj.uerj.br; ajsneto@iprj.br

© Springer Nature Switzerland AG 2020
O. Llanes Santiago et al. (eds.), *Computational Intelligence in Emerging Technologies for Engineering Applications*, Studies in Computational Intelligence 872, https://doi.org/10.1007/978-3-030-34409-2_8

a renewed interest in the last decade, aiming at accurately characterizing the heat fluxes present and other functions in miniaturized electronic chip packaging due to increasing demands of heat dissipation and hot spots remediation [3, 5–7, 13].

Among different methods employed for the formulation and solution of the inverse problem under picture we highlight the use of the Markov Chain Monte Carlo (MCMC) method , implemented within the Bayesian framework, employing prior models derived from the Markov Random Fields theory, as presented by Wang and Zabaras [38] for the estimation of boundary heat flux with timewise or spacewise variations. This subject has also been recently tackled by Orlande et al. [24], aiming at reducing the computational cost required by the MCMC algorithm through the proposition of acceleration strategies such as the approximation error model and the delayed acceptance Metropolis-Hastings.

Other important Bayesian class of methods are the so-called Bayesian filters, which are designed to be employed within online applications mainly, presenting relatively low computational demands if compared to the MCMC algorithm [22]. The most widely known Bayesian filter is the Kalman filter [15, 16, 22]. However, the application of the Kalman filter is limited to linear models with additive Gaussian noises. Extensions of the Kalman filter were developed in the past for less restrictive cases by using linearization techniques [16]. Similarly, Monte Carlo methods have been developed in order to represent the posterior density in terms of random samples and associated weights. Such Monte Carlo methods, usually denoted as particle filters among other designations found in the literature, do not require the restrictive hypotheses of the Kalman filter. Hence, particle filters can be applied to non-linear models with non-Gaussian errors [16]. The estimation of space and time varying boundary heat flux employing the Kalman filter was successfully tackled recently by Pacheco et al. [27].

Taking an example of a timewise varying boundary heat flux reconstruction in a thermally thin plate this work presents a critical comparison of the results obtained with the MCMC algorithm implemented with a total variation prior model [1, 16, 24], against the results obtained with the particle filter known as SIR (Sampling Importance Resampling) [16]. The main contribution of this research is the investigation of combining both methodologies [23]. In this scenario it is considered that the Bayesian filter solution is obtained online in the desired application, i.e., the Bayesian filter solution is constructed, while the experimental data are collected. Next, this solution is set as the initial solutions for the MCMC algorithm, aiming at an offline refinement with quite faster convergence of the Markov chains.

In Sect. 8.2 the inverse problem is formulated through the Bayesian framework, and its solution is discussed first via Markov Chain Monte Carlo methods (MCMC) and, next, via the Sampling Importance Resampling (SIR). In Sect. 8.3 the application chosen as test case for the numerical results is described and the results are presented and critically discussed, demonstrating the advantages of combining the Bayesian filter solution with the MCMC algorithm. Finally, in Sect. 8.4, the main concluding remarks are briefly compiled.

8.2   Inverse Problem Formulation and Solution

The solution of the inverse problem within the Bayesian framework is recast in the form of statistical inference from the *posterior probability density*, which is the statistical model for the conditional probability distribution of the unknown parameters given the measurements [1, 15, 16, 22, 24, 32]. In the present work, two different approaches were addressed: the Markov Chain Monte Carlo Method (MCMC) with the Metropolis-Hastings sampler, and the Bayesian filter known as SIR (Sampling Importance Resampling).

In this context, Bayes' theorem is stated as [11, 16]:

$$\pi_{posterior}(\mathbf{P}) = \pi(\mathbf{P}|\mathbf{Y}) = \frac{\pi_{prior}(\mathbf{P})\pi(\mathbf{Y}|\mathbf{P})}{\pi(\mathbf{Y})} \tag{8.1}$$

where $\pi_{posterior}(\mathbf{P})$ is the posterior probability density, $\pi_{prior}(\mathbf{P})$ is the prior density, modeled with the available prior information regarding the parameters $\mathbf{P}$, $\pi(\mathbf{Y}|\mathbf{P})$ is the likelihood function, and $\pi(\mathbf{Y})$ is the marginal probability density of the measurements, which acts as a normalizing constant.

Assuming the measurement errors are additive and can be modeled as Gaussian random variables, with zero mean and known covariance matrix $\mathbf{W}$, the likelihood function can be expressed as [16]:

$$\pi(\mathbf{Y}|\mathbf{P}) = (2\pi)^{-ND/2}|\mathbf{W}|^{-1/2}\exp\left\{-\frac{1}{2}[\mathbf{Y}-\mathbf{T}(\mathbf{P})]^T\mathbf{W}^{-1}[\mathbf{Y}-\mathbf{T}(\mathbf{P})]\right\} \tag{8.2}$$

where $\mathbf{T}(\mathbf{P})$ is a vector containing the solution of the direct problem at the same location and time instants of the available experimental data $\mathbf{Y}$, employing the parameters values contained in $\mathbf{P}$; and ND is the number of experimental data available, i.e., the dimension of the vector $\mathbf{Y}$. In the application considered in this chapter, there is only one measurement in the spatial domain for each time instant. Therefore, ND corresponds to the number of time instants in which measurements are made. The MCMC algorithm and the SIR filter solution are described next.

8.2.1   The Markov Chain Monte Carlo Method—MCMC

In the MCMC solution the general timewise varying function $f(t)$ is discretized over a regular grid with NP nodes, yielding NP parameters to be estimated:

$$\mathbf{P} = [f_1, f_2, \ldots, f_{NP}]^T \tag{8.3}$$

One way of exploring the posterior probability density is through sampling, for instance, via the Markov Chain Monte Carlo method (MCMC). In this work, the

MCMC is implemented with the Metropolis-Hastings algorithm [15, 16, 22, 23], which is summarized by the following steps:

1. Sample a candidate $\mathbf{P}^*$ from a proposal distribution $p(\mathbf{P}^*, \mathbf{P}^{(n-1)})$.
2. Calculate the acceptance factor:

$$R = \min \left[1, \ \frac{\pi(\mathbf{P}^*|\mathbf{Y}) \, p(\mathbf{P}^{(n-1)}, \mathbf{P}^*)}{\pi(\mathbf{P}^{(n-1)}|\mathbf{Y}) \, p(\mathbf{P}^*, \mathbf{P}^{(n-1)})} \right] \tag{8.4}$$

3. Generate a random value U sampled from a uniform distribution in the interval [0, 1].
4. If $U \leq R$, set $\mathbf{P}^{(n)} = \mathbf{P}^*$. Otherwise, set $\mathbf{P}^{(n)} = \mathbf{P}^{(n-1)}$.
5. Return to step 1.

Hence, a sequence of samples is generated to represent the posterior probability distribution, and inference from this distribution is obtained from the samples $\{\mathbf{P}^{(0)}, \mathbf{P}^{(1)}, \mathbf{P}^{(2)}, \ldots, \mathbf{P}^{(n)}\}$, where the first n_b states before the equilibrium distribution is achieved must be discarded (the burn-in period).

8.2.2 The Sampling Importance Resampling Filter—SIR

The sequential estimation of functions, progressing iteratively as the measurements are acquired, from $k = 1$ up to $k = ND$, is of major interest in several practical applications, especially those demanding on line monitoring, virtual inference, identification, model updating and control [15, 30, 35]. In such problems, the available measured data are used together with prior knowledge about the physical phenomena and the measuring devices, in order to produce estimates of the desired dynamic variables.

In this context, Bayes' theorem given in Eq. (8.1) is rewritten for each time step k, as:

$$\pi_{posterior}(\mathbf{x}_k) = \pi(\mathbf{x}_k|\mathbf{z}_k) = \frac{\pi(\mathbf{x}_k)\pi(\mathbf{z}_k|\mathbf{x}_k)}{\pi(\mathbf{z}_k)} \tag{8.5}$$

where $\mathbf{x}_k$ and $\mathbf{z}_k$ contain, respectively, the sought variables and the measured quantities at the k-th instant. In this scenario, the likelihood function is written as:

$$\pi(\mathbf{z}_k|\hat{\mathbf{z}}_k) = \pi(\mathbf{z}_k|\mathbf{x}_k) = (2\pi)^{-1/2}|\mathbf{W}|^{-1/2} \exp\left\{ -\frac{1}{2}[\mathbf{z}_k - \hat{\mathbf{z}}_k]^T \mathbf{W}^{-1}[\mathbf{z}_k - \hat{\mathbf{z}}_k] \right\} \tag{8.6}$$

where, in the sequential Monte Carlo problem, a set of equations is considered:

(a) *state space model*—represents the dynamic evolution of the state vector $\mathbf{x}_k \in \mathbb{R}^{n_x}$ considering the input vector $\mathbf{u}_k \in \mathbb{R}^m$ and state uncertainty vector $\mathbf{v}_k \in \mathbb{R}^{n_v}$:

$$\mathbf{x}_k = \mathbf{f}_k(\mathbf{x}_{k-1}, \mathbf{u}_{k-1}, \mathbf{v}_{k-1}) \tag{8.7}$$

(b) *observation model*—which provides the solution of the direct problem accounting for the state vector and the measurement uncertainty $\mathbf{n}_k \in \mathbb{R}^{n_z}$:

$$\hat{\mathbf{z}}_k = \mathbf{h}_k(\mathbf{x}_k, \mathbf{n}_k) \tag{8.8}$$

Both functions $\mathbf{f}$ and $\mathbf{h}$ are generally non-linear and the subscript $k = 1, 2, \ldots, NP$ denotes a time instant t_k for a dynamic problem. Where n_x is the dimension of the state and process noise vectors and n_z is the dimension of the measurement vector. The state vector and the input vector, which consider parameters and external signals, contain the variables to be dynamically reconstructed. It can be assumed, without loss of generality, that the state vector is the actual state vector augmented with the input vector, yielding $\mathbf{x} = [\mathbf{x}_{actual}, \mathbf{u}]$.

The evolution and observation models are based on the assumptions that the sequence $\mathbf{x}_k$ depends on the past observations only through its own history, $\pi(\mathbf{x}_k|\mathbf{x}_{k-1})$, and the sequence $\hat{\mathbf{z}}_k$ is a Markovian process with respect to the history of $\mathbf{x}_k$, $\pi(\hat{\mathbf{z}}_k|\mathbf{x}_k)$. For the state and observation noises, it is assumed that the noise vectors $\mathbf{v}_i$ and $\mathbf{v}_j$, as well as $\mathbf{n}_i$ and $\mathbf{n}_j$, are mutually independent, for $i \neq j$, and also mutually independent of the initial state $\mathbf{x}_0$, and the noise vectors $\mathbf{v}_i$ and $\mathbf{n}_j$ are mutually independent for all i and j. Within this framework, different applications can be considered as prediction: filtering, fixed-lag smoothing, and whole-domain smoothing problems [29, 30, 32, 35].

The particle filter method is a Sequential Monte Carlo technique for the solution of the state estimation problem. The key idea is to represent the required posterior density function by a set of N random samples called particles ($\mathbf{x}$) with associated weights ($\mathbf{w}$), given by the set $\{x_k^i, w_k^i\}$ with $i = 1, 2, \ldots, N$ and $k = 1, 2, \ldots, ND$. Therefore, for the Particle Filter, we have $NP = ND$. Then, the estimates can be calculated based on these samples and weights. As the number of samples becomes very large, this Monte Carlo characterization becomes an equivalent representation of the posterior probability function, and the solution approaches the optimal Bayesian estimate. The particle filter algorithms generally make use of an importance density, which is a density proposed to represent the sought posterior density. In this regard, particles are drawn from the importance density instead of the actual density [29, 30, 32, 35].

The sequential application of the particle filter might result in the *degeneracy phenomenon*, where after a few time iterations all but very few particles have negligible weight. The degeneracy implies that a large computational effort is devoted to update particles whose contribution to the approximation of the posterior density function is practically zero. This problem can be overcome with a resampling step in the application of the particle filter. Resampling involves a mapping of the random measure $\{x_k^i, w_k^i\}$ into $\{x_k^{i*}, N^{-1}\}$ with uniform weights; it deals with the elimination of particles originally with low weights and the replication of particles with high weights. Resampling can be performed if the number of effective particles (particles with large weights) falls below a certain threshold number. Alternatively,

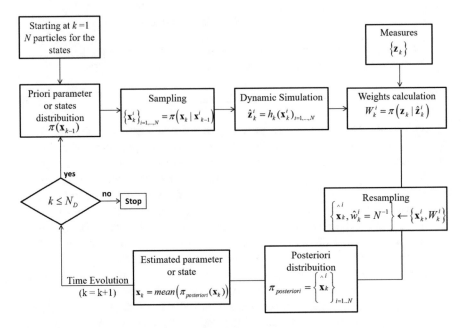

Fig. 8.1 The SIR algorithm (adapted from Ref. [23])

resampling can also be applied indistinctly at every instant t_k, as in the *Sampling Importance Resampling* (SIR) algorithm.

The SIR algorithm is summarized in Fig. 8.1, as applied to the system evolution iteratively from k to $k + 1$. In the first step of the SIR algorithm, it should be noticed that the weights are given directly by the likelihood function $\pi(\mathbf{z}_k | \hat{\mathbf{z}}_k^i)$. Subsequently, in this algorithm, the resampling step is applied at each time instant and then the weights $\hat{w}_k^i = N^{-1}$ are uniform.

8.3 Application

8.3.1 Timewise Varying Heat Flux Estimation

8.3.1.1 Direct Problem

Consider a classical heat conduction problem in a thin plate, with thickness L_{X_1}, to which a time varying heat flux, $q(t)$, is applied at the surface $X_1 = 0$. The opposite surface, $X_1 = L_{X_1}$, exchanges heat with the environment at temperature T_∞ with a heat transfer coefficient H, as schematically represented in Fig. 8.2.

Thus, performing a lumped analysis across the plate thickness, X_1, the heat conduction equation can be simplified to [8, 26]:

Fig. 8.2 Schematic of the
physical problem

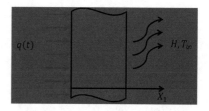

$$C_p \frac{dT(t)}{dt} = \frac{q(t)}{L_{X_1}} - \frac{H}{L_{X_1}}(T(t) - T_\infty), \quad \text{for } t > 0, \tag{8.9a}$$

$$T(0) = T_\infty \tag{8.9b}$$

where T is the plate spatial average temperature and C_p the volumetric heat capacity.

In this work, the parameters H, C_p, and L_{X_1} appearing in the model are considered known, and the goal is to estimate the applied heat flux $q(t)$ employing non-intrusive temperature measurements at the exposed surface, $X_1 = L_{X_1}$.

$$q(t) = \begin{cases} 1000 \, \text{W/m}^2, & \text{for } 1000 \, \text{s} \leq t \leq 2000 \, \text{s} \\ 0 & \text{elsewhere} \end{cases} \tag{8.10}$$

$$q(t) = \begin{cases} (1000 - t) \, \text{W/m}^2, & \text{for } 0 \leq t \leq 1000 \, \text{s} \\ 500 \, \text{W/m}^2 & \text{for } 1500 \, \text{s} \leq t \leq 3000 \, \text{s} \\ 0 & \text{elsewhere} \end{cases} \tag{8.11}$$

8.3.1.2 Inverse Problem

For the inverse problem solution, temperature measurements on the plate's exposed surface are considered to be taken, as obtainable from an infrared thermography system [18], for instance. The synthetic experimental data were simulated by calculating the exact solution for the direct problem, and adding noise with a Gaussian distribution with zero mean and two different standard deviations: $\sigma = 1 \, °C$ and $\sigma = 7 \, °C$. It should be highlighted that although $\sigma = 1 \, °C$ can already be considered a high error for laboratory measurements [18], measurements with $\sigma = 7 \, °C$ are also examined in order to verify the robustness of the methodologies presented.

Considering the Markov Chain Monte Carlo method, the heat flux $q(t)$ was discretized over a grid with 50 nodes, yielding $NP = 50$ parameters to be estimated, $q_k = q(t_k)$, $k = 1, 2, \ldots, 50$. The proposal density was considered as a uniform distribution, $U[P_k^{(n-1)} - \delta, P_k^{(n-1)} + \delta]$, $k = 1, 2, \ldots, NP$, with $\delta = 5$.

For the particle filter, $N = 100$ particles were employed and the boundary heat flux is estimated sequentially according to the number of measurements. The objective of this kind of problem in general includes the estimation of the temperature evolution in addition to the boundary heat flux (the input). Thus, the augmented state vector is given by $\mathbf{x} = [\mathbf{T}, \mathbf{q}]$, where the vectors $\mathbf{T}$ and $\mathbf{q}$ represent, respectively, the timewise variation of the temperature and the heat flux. These variables are related by means of Eq. (8.9), which has to be solved for each sample particle sequentially, for $k = 1, 2, \ldots, ND$.

Hence, for the SIR filter solution the reconstruction of the boundary heat flux is achieved with a discretization equal to the number of measurements ($ND = 300$). In the particle filters framework, two auxiliary models are needed: (1) an evolution model and (2) an observation model. For the temperature evolution model, the spatially averaged temperature was integrated by means of the Runge–Kutta method, to which it was added a Gaussian uncertainty with standard deviation of 1% of the initial temperature, and for the heat flux components evolution model, it was used a random walk, as follows:

$$q_k = q_{k-1} + \sigma_q \varepsilon_q \tag{8.12}$$

where σ_q is the standard deviation related to the model evolution, taken as $300 \, \text{W/m}^2$ (30% of the expected maximum heat flux) in all cases here presented, and ε_q are random numbers with uniform distribution between $[-1, 1]$.

In order to allow for a close comparison regarding the estimates obtained with the different methodologies, it is also calculated the RMS error for the estimated boundary heat flux and the RMS error for the estimated temperatures with respect to the experimental data, defined as:

$$RMS_q = \sqrt{\frac{1}{NP} \sum_{i=1}^{NP} \left[q_i - q_{est,i} \right]^2} \tag{8.13}$$

$$RMS_T = \sqrt{\frac{1}{ND} \sum_{i=1}^{ND} \left[T_{\exp,i} - T_{est,i} \right]^2} \tag{8.14}$$

where q_i corresponds to the discretization of the exact heat flux, $q_{est,i}$ is the discretization of the estimated heat flux, $T_{\exp,i}$ are the experimental data employed, and $T_{est,i}$ are the estimated temperatures, i.e., the calculated temperatures after the solution of the inverse problem.

8.3.1.3 Results

For the computation of the numerical results that are presented in this section, it was considered a plate with the thickness $L_{X_1} = 1.5$ mm made of aluminum with the following properties ($k = 237$ W/mK, $\rho = 2702$ kg/m^3, and $c_p = 903$ J/kg K) [14]. Therefore, $C_p = \rho c_p = 2439$ kJ/m^3 K. The heat transfer coefficient was chosen as $h = 15$ W/m^2 K, typical of natural convection, with initial and room temperatures given by $T(0) = T_\infty = 20\,^\circ$C. The numerical computations were performed on a computer equipped with an Intel CORE i7 processor and 16GB RAM. A final experiment time of 3000 s is considered. Measurements are supposed to be taken at each 10 s, resulting in $NP = 300$ measurements.

In order to model the prior density in MCMC, besides considering a positivity constraint for the heat flux in all cases, a total variation density is employed, as it is expected that the timewise variation of the boundary heat flux be mostly continuous. The total variation prior is given in the form [1, 16, 24]:

$$\pi(\mathbf{P}) \propto \exp\left[-\frac{\gamma}{2}\, TV(\mathbf{P})\right] \tag{8.15}$$

with

$$TV(\mathbf{P}) = \sum_{k=2}^{NP-1} \Delta t\left[|q(t_k) - q(t_{k+1})| + |q(t_k) - q(t_{k-1})|\right] \tag{8.16}$$

where γ is a regularization constant, to be adjusted empirically. A value too small may yield a profile with large fluctuations, whereas the opposite may yield a flat profile. In the total variation prior density it was considered $\gamma = 0.001$ for the cases with a standard deviation of $\sigma = 1\,^\circ$C and $\gamma = 0.00014$ for the cases with a standard deviation of $\sigma = 7\,^\circ$C.

Figure 8.3a, b present the estimated heat flux for the two cases examined in this chapter for the measurements with small standard deviation ($\sigma = 1\,^\circ$C), together with the exact curves, for the three methods described. It is highlighted that the MCMC solution is achieved considering a constant initial state of 500 W/m^2, whereas in the solution obtained with the combination of the SIR filter with the MCMC, the SIR filter solution (supposed to be obtained online) is employed as the initial state for the MCMC method in a refinement stage. For both functional forms of the heat flux, the results show that the SIR filter solution is quite sensitive to the measurement error, as the estimated fluxes present fluctuations of relatively large magnitudes. On the other hand, the MCMC solutions (both employing the

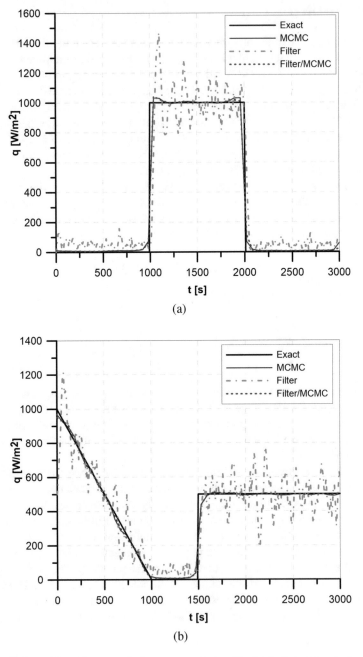

Fig. 8.3 Estimated heat fluxes employing measurements with standard deviation in the experimental error, of 1 °C for (**a**) Case 1; (**b**) Case 2

constant initial state and the SIR filter solution as the initial state) present more accurate results, mainly due to the regularization provided by the total variation prior model. It should also be highlighted that while the MCMC algorithm with constant initial state takes around 7 min to achieve fully converged Markov chains (with around 10,000 states in the worst scenario), the SIR filter solution takes only 30 s to conclude, allowing for online estimation in the problem under picture.

Table 8.1 presents a comparison between the three methodologies employed, confirming that both RMS errors, related to the estimated heat flux and the temperature measurements, are much higher for the estimates obtained with SIR in comparison with the estimates obtained with the MCMC and SIR/MCMC, these two last presenting very similar results in terms of accuracy.

In order to evaluate the gain achieved in employing the SIR filter solution as the initial state for the MCMC algorithm, Fig. 8.4a, b present the Markov chain evolution of a selected parameter in Cases 1 and 2, respectively. For Case 1, presented in Fig. 8.4a, analyzing the parameter q_{25}, which corresponding to the time instant $t = 1500$ s, the gain is slightly higher (4000 states needed in the MCMC solution against 1000 states needed in the SIR/MCMC solution). For Case 2, the parameter q_{20}, corresponding to the time instant $t = 1200$ s, converges after around 2000 states (1.4 min) for the MCMC solution with constant initial state, whereas for the SIR/MCMC solution the same parameter is essentially converged from the very beginning of the iterative process.

Some similar results are now reported for the high noise level case ($\sigma = 7$ °C). Figure 8.5a, b present the estimated heat fluxes together with the exact ones, demonstrating that the MCMC with the total variation prior model remains able to capture the abrupt changes in the heat flux, achieving quite good results, despite the severe conditions of this example. The SIR filter solution, on the other hand, as it is more sensitive to measurement errors, presents some difficulty in capturing the abrupt changes, especially in Case 2. Nonetheless, the SIR solution still provides a good initial state for the MCMC, as it can be observed from the good results achieved with the SIR/MCMC solution [31]. Table 8.2 presents the RMS error for the estimates, allowing for a more objective analysis of this conclusion.

Table 8.1 Calculated parameters of the inverse problem solution with standard deviation of 1 °C

Parameters	MCMC		SIR		SIR/MCMC	
	Case 1	Case 2	Case 1	Case 2	Case 1	Case 2
$RMS_q [\text{W/m}^2]$	19.7	15.86	145.79	110.71	20.8	15.8
$RMS_T [°C]$	1.16099	0.84276	1693.20	1693.61	1.1564	0.84738

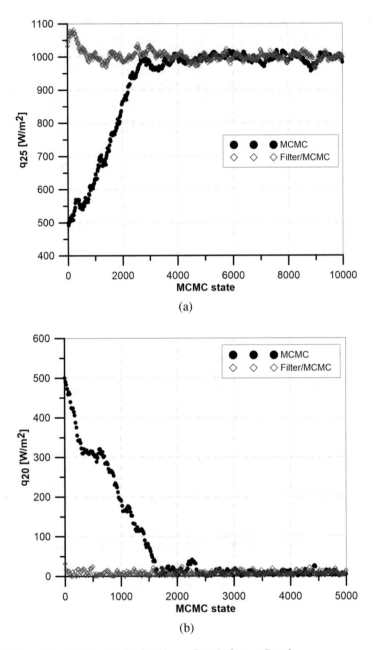

Fig. 8.4 Evolution of Markov chains for (**a**) q_{25}, Case 1; (**b**) q_{20}, Case 2

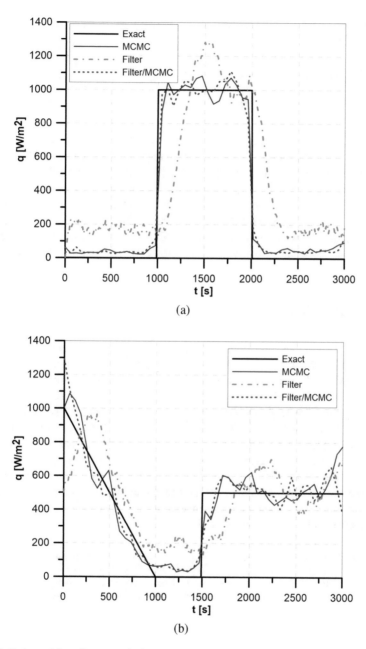

Fig. 8.5 Estimated heat fluxes employing measurements with standard deviation in the experimental error, of 7 °C for (**a**) Case 1; (**b**) Case 2

Table 8.2 Calculated parameters of the inverse problem solution with standard deviation of $7\,°C$

	MCMC		SIR		SIR/MCMC	
Parameters	Case 1	Case 2	Case 1	Case 2	Case 1	Case 2
$RMS_q\,[W/m^2]$	54.4	91.47	320.76	173.62	59.9	86.41
$RMS_T\,[°C]$	7.720801	7.23283	1686.49	1691.72	7.79912	7.2598

Finally, in order to give a whole view of the problem solution, Fig. 8.6 presents the temperature measurements for the case with standard deviation of $7\,°C$, together with the calculated temperatures with MCMC, SIR and SIR/MCMC, for Case 1 in Fig. 8.6a and Case 2 in Fig. 8.6b after the solution of the inverse problem. One may observe that the measurement errors in this case are indeed very high, with the standard deviation representing more than 10% of the maximum temperature span. Regarding the estimated temperatures, the MCMC and SIR/MCMC solutions are quite similar, as expected, while the SIR solution tends to present a lagged behavior.

8.4   Conclusions

This chapter addresses the reconstruction of timewise varying functions, employing the MCMC method, the SIR filter, and the combination of both. For a specific application in a heat conduction problem of heat flux reconstruction, the results showed that the MCMC method with a total variation prior model is very robust, presenting excellent accuracy, even when employing very noisy measurements. The SIR filter presented worse estimations, but with very little computational effort, allowing for online application in the problem considered. Some results were also reported for the combination of both methods, considering an online estimation obtained through the SIR filter, and an offline refinement, in a second stage, employing the MCMC method with the SIR solution as the initial state. This last approach allowed for a faster convergence of the Markov chains in up to 3000 states in comparison with a constant initial state.

Acknowledgements The authors acknowledge the financial support provided by the Brazilian sponsoring agencies FAPERJ, Fundação Carlos Chagas Filho de Amparo à Pesquisa do Estado do Rio de Janeiro, CNPq, Conselho Nacional de Desenvolvimento Científico e Tecnológico and CAPES, Fundação Coordenação de Aperfeiçoamento de pessoal de Nível Superior (Finance Code 001).

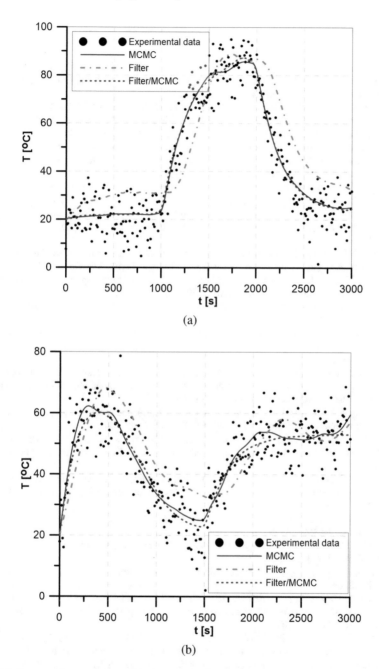

(a)

(b)

Fig. 8.6 Recovered temperature with the estimated heat fluxes employing measurements with standard deviation of 7 °C for (**a**) Case 1; (**b**) Case 2

References

1. Abreu, L.A.S., Orlande, H.R.B., Kaipio, J., Kolehmainen, V., Cotta, R.M., Quaresma, J.N.N.: Identification of contact failures in multilayered composites with the Markov chain Monte Carlo method. J. Heat Transf. **136**(10), 101, 302 (2014)
2. Abreu, L.A., Orlande, H.R., Colaço, M.J., Kaipio, J., Kolehmainen, V., Pacheco, C.C., Cotta, R.M.: Detection of contact failures with the Markov chain Monte Carlo method by using integral transformed measurements. Int. J. Thermal Sci. **132**, 486–497 (2018)
3. Bar-Cohen, A., Wang, P.: Thermal management of on-chip hot spot. In: ASME 2009 Second International Conference on Micro/Nanoscale Heat and Mass Transfer, pp. 553–567. American Society of Mechanical Engineers, New York (2009)
4. Beck, J.V., Woodbury, K.A.: Inverse heat conduction problem: sensitivity coefficient insights, filter coefficients, and intrinsic verification. Int. J. Heat Mass Transf. **97**, 578–588 (2016)
5. Chen, T.C., Hsu, S.J.: Input estimation method in the use of electronic device temperature prediction and heat flux inverse estimation. Numer. Heat Transf. A: Appl. **52**(9), 795–815 (2007)
6. Chen, W.L., Yang, Y.C.: Estimation of the transient heat transfer rate at the boundary of an electronic chip packaging. Numer. Heat Transf. A: Appl. **54**(10), 945–961 (2008)
7. Cheng, J.T., Chen, C.L.: Active thermal management of on-chip hot spots using EWOD-driven droplet microfluidics. Exp. Fluids **49**(6), 1349–1357 (2010)
8. Cotta, R.M., Mikhailov, M.D.: Heat Conduction: Lumped Analysis, Integral Transforms, Symbolic Computation. Wiley, Chichester (1997)
9. Doudard, C., Calloch, S., Hild, F., Roux, S.: Identification of heat source fields from infrared thermography: determination of 'self-heating' in a dual-phase steel by using a dog bone sample. Mech. Mater. **42**(1), 55–62 (2010)
10. Eren, G., Aubreton, O., Meriaudeau, F., Secades, L.S., Fofi, D., Naskali, A.T., Truchetet, F., Ercil, A.: Scanning from heating: 3D shape estimation of transparent objects from local surface heating. Opt. Express **17**(14), 11,457–11,468 (2009)
11. Gamerman, D., Lopes, H.F.: Markov Chain Monte Carlo: Stochastic Simulation for Bayesian Inference, 2nd edn. Chapman and Hall/CRC, Boca Raton (2006)
12. Gu, Y., Wang, L., Chen, W., Zhang, C., He, X.: Application of the meshless generalized finite difference method to inverse heat source problems. Int. J. Heat Mass Transf. **108**, 721–729 (2017)
13. Hetsroni, G., Mosyak, A., Segal, Z.: Nonuniform temperature distribution in electronic devices cooled by flow in parallel microchannels. IEEE Trans. Comput. Packag. Technol. **24**(1), 16–23 (2001)
14. Incropera, F.P., DeWitt, D.P.: Fundamentals of Heat and Mass Transfer. Wiley, New York (2002). OCLC: 439024729
15. Kaipio, J.P., Fox, C.: The Bayesian framework for inverse problems in heat transfer. Heat Transf. Eng. **32**(9), 718–753 (2011)
16. Kaipio, J., Somersalo, E.: Statistical and Computational Inverse Problems, vol. 160. Springer, Berlin (2006)
17. Knupp, D.C., Abreu, L.A.: Explicit boundary heat flux reconstruction employing temperature measurements regularized via truncated eigenfunction expansions. Int. Commun. Heat Mass Transf. **78**, 241–252 (2016)
18. Knupp, D.C., Naveira-Cotta, C.P., Ayres, J.V.C., Orlande, H.R.B., Cotta, R.M.: Space-variable thermophysical properties identification in nanocomposites via integral transforms, Bayesian inference and infrared thermography. Inverse Prob. Sci. Eng. **20**(5), 609–637 (2012)
19. Le Niliot, C., Lefèvre, F.: A method for multiple steady line heat sources identification in a diffusive system: application to an experimental 2D problem. Int. J. Heat Mass Transf. **44**(7), 1425–1438 (2001)
20. Marinetti, S., Vavilov, V.: IR thermographic detection and characterization of hidden corrosion in metals: general analysis. Corros. Sci. **52**(3), 865–872 (2010)

21. Mital, M., Scott, E.P.: Thermal detection of embedded tumors using infrared imaging. J. Biomed. Eng. **129**(1), 33–39 (2007)
22. Orlande, H.R.: Inverse problems in heat transfer: new trends on solution methodologies and applications. J. Heat Transf. **134**(3), 031, 011 (2012)
23. Orlande, H.R.: The use of techniques within the Bayesian framework of statistics for the solution of inverse problems. METTI 6 Advanced School: Thermal Measurements and Inverse Techniques (2015)
24. Orlande, H.R., Dulikravich, G.S., Neumayer, M., Watzenig, D., Colaço, M.J.: Accelerated Bayesian inference for the estimation of spatially varying heat flux in a heat conduction problem. Numer. Heat Transf. A: Appl. **65**(1), 1–25 (2014)
25. Ozisik, M.N.: Inverse Heat Transfer: Fundamentals and Applications. Taylor e Francis, Oxford (2018)
26. Özişik, M.N., Orlande, H.R., Colaço, M.J., Cotta, R.M.: Finite difference methods in heat transfer. CRC Press, Boca Raton (2017)
27. Pacheco, C.C., Orlande, H.R.B., Colaço, M.J., Dulikravich, G.S.: Estimation of a location and time-dependent high-magnitude heat flux in a heat conduction problem using the Kalman filter and the approximation error model. Numer. Heat Transf. A: Appl. **68**(11), 1198–1219 (2015)
28. Pradere, C., Joanicot, M., Batsale, J.C., Toutain, J., Gourdon, C.: Processing of temperature field in chemical microreactors with infrared thermography. Quant. InfraRed Thermogr. J. **3**(1), 117–135 (2006)
29. Ristic, B., Arulampalam, S., Gordon, N.: Beyond the Kalman filter. IEEE Aerosp. Electron. Syst. Mag. **19**(7), 37–38 (2004)
30. Silva, W.B., Rochoux, M., Orlande, H.R.B., Colaço, M.J., Fudym, O., El Hafi, M., Cuenot, B., Ricci, S.: Application of particle filters to regional-scale wildfire spread. High Temp. High Pressures **43**, 415–440 (2014)
31. Silva, W.B., Dutra, J.C.S., Abreu, L.A.S., Knupp, D.C., Silva Neto, A.J.: Estimation of timewise varying boundary heat flux via Bayesian filters and Markov Chain Monte Carlo method. In: 2nd International Symposium of Modeling Applied to Engineering (MAI2016), 18 Convención Científica de Ingeniería y Arquitectura. Havana, Cuba (2016)
32. Silva, W.B., Dutra, J.C.S., Costa, J.M.J., Abreu, L.A.S., Knupp, D.C., Silva Neto, A.J.: A hybrid estimation scheme based on the sequential importance resampling particle filter and the particle swarm optimization (PSO-SIR). In: Computational Intelligence, Optimization and Inverse Problems with Applications in Engineering, pp. 247–261. Springer, Berlin (2019)
33. Silva Neto, A.J., Ozisik, M.N.: Simultaneous estimation of location and timewise-varying strength of a plane heat source. Numer. Heat Transf. A: Appl. **24**(4), 467–477 (1993)
34. Silva Neto, A.J., Ozisik, M.N.: The estimation of space and time dependent strength of a volumetric heat source in a one-dimensional plate. Int. J. Heat Mass Transf. **37**(6), 909–915 (1994)
35. Smith, A.: Sequential Monte Carlo Methods in Practice. Springer, Berlin (2013)
36. Su, J., Hewitt, G.F.: Inverse heat conduction problem of estimating time varying heat transfer coefficient. Numer. Heat Transf.: A: Appl. **45**(8), 777–789 (2004)
37. Su, J., Silva Neto, A.J.: Two-dimensional inverse heat conduction problem of source strength estimation in cylindrical rods. Appl. Math. Model. **25**(10), 861–872 (2001)
38. Wang, J., Zabaras, N.: A Bayesian inference approach to the inverse heat conduction problem. Int. J. Heat Mass Transf. **47**(17), 3927–3941 (2004)

Chapter 9
Health Monitoring of Automotive Batteries in Fast-Charging Conditions Through a Fuzzy Model of the Incremental Capacity

Luciano Sánchez, José Otero, Inés Couso, and David Anseán

Abstract A method for monitoring the condition of lithium iron phosphate (LFP) rechargeable automotive batteries under fast-charging conditions is proposed. A learning fuzzy dynamic model is used for expressing the battery voltage as a sum of the open circuit voltage and the overpotential. The open circuit voltage term depends on a fuzzy rule-based model of the incremental capacity curve, whose linguistic expression is linked to the battery condition. The parameters of the model are learnt from samples of current and voltage obtained in charge and discharge cycles. The performance of the model is assessed in batteries with different states of health.

Keywords Condition monitoring · Soft sensors · Battery health · Fractional order fuzzy models

9.1 Introduction

An accurate data-driven prediction of the state of health (SoH) of the rechargeable batteries of electrical vehicles has undeniable advantages in both daily operation and maintenance routines. There is a growing interest in predicting capacity losses (that limit the range of the vehicle), changes in internal resistance (that limits the

L. Sánchez (✉) · J. Otero
Departamento de Informática, Universidad de Oviedo, Gijón, Spain
e-mail: luciano@uniovi.es; jotero@uniovi.es

I. Couso
Departamento de Estadística e I. O., Universidad de Oviedo, Gijón, Spain
e-mail: couso@uniovi.es

D. Anseán
Departamento de Ingeniería Eléctrica, Electrónica, C. y S., Universidad de Oviedo, Gijón, Spain
e-mail: anseandavid@uniovi.es

© Springer Nature Switzerland AG 2020
O. Llanes Santiago et al. (eds.), *Computational Intelligence in Emerging Technologies for Engineering Applications*, Studies in Computational Intelligence 872, https://doi.org/10.1007/978-3-030-34409-2_9

available power) or abnormal electrochemical reactions (that might compromise the security of the vehicle) [1].

Many deteriorations are gradual and can be related to the capacity degradation, thus they can be assessed by monitoring the discharge capacity. In recent works [2], data-driven methods have been proposed that can predict the cycle life of lithium iron phosphate (LFP) cells using early-cycle data, and it was concluded that the capacity loss during the first 100 cycles is highly correlated with battery life (the number of cycles until 80% of nominal capacity). There are also "silent" deteriorations that do no manifest themselves as a gradual loss of discharge capacity but are detectable in the voltage curves. Differential methods, such as the analysis of dQ/dV (the so-called incremental capacity analysis or ICA), allow to diagnose the causes of the degradation [3, 4]. However, differential methods require that the battery is discharged at low currents [5], or else the kinetic effects mask the changes in the voltage curves on which these techniques are based. As a consequence of this, there is not a widely accepted method for diagnosing silent deteriorations in cells that are cycled under fast-charging conditions.

In this paper a method is proposed for approximating ICA curves in fast-charging conditions. This method relies on a dynamic fuzzy model of the battery voltage as a function of the charging current. The model is learnt with high currents but can extrapolate the voltage curve of the battery to low currents, from which an approximation to the ICA curve can be computed.

This document consists of the following parts: Sect. 9.2 introduces ICA curves and contains a description of the proposed model. Section 9.3 describes an empirical validation on batteries with different degree of deterioration. Section 9.4 concludes this document and suggests lines of future work. Final section is a glossary of terms.

9.2 Model-Based Differential Methods

Incremental capacity analysis (ICA) [5] relates certain battery degradations to changes in the derivative of the charge stored in the battery with respect to the voltage at its terminals (dQ/dV). ICA consists in locating the transitions that occur at certain characteristic voltages while the battery is being discharged. These transitions match the extrema in the IC curve (see Fig. 9.1, left side). Five characteristic peaks are used in this study (numbered 1–5 in the figure). The shapes and locations of these extrema change when the battery degrades, and these changes can be mapped to different deterioration mechanisms [5].

Data for determining IC curves must be captured while the battery is being discharged at low current (discharges from 10 to 25 h). If discharged at a higher rate, the kinetics of the battery mask the transitions in the battery voltage (see Fig. 9.1, right side) and cannot be used for diagnosing purposes.

The main purpose of this contribution is to define a fuzzy model of the derivative of the stored charge with respect to the voltage that can be fitted to data from fast

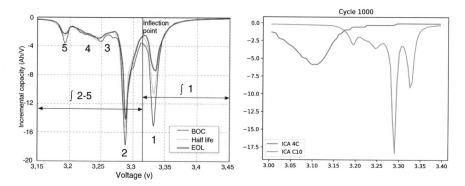

Fig. 9.1 Left: IC curve and characteristic points for new batteries, half-life and end of life. Right: IC curves computed from slow (C10, 10 h, orange line) and fast (4C, 15 min, blue line) discharges. The 4C curve cannot be used for diagnosing the degradation because the characteristic points are masked by the kinetic effects. Curves computed with publicly available data taken from reference [2]

charges and discharges. The model contains a fuzzy knowledge base, where the antecedents of the rules match the extrema of the IC curve and the consequents correspond to the heights of the same curve, so the SoH of the battery can be obtained after a simple exploration of the learned rules. A detailed description of such knowledge base is given in Sect. 9.2.1.

It is remarkable that the proposed model cannot be fitted to data by means of standard regression techniques because the characteristic points of dV/dQ are smoothed out at fast-discharge rates, as mentioned before. On the contrary, the model has to be embedded in a set of differential equations, and fitted to data as explained in Sect. 9.2.2.

9.2.1 Fuzzy Knowledge Base

IC curves are modelled with TSK fuzzy systems of order 0 [6], comprising five rules each (a rule for each characteristic point). Rules have the following form:

Rule i : If V is $\mu(V, c_i, \sigma_i)$ **then** C is h_i,

where the membership function of the antecedent is the derivative of the softsign function [7]

$$\text{softsign}(x) = \frac{x}{1 + |x|} \qquad (9.1)$$

that is

$$\text{softsign}'(x) = \frac{1}{(1 + |x|)^2} \tag{9.2}$$

whose shape is closer to the peaks of the IC curve than the most usual memberships in the literature (i.e., triangular or Gaussian). The antecedent of the i-th rule depends on V, has centre c_i and width σ_i:

$$\mu(V, c_i, \sigma_i) = \text{softsign}' \left(\frac{V - c_i}{\sigma_i} \right) = \left(1 + \left| \frac{V - c_i}{\sigma_i} \right| \right)^{-2}. \tag{9.3}$$

The output of this fuzzy system is

$$\text{ICA(V)} = \sum_{i=1}^{5} \mu(V, c_i, \sigma_i) \cdot h_i. \tag{9.4}$$

9.2.2 Dynamic Behaviour of the Battery

Pseudo-OCV IC curves depend on the open circuit voltage (OCV) of the battery:

$$\text{ICA(OCV)} = \frac{dQ}{dOCV}. \tag{9.5}$$

The open circuit voltage is the voltage at the battery terminals when the charge or discharge current is near zero. The battery voltage V is lower than OCV when the battery is being discharged, and larger when the battery is being charged. The difference between the battery voltage and the open circuit voltage is called "overpotential" (OVP) [8]:

$$\text{OVP}(t) = V(t) - \text{OCV}(t). \tag{9.6}$$

The overpotential has a nonlinear dependence with respect to the current. Assuming that the charge profile is stable (i.e. the charge or discharge current is constant and the battery is not nearly full or empty), this dependence can be modelled in steady state by means of the Tafel equation [9]: resistive behaviour for small currents and a non-linear dependence for large currents:

$$\text{OVP(I)} = \begin{cases} I \cdot \frac{a + b \log(I_0)}{I_0} & \text{if } I \leq I_0 \\ a + b \log(I) & \text{if } I > I_0, \end{cases} \tag{9.7}$$

where b is the Tafel slope, a is introduced for compensating the exchange current density and I_0 is the threshold for "small currents": the overpotential is linear for currents under I_0 and logarithmic for currents higher than I_0.

Recalling Eq. (9.5),

$$\text{ICA(OCV)} = \frac{dQ(t)/dt}{d\text{OCV}(t)/dt} = \frac{I(t)}{d\text{OCV}(t)/dt} \tag{9.8}$$

and

$$\frac{d\text{OCV}(t)}{dt} = \frac{I(t)}{\text{ICA(OCV}(t))} \tag{9.9}$$

therefore

$$\text{OCV}(t) = \text{OCV}(0) + \int_0^t \frac{I(\tau)}{\text{ICA(OCV}(\tau))} d\tau, \tag{9.10}$$

thus the output $\widetilde{V}$ of the dynamic model is

$$\widetilde{V}(t) = \text{OVP}(I(t)) + \text{OCV}(0) + \int_0^t \frac{I(\tau)}{\text{ICA(OCV}(\tau))} d\tau. \tag{9.11}$$

Let C be the capacity of the battery and let also

$$Q_i = Q(t_i) = \frac{1}{C} \int_0^{t_i} I(\tau) d\tau \tag{9.12}$$

be the normalized state of charge. Given a set of data comprising N samples of current $I_i = I(t_i)$, voltage $V_i = V(t_i)$ and normalized charge Q_i, for $i = 1, \ldots, N$, the approximation error of a fuzzy model depends on 18 parameters, namely a, b, I_0 and $c_1, \ldots, c_5, h_1, \ldots, h_5, \sigma_1, \ldots, \sigma_5$: is as follows:

$$\text{err} = \sum \left\{ (\widetilde{V}(t_i) - V_i)^2 \mid i = 1, \ldots, N, \ 0.05 \le Q_i \le 0.95 \right\}, \tag{9.13}$$

where the output of the model when the battery is near empty or full is ignored $(0.05 \le Q_i \le 0.95)$ for Eq. (9.7) being a valid approximation to OVP. This expression has a large computational cost because of the numerical integration in Eq. (9.11), that can be alleviated by means of a change of variables. Observe that the expression of the fuzzy system (Eq. (9.4)) has the following property:

$$\text{OCV}^{-1}(V) = \int_0^V \frac{d\text{OCV}^{-1}}{dv} dv = \int_0^V \text{ICA}(V) dv$$

$$= \sum_{i=1}^5 h_i \sigma_i \left(\text{softsign}\left(\frac{V - c_i}{\sigma_i}\right) - \text{softsign}\left(\frac{-c_i}{\sigma_i}\right) \right), \tag{9.14}$$

where ICA is the pseudo-OCV IC curve (Eq. (9.5)) and OCV^{-1} is the inverse of the OCV function,

$$\text{OCV}^{-1}(\text{OCV}(Q)) = Q \tag{9.15}$$

and the approximation error can be rewritten as

$$\text{err} = \sum \left\{ (\tilde{V}(Q_i) - V_i)^2 \mid i = 1, \ldots, N, \ 0.05 \leq Q_i \leq 0.95 \right\}, \tag{9.16}$$

where

$$\tilde{V}(Q_i) = \text{OVP}(I_i) + \text{OCV}(Q_i) \tag{9.17}$$

and $\text{OCV}(Q_i)$ is the solution of the following equation (recall the definition of $\text{OCV}^{-1}(\cdot)$ in Eq. (9.14)),

$$\text{OCV}^{-1}(v) = Q_i. \tag{9.18}$$

Since solving this equation is much faster than computing the integral in Eqs. (9.11) and (9.16) is the preferred implementation for the approximation error of the model.

9.3 Empirical Study

The empirical study in this section compares IC curves arising from the proposed fuzzy model with IC curves measured at the laboratory. The experimental setup is described in Sect. 9.3.1.

9.3.1 Experimental Setup

In order to fit the model to fast discharges, an instrumental model is fitted first to a C25 discharge. The locations of the characteristic points of the ICA curve (parameters $c_1, \ldots, c_5$) are obtained in this initial learning. The instrumental model is the starting point for a greedy optimization algorithm that learns a second model of the fast charge/discharge data. In other words, a greedy algorithm is launched whose starting point is the set of 18 parameters fitted to the slow discharge data and only 13 parameters are optimized for fast-charging conditions (i.e. all parameters but $c_1, \ldots, c_5$).

The data for the experimentation was obtained with a battery analysis equipment of the Arbin brand, model BT-2000, and a Memmert environmental chamber (see Fig. 9.2). The ambient temperature is 23 °C. Cylindrical batteries of the LFP type

Fig. 9.2 Image of the measurement equipment used in the experimentation

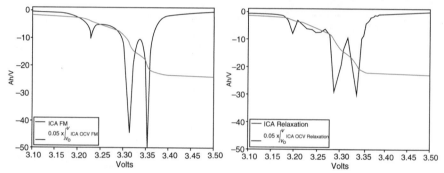

Fig. 9.3 Left: indirect estimation (FM, fuzzy model) of the IC curve at the beginning of the battery life. Right: laboratory estimation (ground truth)

(A123 Systems brand) of 2.3 Ah capacity, commonly used in electric cars, have been used. Batteries have been tested at the beginning of its life, at half-life (3000 cycles) and at the end of his life (6000 cycles).

All methods were coded in Python. Python packages Keras 1.0.4, TensorFlow 1.10.1 and PyFlux 0.4.15 were used. Experiments were carried on a workstation that is based on an Intel Xeon E3-1226 processor, has 16Gb of RAM and a Tesla K40c GPU accelerator.

9.3.2 Compared Diagnosis of Battery Deterioration

Fuzzy model-based (FM) IC curves of batteries at the beginning of its life, at 3000 and 6000 cycles are shown in the left side of Figs. 9.3, 9.4 and 9.5, respectively. The integral of that curve (OCV^{-1}) is plotted in the same figures. Batteries were discharged at C1 (2.3 A).

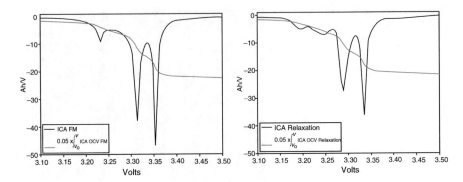

Fig. 9.4 Left: indirect estimation (FM, fuzzy model) of the IC curve at the middle of the battery life. Right: laboratory estimation (ground truth)

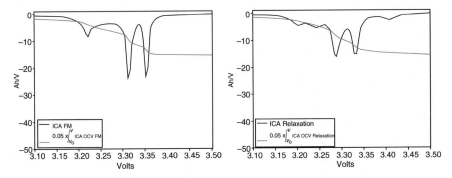

Fig. 9.5 Left: indirect estimation (FM, fuzzy model) of the IC curve at the end of the battery life. Right: laboratory estimation (ground truth)

The evolution of the area of the IC curve to the left and to the right of the minimum between peaks 1 and 2 is used to diagnose different types of deterioration. In the right side of the same figures, the "ground truth" curves, as measured in the laboratory for a slow discharge (C25), are plotted.

The relative differences between these areas are coherent with the laboratory estimations. Note that the deterioration between cycles 3000 and 6000 is noticeable to the naked eye (height difference between Figs. 9.4 and 9.5) but the anomalous peak due to electrodeposition (deformation in 3.42 V, right side of Fig. 9.5) was not detected by the diagnostic in fast-charging conditions.

Table 9.1 includes a numerical comparison between proposed model and the reference diagnosis. Following [5], the areas described in Table 9.1 were measured when the batteries were new, at half-life and at the end of life. The results of the fuzzy model are boldfaced. The fact that the values of the fuzzy estimations of these areas are similar to the ground truth validates the hypothesis that the model can extrapolate measurements taken at high currents and recover information that is only detectable at low currents.

Table 9.1 Areas (in Ah) between the different extrema of the incremental capacity curve of the battery. The results of the fuzzy model are boldfaced

	Method	New battery	Half-life	End of life
Marker #1	Laboratory	147	151	102
Marker #1	Fuzzy model	**158**	**166**	**81**
Markers #2–#5	Laboratory	312	281	211
Markers #2–#5	Fuzzy model	**332**	**292**	**230**

The diagnosis made in the laboratory coincides with the fuzzy estimation: up to half of the battery's life, peak area 1 is maintained and area 2–5 decreases slightly. Between the middle and the end of life both area 1 and area 2–5 decrease: there are effects of type "Loss of Lithium Inventory" and "Loss of Active Material in the negative electrode"

In both cases, at half-life, peak 1 area is maintained and area 2–5 decreases slightly. Between the middle and the end of life, both area 1 and area 2–5 decrease. This is compatible with LLI (loss of lithium inventory) and LAM (NE) (loss of active material—negative electrode) deteriorations [5].

9.4 Concluding Remarks

This study presents a method for estimating ICA curves from fast-discharge data. It has been concluded that the method can detect certain types of deterioration, such as loss of inventory of lithium and active material in the electrodes, but the kinetic effects mask other degradations that can only be detected with differential methods when the currents are low.

In future work we will develop a dynamical model of the overpotential for high currents and introduce the temperature in the model. The temperature can be safely discarded in the tests at low current, but it is a relevant factor that influences the voltage curves in fast charges and discharges. Given that there is a low dependence between the OCV and the temperature in LFP cells, it is expected that pseudo-OCV ICA curves are also independent of the temperature, but the overpotential is heavily affected and the model-based estimation of the OCV may lead to inconclusive results if the temperature of the battery was not properly controlled or contemplated in the model.

It is also remarked that the overpotential model used in this paper is equivalent to a non-linear resistance, which is a memoryless model. However, there are cases where this simplification is not adequate. In particular, if a fast discharge is performed in the latter cycles of the battery life, when the battery is almost depleted the overpotential grows much faster that it would do for batteries in a good condition. The resistive analogy is not adequate for this case, and it would be needed that the parameters of the resistance depend on the stored charge or else an impedance with memory is adopted. This is a limitation of this method that should be studied in detail in further studies.

Glossary of Terms

C Capacity of the battery
c_i Parameter of the antecedent of a fuzzy rule (centre)
h_i Parameter of the consequent of a fuzzy rule (height)
I Charge or discharge current
IC Incremental capacity
μ Membership function
OCV Open circuit voltage (voltage of the battery when $I = 0$)
OVP Overpotential (difference between V and OCV)
Q Battery charge
σ_i Parameter of the antecedent of a fuzzy rule (width)
V Battery voltage
$\widetilde{V}$ Output of the voltage model

Acknowledgements This work has been funded by the Ministry of Science and Innovation of the Government of Spain, Regional Ministry of the Principado de Asturias and by the FEDER funds of the European Community through projects with codes FC-IDI/2018/000226, TEC2016-80700-R (AEI / FEDER, UE) and TIN2017-84804-R.

References

1. Vetter, J., Novák, P., Wagner, M.R., Veit, C., Möller, K.C., Besenhard, J., Winter, M., Wohlfahrt-Mehrens, M., Vogler, C., Hammouche, A.: J. Power Sources **147**(1-2), 269 (2005)
2. Severson, K.A., Attia, P.M., Jin, N., Perkins, N., Jiang, B., Yang, Z., Chen, M.H., Aykol, M., Herring, P.K., Fraggedakis, D., Bazant, M.Z., Harris, S.J., Chueh, W.C., Braatz, R.D.: Data-driven prediction of battery cycle life before capacity degradation. Nat. Energy. **4**(5), 383 (2019)
3. Sánchez, L., Blanco, C., Antón, J.C., García, V., González, M., Viera, J.C.: IEEE Trans. Ind. Electr. **62**(1), 555 (2015)
4. Sánchez, L., Couso, I., Otero, J., Echevarría, Y., Anseán, D.: A model-based virtual sensor for condition monitoring of Li-Ion batteries in cyber-physical vehicle systems. In: J. Sens. **2017**, Article ID 9643279, 12 (2017) https://doi.org/10.1155/2017/9643279
5. Dubarry, M., Truchot, C., Liaw, B.Y.: J. Power Sources **219**, 204 (2012)
6. Takagi, T., Sugeno, M.: IEEE Trans. Syst. Man, Cybern. **1**, 116 (1985)
7. Glorot, X., Bengio, Y.: In: Proceedings of the Thirteenth International Conference on Artificial Intelligence and Statistics, pp. 249–256 (2010)
8. Bard, A.J., Faulkner, L.R., Leddy, J., Zoski, C.G.: Electrochemical Methods: Fundamentals and Applications, vol. 2. Wiley, New York (1980)
9. Heubner, C., Schneider, M., Michaelis, A.: J. Power Sources **288**, 115 (2015)

Chapter 10
Fault Detection and Isolation in Smart Grid Devices Using Probabilistic Boolean Networks

Pedro J. Rivera-Torres and Orestes Llanes Santiago (iD)

Abstract The area of smart power systems needs continuous improvement of its efficiency and reliability, to produce power with optimal quality in a resilient, fault-tolerant grid. Components must be highly reliable, properly maintained, and the occurrence of faults and failures has to be studied. Guaranteeing correct system operation to performance specifications involving the aforementioned activities is an active research area that applies novel methodology to the detection, classification, and isolation of faults and failures, modeling and simulating processes using predictive algorithms, with innovative AI techniques. To maintain complex power grids, predictive analytics is necessary, as employing it to plan and perform activities lowers maintenance costs and minimizes downtime. Detecting multiple faults in dynamic systems is a difficult task. Biomimetic methodologies have been applied widely in engineering systems to solve many complex problems of this field. This contribution presents a complex-adaptive bioinformatic, self-organizing framework, probabilistic Boolean networks (PBN), as a means to understand the rules that govern dynamic power systems, and to model and analyze their behavior. They are used to describe Gene Regulatory Networks, but have been recently expanding to other fields. PBNs can model system behavior, and with model checking and formal logic, assure the process' mathematical correct-ness. They enable designers, reliability and electrical engineers, and other experts to make intelligent decisions, since PBNs self-organize into attractors that model the system's operating modes, permit design for reliability, create intelligent fault diagnosis systems, assist the reliability engineering design process, and use data to analyze a system's behavior, achieving predictive maintenance.

P. J. Rivera-Torres (✉)
Department of Computer Science-School of Natural Sciences, University of Puerto Rico at Río Piedras, San Juan, PR, USA
e-mail: pedro.rivera1@upr.edu

O. Llanes Santiago
Department of Automation and Computing, Universidad Tecnológica de La Habana José Antonio Echeverría, CUJAE, La Habana, Cuba
e-mail: orestes@tesla.cujae.edu.cu

© Springer Nature Switzerland AG 2020   165
O. Llanes Santiago et al. (eds.), *Computational Intelligence in Emerging Technologies for Engineering Applications*, Studies in Computational Intelligence 872, https://doi.org/10.1007/978-3-030-34409-2_10

Keywords Smart power systems · Fault detection and isolation · Probabilistic
Boolean networks · Reliability · Failure modes

10.1 Introduction

Electrical power distribution systems (EPDSs) are complex systems that can be
affected by a wide range of events that can cause faults. Restoration procedures may
take anywhere from minutes to hours. One way to accelerate restoration procedures
is by inserting smart devices in the EPDSs, intended to automatically avoid failure
or coordinate actions for a prompt recuperation. One of these mechanisms is
the intelligent power router (IPR). An IPR is a device inspired on data routers,
which has embedded intelligence that allows switching power lines and shedding
loads. In the event of a power or component failure, each IPR makes a local
decision and coordinates actions with routers in its vicinity to bring the system
back into operation, realizing thus, a peer-to-peer mesh automatic decision making
architecture. Figure 10.1 depicts the basic elements of an IPR unit.

For an EPDS endowed with IPRs (EPDS-IPRs) to have the capacity of restoring
service automatically, a network of IPRs must be strategically deployed over the
power grid, and be programmed to exchange information and reconfigure the
network following a set of pre-specified rules, whenever a perturbation occur. The
central hypothesis behind EPDS-IPR systems design is:

> If IPRs are properly distributed over the network, and are endowed with the proper
> intelligence and control functions, EPDSs survivability can be achieved for a predefined set
> of system perturbation events, while maintaining an optimal use of the system's resources

The design of an EPDS-IPR capable of satisfying this hypothesis is a very
complex endeavor. No explicit mathematical model is available to guide this design.
The main contribution of this chapter is to provide a methodology based on a
probabilistic Boolean network (PBN) model of IPRs, which is formally testable
with model checking techniques, and allows detection of multiple faults. This model
allows the simulation of several scenarios for an IPR, and as such, it has the
potential to provide enough insights for the allocation, design, and programming
of the nodes of an IPR network. The ultimate aim is to endow the EPDS-IPR

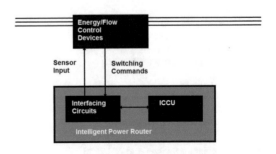

Fig. 10.1 Basic elements of an IPR

with enough intelligence to survive a fairly large set of different disturbing events. EPDSs need to comply with laws and regulations (environmental, safety, health, etc.), to produce electrical power while maintaining high standards, efficiency, and committing to continuous improvement. It is not trivial to stop power generation because of the importance of this service's continuity in modern society, but there may be unforeseen events that affect negatively the availability of systems, and the provision of power to clients. EPDSs have to be operated properly, and this task compensates the effects that disturbances and changes can have, assuring the continuous operation within performance specifications. To achieve this goal, faults and failures have to be detected, isolated, and eliminated; all of which are tasks related to fault detection and isolation (FDI) [22]. FDI methods are mainly divided in two categories, which are model-based and process-history-based [40–42]. Model-based methods make use of either an analytical or computational model of the systems. A varied spectrum of the reviewed model-based methods are based on the parity space, observer approach and parameters identification or estimation approach concepts [16, 20, 21, 45]. For models that are non-linear in nature, the complexity on the observer design method increases, whereas a precise system model is needed for the parity space approach [16]. To overcome these problems a more recent approach based in the solution of an inverse problem using computational intelligence tools has been presented [1, 6, 8]. In general, the developed researches have been limited to the diagnosis of independently occurring faults.

Diagnosing simultaneous faults is an area not sufficiently addressed in scientific literature. Multiple fault detection in dynamic systems can be challenging, because the effects of a fault may hide or be compensated with the effects of different type of faults, and because equal types of multiple faults can manifest themselves in different forms, considering their order of occurrence. The computational intelligence tools have been the most used to address this area [7, 33, 37, 43]. In this sense, research has focused on static systems [38], solutions to the multiple faults problem through observations on imperfect tests as in [34], to determine the closest evolution relative to the state of the fault. The authors in [44] postulate an algorithm-based pattern recognition method for diagnostics, which resulted in high efficiency and precision, but with cases in experimental data where particular fault tests did not have a solution. Other developments include SLAT patterns for multiple fault diagnosis [5], and model-based methods for describing multiple faults in rotor systems [3]. However, multiple fault diagnosis is a current research area which demands the development of novel strategies for improving the performances of the fault diagnosis systems.

Biomimetic methodologies, or those that mimic the behavior of biological systems, are widely used in power system design and analysis for the solution of many complex problems. Qualitative frameworks such as PBNs allow describing large biological networks without loss of important system properties, and allowing the representation of complex behavior, such as self-organization. PBNs are used to model gene regulatory networks (GRNs); collections of DNA segments inside a cell that interact indirectly with other segments and substances in it to regulate/govern the expression level of genes [27, 35]. They are used to understand the general rules that govern gene regulation in genomic DNA. PBNs are transition systems that satisfy the Markov property (memoryless, not dependent on the history of the system). Proposed by Shmulevich [35] by extending Kauffman's Boolean network (BN) concept, they combine the rule-based modeling of Boolean networks with uncertainty principles. These probabilistic Boolean networks consist of a series of constituent BNs that have assigned selection probabilities, where each BN may be considered a "context." Data for each cell comes from dissimilar sources; where each one represents a cell context. In each given point in time t, a system can be governed by one of these constituent BNs, and the system switches to another constituent BN at another time, with a given switching probability. PBNs for manufacturing systems were introduced in [31] and further developed in [29, 30, 32].

In genomic research, one of the central focuses is to understand the way in which cells control and execute a large number of operations that are required for their function. Such systems are parallel and remarkably integrated, and an approach that considers a perspective superior to that of a single gene is necessary in order to gain a better insight into biological processes. Genes, cells, and molecules are a networked system that needs to be understood in order to create better medicines and medicine delivery mechanisms for correcting human disease. That requires computational methods and formal verification to process large amounts of data, understand the principles that rule the system, and make meaningful predictions about the behavior of the system given a set of conditions. EPDSs are similar to gene regulatory networks because in order to better understand the general rules that govern the system, and to make predictions on how the system will behave, endure, or fail under a set of circumstances, a model that accurately describes the system and its behavior is needed. The coordinated interaction and regulation between genes and their products form these networks, in which gene expression is key. Several mathematical constructs have been proposed for modeling GRNs, including Boolean networks (BNs), Bayesian networks, regression, and Markov chains.

Here, the use of PBNs, which has already been applied successfully in manufacturing systems, will be expanded to allow the analysis of IPR reliability, and the consideration of faults that may lead to catastrophic failure, being this a first contribution of this research. The proposed model allows detection and classification of single and multiple faults which constitute another contribution of the proposal. It allows identification of fault states in which it is possible to continue operation, and those where it is not possible to continue (failure). It also allows to forecast a time in hours by which the fault or failure will be imminent. As a final contribution, the system provides information about the maximum probability of fault and failure occurrence, which allows better maintenance planning. This chapter is organized in the following manner: Sect. 10.2 discusses probabilistic Boolean networks and their use in systems modeling, Sect. 10.3 presents how these PBNs can be used for FDI in these systems. Section 10.4 discusses the experimental results. Finally, the conclusions of this research and future works are presented.

10.2 Probabilistic Boolean Networks and Model Checking

Boolean networks [23, 24] and probabilistic Boolean networks [35, 36] have utilized for modeling systems and process' dynamics, and predict their future behaviors with statistical analysis and discrete event simulation. This use has been very well documented in literature, for modeling biological systems [2, 4, 9, 14, 17], and for modeling GRNs [10–13, 17, 25, 39]. The mechanism of intervention [36] is used to steer the evolution of the network and guide it away from undesired states, such as those associated with disease.

BNs are a finite set of Boolean nodes, with states quantized to 0 or 1 (they allow for different quantizations), for which their state is determined by the current state of other nodes in the BN. Their finite set of input nodes are called regulatory nodes, and there is a set of Boolean functions that are known as predictors, that regulate the value of a target node. If the set of nodes and their corresponding functions are defined, the BN is defined. PBNs are basically a collection or tree of BNs for which at any discrete time interval, the node state vector transitions are based on one of the rules of the constituent BNs. These context-sensitive, dynamical, and probabilistic BNs satisfy the Markov property.

In [31], the authors demonstrated that PBNs are valid for modeling systems outside the biological realm; such as manufacturing and industrial systems, by establishing a system model, validating it through model checking, and comparing the results obtained through simulation with actual machine data. In [32], the authors used the same methodology applied to a manufacturing process to obtain quantitative occurrence data for DFMEA. In [30], the authors expanded the application of PBNs in industrial manufacturing systems by incorporating the intervention mechanism to guide a modeled manufacturing system away from possible failure modes, thus delaying eventual failure of the system.

A BN as defined in [36, 37] is a graph $G(V, F)$ consisting of a set of nodes $V = \{v_1, v_2, ..., v_n\}$ and a set of Boolean functions $F = \{f_1, f_2, ..., f_n\}$. Each node $v_i \in \{0, 1\}$, $i = 1, ..., n$ is a binary variable. The value of each node is determined by the value of nodes $v_{j1}, v_{j2}, ..., v_{jk_i}$ at time t, by a function $f_i : \{0, 1\}^{k_i} \rightarrow \{0, 1\}$. In other words, there exist k_i regulatory nodes assigned to v_i. These regulatory nodes determine the value of v_i. Thus, in time $t + 1$, $v_i(t + 1) = f_i(v_{j1}(t), v_{j2}(t), ..., v_{jk_i}(t))$. There are 2^n possible states that the network can assume. $v_i(t)$ defines the state of the node at a time t, and the Boolean functions dictate the regulatory interaction between nodes. $v(t) = (v_1(t), v_2(t), ..., v_n(t))^T$ is called the GAP (gene activity profile) and gives us a map of the state of all the nodes in a given time.

PBNs extend BNs by incorporating stochasticity. A PBN is, according to [4], a collection of BNs. One of the constituent networks rules node activity for a random period, before another one is chosen randomly to govern the activity in response to an (random) event. PBNs are, like BNs, graphs, with a set of nodes $V = \{v_1, v_2, ..., v_n\}$ where $v_i \in \{0, 1\}$, $i = 1, 2, .., n$ and a sequence $\{\mathbf{f}_l\}$, $l = 1, ..., m$ of vector-valued functions that define the networks. Each of these functions in the form of $\mathbf{f_l} = (f_l^{(1)}, f_l^{(2)}, ..., f_l^{(n)})$ determines a context, or instance of a constituent network (BN) of the PBN. Instead of a single Boolean function, there are several predictor functions $f_l^{(i)} : \{0, 1\}^n \to \{0, 1\}$, where $l = 1, ..., l(i)$ and $l(i) \leq 2^{2^n}$ (the number of possible BNs in node v_i). The selection probability in a PBN is given as $\{c_l\}_{l=1}^m$, $\sum_{i=1}^n c_{ji}^{(i)} = 1$.

Model checking is performed on transition systems that admit a representation as a Kripke structure. A Kripke structure is a quadruple (S, S_0, R, L), where S is a finite set of states, S_0 a subset of S consisting of the initial states, R is a transition relation defined on all pairs of states, and L is a mapping that associates to each state s a set of atomic logical propositions describing the characteristics of s that are of interest for the system. The model-checking problem is the following system's property verification problem:

Given a Kripke structure $M = (S, S_0, R, L)$ and a logic formula Ψ expressing a property of the system modeled by M, find all states in S that satisfy Ψ

In this chapter, Ψ shall be expressed as a computation tree logic (CTL) formula. CTL has been developed to describe properties on computation trees. These are trees whose nodes represent states and whose edges, the transitions between the states. A CTL formula is a well-posed logical expression composed by the usual atomic propositions and Boolean logic connectives, but with the addition of temporal operators and path quantifiers. There are four temporal operators for describing the validity of a proposition p along a path. A path in a Kripke structure is a sequence of states: $s_0, s_1, ..., s_n$, where s_0 is the initial state. The temporal operators are

- X_p which means that p holds in the *next state* in the path;
- F_p which means that p holds at *some future state* along the path;
- G_p which means that p *holds always on the path*;
- $p \bigcup q$ which means that p holds until q holds.

Each temporal operator must be immediately preceded by either of the path quantifiers

- A, which means for all paths or
- E, which means for all paths.

As an example, consider the formula:

$$AG(Shut\ Down\ Request \to AX\ Shut\ Down\ Execute)$$

which means that wherever a shutdown is requested, the shutdown is executed in the next time step.

A model checker is an algorithm for solving the model-checking problem. These algorithms have encoded the Kripke structure representing the system under study, and receive as input a CTL formula Ψ. The method works by recursively labeling the state graph with the sub-formulas of Ψ and then parsing the graph to compute the truth-value of each sub-formula. The output could be either that Ψ is true or a counterexample trace showing why property Ψ is false.

Automated verification techniques, such as model checking, provide powerful methods for rigorously analyzing the correctness of systems. Increasingly, this analysis takes into account quantitative aspects of the systems being verified, including real-time and probabilistic behavior. Systems often exhibit stochastic behavior due, for example, to component failures, unreliable communication media, or the use of randomization. But yet another vital ingredient in system modeling is non-determinism, which is often used to capture concurrency between parallel components and to under-specify or abstract certain aspects of a system. The need for automated formal verification techniques in all these domains is clear. Thus, while a Kripke structure is a basic abstraction for capturing the elements that are common to all systems that can be checked with a model checker, several extensions of the Kripke structure and of the CTL language have been proposed to tailor the model to the particularities of the system under study. The best-known extensions are:

1. Discrete-time Markov chains (DTMCs)
2. Continuous-time Markov chains (CTMCs)
3. Markov decision processes (MDPs)
4. Probabilistic automata (PAs)
5. Probabilistic timed automata (PTAs)

The verification of reliability properties of an EPDS-IPR system shall be done with the MDP paradigm. Examples of these properties are TCL statements for:

1. *The maximum probability of breaker b_j to complete an action within 0.02 s or*
2. *The minimum probability that software s_j correctly answer to a signal within 1 s.*

The MDP model is very close to the original Kripke structure. The main changes are the inclusion of probabilities in the transitions and the extension of the CTL language to the continuous stochastic logic (CSL). This logic allows for the formulation of time-bounded statements and the quantification of the probabilities for their occurrence. For example, statements of the form: "Ψ is true with probability P within 10 h" or even, "State s is true after 10 h with probability P." Property specifications are expressed in an extension of CTL called probabilistic continuous time logic (PCTL). PTAs models are much harder to construct and much more computer intensive. However, the need for automated formal verification tools for PTAs is clear and demand for software tools is increasing.

10.3   EPDS, Intelligent Power Routers, and PBN Modeling

Electricity generated at power plants is the result of transforming different energy sources, such as nuclear, geothermal, solar, and others, into raw energy. The management of this energy is guided by the need to ensure a permanent supply of high quality electrical power, important pillar of the success of modern economies. Power plants have grown in size and generation capacity since they were devised more than a 150 years ago. Generation generally occurs far away from where the power is ultimately consumed, and in such scheme, consumers may be separated from plants by thousands of kilometers.

There are two distinguishable networks that connect end users to generation sites. These are:

- *Transmission or transport networks*: they cover a rather large area, making sure that all provinces/areas of a given country are covered. These run at rather high voltages (230 kV or 138 kV). High voltages at transmission allow the network to minimize losses. Different lines are joined at transmission substations, and these networks eventually connect and feed electricity to the lower voltage distribution network, that reaches the end-customer.
- *Distribution networks*: Distribution networks are designed to cover a smaller area and run at a lower voltage than transmission networks, but they are also denser, as they deliver electrical power to the end users. They use lower voltages for several reasons, including security, and installation costs. They also have the ability to provide different voltage levels to different types of consumers, through the use of transformers.

Industrial and residential customers need to receive reliable power. There are several factors, natural and induced, in the generation, transportation, and distribution of electrical power that can cause equipment damage. Some of these are wind, icing, lightning, hurricanes, tree growth that can cause short-circuits, and other natural disasters, as well as intentional sabotage. Some of these factors are unpredictable and have to be dealt with as they occur (in real-time). There are also other frequent factors that can contribute to network state unbalance. Temperature fluctuations, for example, can cause changes of load, and the overall demand for power may vary according to the time of day, week, the season, weather, and others. Some of the aforementioned factors affect the quality of the power supplied, while others create emergency situations forcing the network operators to cut power to areas that are causing problems, to prevent chain reactions. More serious situations can cause power station disconnections and imbalance in the network's power. Intentional power cuts need to be controlled and kept to a minimum.

Electrical power networks are usually managed from a control room. Some networks are telemetered, so that the control engineers have on-line information about the network's status. These may have protection equipment that can be operated from the control room, so the engineers are able to operate said equipment to prevent larger failures. In other instances telemetry may prove to be expensive, and abnormal network states may be reported by field engineers, crew, or customer phone calls. Repair and maintenance work may be performed manually also by field engineers.

A substation may host several busbars, where two of these can be connected to each other by a switch or a conductor line. Both edges of a line are connected to a circuit breaker. Breakers are standard protection mechanisms that have a relay that automatically opens the breaker in case of a short circuit, making it possible to disconnect every single or a busbar from the remaining network. Alarm messages to the control room may be generated as well. From there, the engineers may have the possibility of controlling the state of the breakers. The objective of fault management in an electrical power network is restoring energy supply as fast as possible to as many customers as possible. Since in a network there may be different possible routes through which electricity can be delivered, the network can be switched to establish an alternate route through the operation of breakers and switches that skip the devices causing problems. The need to identify any malfunction in protective equipment and/or switches generates a correct diagnosis out of the alarm messages that have been received, and proceeds to create a safe switching plan to efficiently restore power to the maximum number of customers arises.

EPDSs are a group of one or more generators and transmission lines that operate under a common management or supervision in order to supply customers. Power delivery systems are formed by the juncture of distribution and transmission systems. Transmission systems transport high voltage electricity over long distances. The high voltage loads are reduced at major load centers and then sent to final users. Distribution systems transport electrical power from the transmission system to the final users. EPDSs are found everywhere, from ships to data centers. For the scope of this chapter, generation and transmission systems, and not distribution systems are considered.

IPRs [19] are the main components of an electrical power smart-grid that was proposed as a distributed framework for decentralized control and communication between system components. Intelligence used for control and coordination of system operations is embedded into a series of computing devices that are connected to generators and power lines, enabling them to have knowledge of current network conditions, allocate resources in response to component failures or demand, activate breaker banks, etc. IPRs are based on a peer-to-peer (P2P) mesh architecture, and in the event of a system or component failure, the IPR can make local decisions and coordinate with routers in its vicinity to bring the system back into operation.

Command and control of the process of generating and distributing power, even with redundant generators and lines, is still done in a centralized manner. It is desirable that future EPDS are capable of distributing coordination and control of generation and distribution tasks when there are contingencies or emergency situations. IPRs are designed to provide survivability, fault tolerance, scalability, cost-effectiveness, and unattended 24/7 operation. An IPR is at its core an energy flow controller with embedded intelligence, composed of two main components: interfacing circuits (ICKT) and an integrated control and communications unit (ICCU). The interfacing circuits operate energy flow control and sensing devices, such as circuit breakers, series compensation capacitors, phase-shifting transformers, etc. It also receives information about the current state of the network from sensors and dynamic system monitors. It operates in direct control of the ICCU, which has logic and software to figure out how to route power, change loads, and take any corrective or preventive actions that enhance system stability and security.

The network architecture and communication protocols are similar to local area IP networks. Loads can be assigned a priority, and contrary to ordinary power networks, upon a failure of a source the ICCU of an IPR can react to source failure by reconfiguring the system in order to serve the load with the highest priority. In order to calculate the reliability of individual IPRs, they have been divided into three subsystems: power hardware (circuit breakers or other power switching elements), computer hardware (for data routing, namely IPR-IPR communications and CPU functions), and software (intelligence). Reliability estimates of each subsystem are provided in [19]. The reliability of a circuit breaker is obtained from literature as 0.99330 [19]. Each IPR requires a minimum of two circuit breakers, one for an input line and another for an outgoing line. For a data router, the reliability is estimated at 0.9009 for a 1-year period, for a router with a mean time between failure (MTBF) of 9.5 years. And finally, the reliability of software is estimated at 0.99 [19].

PBNs can accurately model an EPDS with IPRs since our system has a lot of similarities with gene regulatory networks that are modeled using BNs and PBNs. First the PBN representing our EPDS model is constructed. Each node of the system is treated analogous to the gene abstraction in a GRN using binary quantization; 0 means the IPR of the observed device is OFF and 1 meaning it is ON. For each node, an attempt to determine the binary relationships that determine the state of its IPR in time $t + 1$ given the state of the system's nodes in time t is made.

Next, a matrix for each node in order to construct the node's Boolean function or predictor is constructed. In the calculation of the predictors, nodes that directly affect the status of the node are considered. Non-relevant nodes are ignored. In reference to the connections between the relevant nodes and the observed nodes, the logical equation (built with basic Boolean operators, AND, OR, and NOT) that determines the state of each node is postulated. Each node has its own predictor function or set of functions. For those statements, the symbol "+" stands for the Boolean OR operator, "&" for AND, and "!" for NOT. Predictors are determined from the examination of each node and the relationship between it and its relevant nodes. All the possible states of all relevant nodes are considered, and an assessment is made about the next state of the node in time $t + 1$, given the state of all relevant nodes in time t.

The method proposed in this chapter adapts the FDI scheme described in [28], and shown in Fig. 10.2, where a model is used for normal operation of the process and another model is used for each one of the different faults.

PBNs self-organize into attractor states, and these states are related to the different failure modes that the system experiences. Model construction and semantics are similar to [31]. Through characterization of the failure modes, the models can, with property verification, characterize the state of their relevant components to determine which component failures correlate to machine and/or system fault conditions. The method is very flexible, and the design of the PBN and its state transitions depends on the amount of resolution that the experts need, based on design specifications. The system can grow in complexity and expression depending on the needs of the experts.

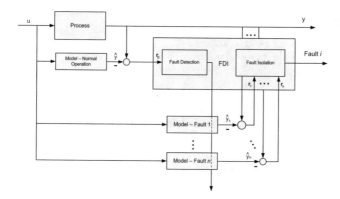

Fig. 10.2 Fault detection and isolation method from Mendonça et al. [28]

Operation is modeled through simulation of the system's components, based on the reliability analysis performed in [19]. This can be modeled for the IPR, or for a test system, such as the 9 bus WSCC, or the IEEE RTS 1979, through simulation of their relevant components, based on each of the component's mean time between failures (MTBF) data. Each of the IPR's faults is modeled based of the reliability analysis conducted in [19]. Therefore, the model is able to detect and isolate single and multiple IPR faults.

With this structure, faults and failure modes can be classified for an individual IPR, or per IPR on a network and EPDS system faults and failure modes. The authors propose the establishment of the model using the PRISM model checker [26], in order to validate its use and check its formal correctness using probabilistic computational tree logic (PCTL). The model is composed of an input module, which uses PRISM's local non-determinism to provide the input to the PBNs. A module for the PBN model of the IPR is then built and a fourth system PBN module modules the behavior of the whole process. An output module produces the system state based on the state of the individual modules. This way, given the different faults and failure modes of the IPRs (which are based on the possible fault conditions of their components) the model produces the failure modes corresponding to the system.

10.4   Experimental Results

This section details the experimental results of the tests performed to validate the adequacy of the proposed model. PRISM was employed to validate the model quantitatively. Experiments were conducted using a PBN model representing the IPR. The IPRs major subsystems (data router, software, breakers) are modeled using PRISM and their relationship expressed as predictors. These subsystems are considered the PBNs genes or nodes. These nodes give the overall state of the IPR. Each subsystem can be in one of two different states:

- Breakers
    - 0—the breakers switch properly
    - 1—the breakers do not close
- Router
    - 0—the router communicates properly
    - 1—the router does not communicate/send any information
- Software
    - 0—the software works properly
    - 1—the software takes an incorrect decision

The state of the IPR is therefore a set of states of its subsystems (RSBB). The IPR can be in $(2 \times 2 \times 2 \times 2) = 2^4 = 16$ states that go from all subsystems operational (0 0 0 0) to all subsystems in failure (1 1 1 1). Some of these states, such as the failure of a single breaker, are identical, and therefore can be merged, resulting in 12 unique states. These states can be classified into categories that describe the IPR failure modes, which are:

- Category 1: If there is an active switching event/signal (AS), the IPR works properly. If there is an inactive switching event/signal (IS), the IPR works unnecessarily.
- Category 2: There is an AS, and the IPR works properly. If there is an IS, the IPR does not switch (works properly).
- Category 3: If there is an AS, the IPR does not work properly. If there is an IS, the IPR works unnecessarily.
- Category 4: If there is an AS, the IPR does not work properly. If there is an AS, the IPR does not work properly. If there is an IS, the IPR does not switch (works properly).

It is assumed that the probability of failure of each component is independent of each other. Reliability estimates for each of the components are needed. The reliability of circuit breakers above 63 kV is obtained from [18]. For a 1-year period, the breakers' reliability is 0.99330 (99.330% of confidence). For routers, reliability is 0.90009 per year, for a 9.5 MTBF router, from [15]. Estimating software reliability is an extremely difficult task in itself. For IPRs we assume the value used in [19], which is 0.99. These have to be changed to yearly figures. For the router, the failure rate per year is 0.000288392. For the software, 0.00271232877, and for the breakers, 0.0000184110 [19].

As an example, the relevant nodes of the IPR's PBN are the router, software, and the input and output breakers of the device. On the IPR devices, the state of these components determine the failure mode on which the device is currently in, given by the following categories. Category 1 describes the first type of fault, where the device acts properly and switches the breakers on an active signal, but may switch them also when there is no need to switch. Category 2 describes the normal operation of the device. Category 3 describes a complete failure of the IPR device.

Table 10.1 Predictors and selection probability, IPR PBN

Component	Predictor	Selection probability
x_1, Software	$x_1(t+1) = x_1(t)$	1
x_2, Router	$x_2(t+1) = x_1(t) \& x_2(t)$	0.9611
	$x_2(t+1) = x_1(t) \mid x_2(t)$	0.0389
x_3, Input breaker	$x_3(t+1) = x_1(t) \& x_2(t) \& x_3(t)$	0.9611
	$x_3(t+1) = x_1(t) \mid x_2(t) \mid x_3(t)$	0.0389
x_4, Output breaker	$x_4(t+1) = x_1(t) \& x_2(t) \& x_4(t)$	0.9611
	$x_4(t+1) = x_1(t) \mid x_2(t) \mid x_4(t)$	0.0389

Lastly, Category 4 describes a fault condition on which the device does not act upon an active event, and may also switch the breakers unnecessarily when there is no AS.

Table 10.1 presents the predictor functions for each of the IPRs subsystems, based on their configuration.

The models presented in [29–32] were expanded to include fault conditions that may lead to failure on the individual IPR or an EPDS. This allows the prediction of conditions that may not cause complete failure, but rather failure modes that may lead to situations where the system continues its operation, but cannot perform the required task to specifications. These constitute unhealthy system states, where a fault condition can be treated or lead to failure. For an IPR, the failure modes described in [19] were used, and a determination was made of which system components and failure modes can produce a failure or a fault.

Three modules constitute the complete models in PRISM, an input module, a module for the PBN, and an output module. The current state of the PBN's components is in module input. The PBN module uses the state of the input variables and applies the corresponding predictors to transition to the next state. Based on the values of these variables, and the fault conditions, the state a global IPR variable is changed, giving us the current state of the device. In these experiments, time is expressed in hours (h).

Property verification in PRISM was employed for determining the maximum probability of occurrence of any of the failure modes that could lead to fault. From an initial state for the device, such as Category 2, a determination is made about the maximum probability of reaching one of the different identified fault conditions. Property verification in PRISM not only allows us to verify the models, they also allow, through experiments, to reach an estimate in time about when fault occurrence is certain.

Detection The models are able to detect faults and failures, based on the application of the PBN. Given the current state of the network genes, the PBN will select an appropriate context and self-organize into one of the attractor states of their constituent Boolean networks. As an example, in Table 10.1 the predictors and selection probability of each predictor are given for the IPR. The authors in [32] equated the context to the different failure modes that can occur. The input module of the model randomizes the current state of the device, and based on the current state, the PBN module will apply the predictors and select a constituent BN. The output model contains all of the identified fault conditions/failure modes of a device, and after the application of the predictors evaluates the state of the device's components, and makes a determination of the state of the device as a whole. The device can be in a complete failure condition, or in a fault condition, that can be specifically described based on the condition of the components, allowing detection and isolation of individual or combined faults.

The first test conducted was to determine the maximum probability of reaching any of the failure modes leading to fault of the IPR through verification of the probabilistic real time computational tree logic (PCTL) property:

$$Pmax = ?[F <= time(r1 = true\&s1 = true\&b1a = true\&b1b = false)|(r1 =$$
$$true\&s1 = true\&b1a = false\&b1b = true)|(r1 = true\&s1 =$$
$$true\&b1a = false\&b1b = false)|(r1 = true\&s1 = false\&b1a =$$
$$true\&b1b = true)|(r1 = true\&s1 = false\&b1a = true\&b1b =$$
$$false)|(r1 = true\&s1 = false\&b1a = false\&b1b = true)|(r1 =$$
$$true\&s1 = false\&b1a = false\&b1b = false)|(r1 = false\&s1 =$$
$$true\&b1a = true\&b1b = true)|(r1 = false\&s1 = true\&b1a =$$
$$true\&b1b = false)|(r1 = false\&s1 = true\&b1a = false\&b1b =$$
$$true)|(r1 = false\&s1 = true\&b1a = false\&b1b = false)|(r1 =$$
$$false\&s1 = false\&b1a = true\&b1b = true)|(r1 = false\&s1 =$$
$$false\&b1a = true\&b1b = false)|(r1 = false\&s1 = false\&b1a =$$
$$false\&b1b = true)|(r1 = false\&s1 = false\&b1a = false\&b1b =$$
$$false)]$$

This property was tested for IPR's PBN model and a control group consisting of the MTBF for the relevant nodes of the system. Two sample T-tests were performed using Minitab 16 to look for statistically significant differences among the group means. The null hypothesis states that there is no difference between the control and PBN groups, or $H_0 : \mu$ Control $= \mu$ PBN. The alternative hypothesis would be finding differences between the control and PBN model groups, or $H_0 : \mu$ control $\neq \mu$ PBN. For an α-level of 0.05 for the test, the conclusion is that for the IPR, there are no statistically significant differences between the groups (p-value > 0.05). This means that there is no difference between both groups. Figure 10.3 shows the result of the first experiment, the occurrence of any of the failure modes, Categories 1, 3, and 4. Figure 10.4 presents the result of the experiment for the maximum probability of occurrence of Category 1, of the first type of fault. Figure 10.5 presents the results for Category 3, failure, and Fig. 10.6 for Category 4, the second type of fault. In these graphs, time is represented in hours (h).

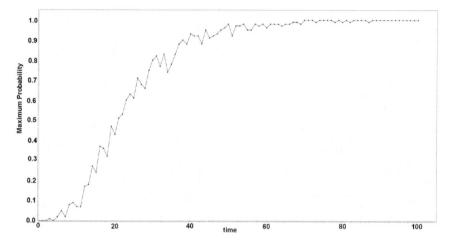

Fig. 10.3 Maximum probability of occurrence of Category 1, 2, or 3 faults

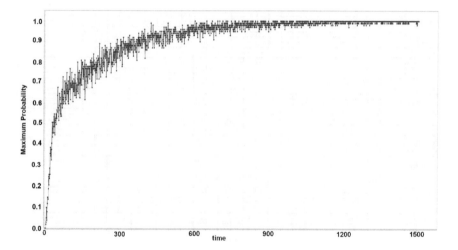

Fig. 10.4 Maximum probability of Category 1 failure mode occurrence

Diagnosis Labels in PRISM can be used to single-out specific states, or sets of states. They can be used to single-out single faults, or combinations of faults. When the PBN is applied and a constituent BN is selected, these labels provide a way of filtering which fault is occurring, or if the device is operating correctly. Within the output module, all of the possible failure and fault conditions on the device caused by the components that have been identified are expressed, and this allows to determine its future state. This allows not only to discern which specific fault or combination of faults is occurring, but through property verification we can make use of these labels to produce a prognosis, an estimate in time of when the fault is expected to occur. Knowing the probability of fault and failure occurrence allows the

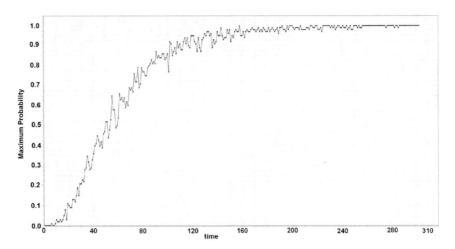

Fig. 10.5 Maximum probability of Category 3 failure mode occurrence

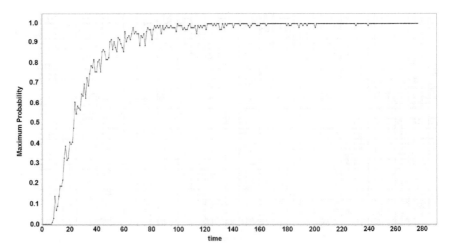

Fig. 10.6 Maximum probability of Category 4 failure mode occurrence

system designers to make decisions about the interventions needed for the system or machine and minimize the downtime needed for maintenance. For example, $Pmax = ? [F \leq time \text{"} class2\text{"}]$ verifies the maximum probability of occurrence of a Category 3 fault, which is a catastrophic failure of the IPR. There are 11 IPR states that are divided into four types of modes that define a normal operating state, two fault conditions and a failure condition, based on the previously shown categories.

The system is also capable of detecting multiple simultaneous faults. Experiments were performed to verify the capability of detecting multiple faults of the device's components. Through property verification, the system is able not only to detect these simultaneous faults, but is also able to tell when the fault is imminent. Figure 10.7 shows a simultaneous fault on the IPR where both breakers fail, and shows that the faults will manifest at around 7000 h of continuous operation.

In Fig. 10.8, diagnosis of faults and failures is illustrated. States of the IPR's components have been individually identified. This means that all faults, single or multiple, can be singled out specifically. With simulations in PRISM a time estimate of when the failure of particular components will occur can be graphed. This figure presents a simulation of 1000 h of operation of the device, and the amount of software failures that can occur within that period.

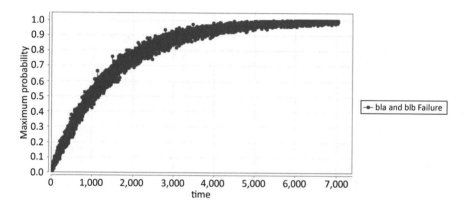

Fig. 10.7 Simultaneous failure of the input and output breakers

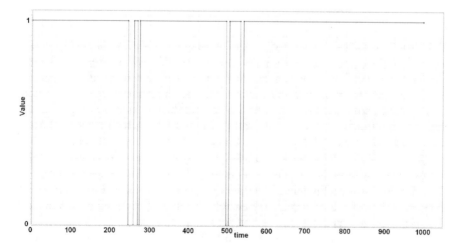

Fig. 10.8 Software failures in 1000 h of device operation

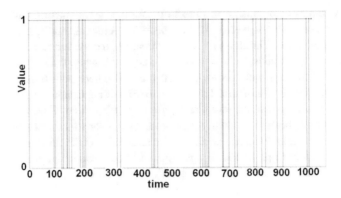

Fig. 10.9 Failures in 1000 h of device operation of the input and output breakers

Figure 10.9 illustrates the combined failures of the input and output breakers.

This novel method of modeling provides a simple, straightforward mechanism of detecting, isolating, and classifying single and multiple faults. Through the obtained experimental results, it is statistically demonstrated that model-based FDI performs satisfactorily when examining a device's possible behavior, for detection, classification, and isolation of single and multiple faults. Given an operational state, and a set of possible fault states, the model is able to characterize possible failure modes in which the device may be, given the state of its variables. Model-based approaches depend on the knowledge of experts that classify the different fault conditions and failure modes occur.

10.5   Conclusions

This chapter presents a bioinspired, complex-adaptive modeling methodology that allows modeling single and multiple faults on smart grid devices using probabilistic Boolean networks. The modifications proposed in this chapter to the aforementioned architecture and to this new method allowed the classification of single and multiple faults. These permit the FDI scheme proposed in [28], and shown in Fig. 10.2, the detection and isolation of single and multiple faults, along with an estimate of when these faults will present themselves. Statistical tests performed of this data validate the proposed approach for future use and further development. Since these models are based on the definition of the PBNs derived from regulating genes/nodes, this discretization creates a limitation in terms of the possible states that it can represent, but greatly simplifies the analysis. The authors are currently working on new models that may allow new faults to be detected, for further analysis.

For future research, an interesting idea is to design a fault diagnosis system based in historical data of the process with the ability to detect and classify multiple and novel faults. Expanding the use of non-binary quantized PBNs will also allow in the future a more rich mechanism of expressing fault conditions and failure modes. Another possible avenue of development is the use of the intervention mechanism in FDI-enabled PBN models.

Acknowledgements The authors would like to acknowledge the support of University of Puerto Rico-Río Piedras, and Universidad Tecnológica de la Habana José Antonio Echeverría, CUJAE.

References

1. Acosta Diaz, C., Camps Echevarria, L., Prieto Moreno, A., Silva Neto, A.J., Llanes-Santiago, O.: A model-based fault diagnosis in a nonlinear bioreactor using an inverse problem approach. Chem. Eng. Res. Des. **114**, 18–29 (2016)
2. Arnosti, D.N., Ay, A.: Boolean modeling of gene regulatory networks: Driesch redux. Proc. Natl. Acad. Sci. **109**(45), 18239–18240 (2012). https://doi.org/10.1073/pnas.1215732109
3. Bachschmid, N., Pennacchi, P., Vania, A.: Identification of multiple faults in rotor systems. J. Sound Vib. **254**, 327–366 (2002)
4. Bane, V., Ravanmehr, V., Krishnan, A.R.: An information theoretic approach to constructing robust Boolean gene regulatory networks. IEEE/ACM Trans. Comput. Biol. Bioinform. **9**(1), 52–65 (2012)
5. Bartenstein, T., Heaberlin, D., Huisman, D., Sliwinski, D.: Diagnosing combinational logic designs using the single location at-a-time (slat) paradigm. In: Proceedings of IEEE International Test Conference (ITC), pp. 287–296. IEEE, Piscataway (2001). https://doi.org/10.1109/TEST.2001.966644
6. Camps Echevarría, L., Silva Neto, A.J., Llanes-Santiago, O., Hernández Fajardo, J.A., Jiménez Sánchez, D.: A variant of the particle swarm optimization for the improvement of fault diagnosis in industrial systems via faults estimation. Eng. Appl. Artif. Intell. **28**, 36–51 (2014)
7. Camps Echevarría, L., Campos Velho, H.F., Becceneri, J.C., Silva Neto, A.J., Llanes-Santiago, O.: The fault diagnosis inverse problem with ant colony optimization and ant colony optimization with dispersion. Appl. Math. Comput. **227**(15), 687–700 (2014)
8. Camps Echevarría, L., Llanes-Santiago, O., Fraga de Campos Velho, H., Silva Neto, A.J.: Fault Diagnosis Inverse Problems: Solution with Metaheuristics. Springer, New York (2019). https://doi.org/10.1007/978-3-319-89978-7
9. Chaouiya, C., Ourrad, O., Lima, R.: Majority rules with random tie-breaking in Boolean gene regulatory networks. PLOS One **8**(7), e69626 (2013). https://doi.org/10.1371/journal.pone.0069626
10. Chen, H., Sun, J.: Stability and stabilisation of context-sensitive probabilistic Boolean networks. IET Control Theory Appl. **8**(17), 2115–2121 (2014)
11. Chen, X., Jiang, H., Ching, W.: On construction of sparse probabilistic Boolean networks. East Asian J. Appl. Math. **2**(1), 1–18 (2012). https://doi.org/10.4208/eajam.030511.060911a
12. Ching, W., Chen, X., Tsing, N.: Generating probabilistic Boolean networks from a prescribed transition probability matrix. IET Syst. Biol. **3**, 453–464 (2009)
13. Ching, W., Zhang, S., Jiao, Y., Akutsu, T., Tsing, N., Wong, A.: Optimal control policy for probabilistic Boolean networks with hard constraints. IET Syst. Biol. **3**(2), 90–99 (2009)
14. Didier, G., Remy, E.: Relations between gene regulatory networks and cell dynamics in Boolean models. Discret. Appl. Math. **160**(15), 2147–2157 (2012). https://doi.org/10.1002/asjc.1722
15. Ebeling, C.E.: An Introduction to Reliability and Maintainability Engineering. McGraw-Hill, New York (1997)

16. Frank, P.M.: Analytical and qualitative model-based fault diagnosis - a survey and some new results. Eur. J. Control **2**(1), 6–28 (1996)
17. Gao, Y., Xu, P., Wang, X., Liu, W.: The complex fluctuations of probabilistic Boolean networks. BioSystems **114**, 78–84 (2013). https://doi.org/10.1016/j.biosystems.2013.07.008
18. Heising, C., Janssen, A.L.J., Lanz, W., Colombo, E., Dialynas, E.N.: Summary of CIGRE 13.06 working group world wide reliability data and maintenance cost data on high voltage circuit breakers above 63kv. In: Industry Applications Society Annual Meeting, vol. 3, pp. 2226–2234 (1994)
19. Irizarry-Rivera, A.A., Rodríguez-Martínez, M., Vélez, B., Vélez-Reyes, M., Ramirez-Orquin, A.R., O'neill-Carrillo, E., Cedeño, J.R.: Operation and Control of Electric Energy Processing Systems, chap. Intelligent Power Routers: Distributed Coordination for Electric Energy Processing Networks, pp. 47–85. Springer, Berlin (2010). https://doi.org/10.1002/9780470602782.ch3
20. Isermann, R.: Process fault detection based on modelling and estimation methods - a survey. Automatica **20**(4), 387–404 (1984). https://doi.org/10.1016/0005-1098(84)90098-0
21. Isermann, R.: Model based fault detection and diagnosis. Status and applications. Ann. Rev. Control **29**(1), 71–85 (2005)
22. Isermann, R.: Fault-Diagnosis Applications: Model-Based Condition Monitoring: Actuators, Drives, Machinery, Plants, Sensors, and Fault-tolerant Systems, vol. 24. Springer, Berlin (2011). https://doi.org/10.1002/rnc.3142
23. Kauffman, S.A.: Homeostasis and differentiation in random genetic control networks. Nature **224**, 177–178 (1969). https://doi.org/10.1038/224177a0
24. Kauffman, S.A.: Metabolic stability and epigenesis in randomly constructed genetic nets. J. Theor. Biol. **22**, 437–467 (1969). https://doi.org/10.1016/0022-5193(69)90015-0
25. Kobayashi, K., Hiraishi, K.: Reachability analysis of probabilistic Boolean networks using model checking. In: Proceedings of SICE Annual Conference 2010, vol. 2014, pp. 1–8 (2010). https://doi.org/10.1155/2014/968341
26. Kwiatkowska, M.Z., Norman, G., Parker, D.: Prism 4.0: verification of probabilistic real-time systems. In: G. Gopalakrishnan, S. Qadeer (eds.) Computer Aided Verification. Lecture Notes in Computer Science, vol. 6806, pp. 585–591. Springer, Berlin (2011). https://doi.org/10.1007/978-3-642-22110-1_47
27. Liu, Q., Zeng, Q., Huang, J., Li, D.: Optimal intervention in semi-Markov-based asynchronous probabilistic Boolean networks. Complexity **2018**(ID 8983670) (2018). https://doi.org/10.1155/2018/8983670
28. Mendonça, L., Sousa, J., Sá da Costa, J.: An architecture for fault detection and isolation based on fuzzy methods. Expert Syst. Appl. **36**, 1092–1104 (2009). https://doi.org/10.1016/j.eswa.2007.11.009
29. Rivera Torres, P., Serrano Mercado, E.: Probabilistic Boolean network modeling as an aid for DFMEA in manufacturing systems. In: Proceedings of XVIII Scientific Convention in Engineering and Architecture (CCIA 2016). La Habana, Cuba (2016)
30. Rivera Torres, P., Serrano Mercado, E., Llanes-Santiago, O., Anido Rifón, L.: Modeling preventive maintenance of manufacturing processes with probabilistic Boolean networks with interventions. J. Intell. Manuf. **29**(8), 1941–1952 (2018). https://doi.org/10.1007/s10845-017-1321-7
31. Rivera Torres, P.J., Serrano Mercado, E., Anido, R.L.: Probabilistic Boolean network modeling of an industrial machine. J. Intell. Manuf. **29**(4), 875–890 (2018). https://doi.org/10.1007/s10845-015-1143-4
32. Rivera Torres, P.J., Serrano Mercado, E., Anido Rifón, L.: Probabilistic Boolean network modeling and model checking as an approach for DFMEA for manufacturing systems. J. Intell. Manuf. **29**(6), 1393–1413 (2018). https://doi.org/10.1007/s10845-015-1183-9
33. Rodríguez Ramos, A., Domínguez Acosta, C., Rivera Torres, P.J., Serrano Mercado, E.I., Beauchamp Báez, G., Anido Rifón, L., Llanes-Santiago, O.: An approach to multiple fault diagnosis using fuzzy logic. J. Int. Manag. (2016). https://doi.org/10.1007/s10845-016-1256-4

34. Ruan, S., Zhou, Y., Feili, Y., Pattipati, K., Willett, P., Patterson-Hine, A.: Dynamic multiple-fault diagnosis with imperfect tests. IEEE Trans. Syst. Man Cybern. A: Syst. Humans **39**, 1224–1236 (2009). https://doi.org/10.1109/tsmca.2009.2025572
35. Shmulevich, I., Dougherty, E., Kim, S.: Probabilistic Boolean networks: a rule-based uncertainty model for gene regulatory networks. Bioinformatics **18**(2), 261–274 (2002). https://doi.org/10.1093/bioinformatics/18.2.261
36. Shmulevich, I., Dougherty, E.R.: Probabilistic Boolean Networks: Modeling and Control of Gene Regulatory Networks. SIAM, Philadelphia (2010). https://doi.org/10.1137/1.9780898717631
37. Simani, S., Farsoni, S., Castaldi, P.: Wind turbine simulator fault diagnosis via fuzzy modelling and identification techniques. Sustainable Energy, Grids Netw. **1**, 45–52 (2015). https://doi.org/10.1016/j.segan.2014.12.001
38. Sobhani-Tehrani, E., Talebi, H., Khorasani, K.: Hybrid fault diagnosis of nonlinear systems using neural parameter estimators. Neural **50**, 12–32 (2014). https://doi.org/10.1016/j.neunet.2013.10.005
39. Trairatphisan, P., Mizera, A., Pang, J., Tantar, A.A., Schneider, J., Sauter, T.: Recent development and biomedical applications of probabilistic Boolean networks. Cell Commun. Signal **11**, 46 (2013). https://doi.org/10.1186/1478-811x-11-46
40. Venkatasubramanian, V., Rengaswamy, R., Yin, K., Kavuri, S.N.: A review of process fault detection and diagnosis-part I: quantitative model-based methods. Comput. Chem. Eng. **27**(3), 293–311 (2003). https://doi.org/10.1016/s0098-1354(02)00161-8
41. Venkatasubramanian, V., Rengaswamy, R., Yin, K., Kavuri, S.N.: A review of process fault detection and diagnosis-part II: qualitative model-based methods and search strategies. Comput. Chem. Eng. **27**(3), 313–326 (2003). https://doi.org/10.1016/s0098-1354(02)00161-8
42. Venkatasubramanian, V., Rengaswamy, R., Yin, K., Kavuri, S.N.: A review of process fault detection and diagnosis-part III: process history based methods. Comput. Chem. Eng. **27**(3), 327–346 (2003). https://doi.org/10.1016/s0098-1354(02)00161-8
43. Vong, C., Wong, P., Wong, K.: Simultaneous-fault detection based on qualitative symptom descriptions for automotive engine diagnosis. Appl. Soft Comput. **22**, 238–248 (2014). https://doi.org/10.1016/j.asoc.2014.05.014
44. Wang, Z., Marek'Sadowska, M., Tsai, K., Rajski, J.: Analysis and methodology for multiple fault diagnosis. IEEE Trans. Comput.-Aided Des. Integr. Circuits Syst. **25**, 558–575 (2006)
45. Witczak, M.: Modelling and Estimation Strategies for Fault Diagnosis of Non-Linear Systems From Analytical to Soft Computing Approaches, vol. 354. Springer, Berlin (2007). https://doi.org/10.1007/978-3-540-71116-2

Chapter 11
Evaluating Automated Machine Learning on Supervised Regression Traffic Forecasting Problems

Juan S. Angarita-Zapata ⓘ, Antonio D. Masegosa ⓘ, and Isaac Triguero ⓘ

Abstract Traffic forecasting is a well-known strategy that supports road users and decision-makers to plan their movements on the roads and to improve the management of traffic, respectively. Current data availability and growing computational capacities have increased the use of machine learning methods to tackle traffic forecasting, which is mostly modelled as a supervised regression problem. Despite the broad range of machine learning algorithms, there are no baselines to determine what are the most suitable methods and their hyper-parameters configurations to approach the different traffic forecasting regression problems reported in the literature. In machine learning, this is known as the model selection problem, and although automated machine learning methods have proved successful dealing with this problem in other areas, it has hardly been explored in traffic forecasting. In this work, we go deeply into the benefits of automated machine learning in the aforementioned field. To this end, we use Auto-WEKA, a well-known AutoML method, on a subset of families of traffic forecasting regression problems characterised by having loop detectors, as traffic data source, and scales of predictions focused on the point and the road segment levels within freeway and urban environments. The experiments include data from the Caltrans Performance Measurement System and the Madrid City Council. The results show that AutoML methods can provide competitive results for TF with low human intervention.

J. S. Angarita-Zapata (✉)
DeustoTech, Faculty of Engineering, University of Deusto, Bilbao, Spain
e-mail: js.angarita@deusto.es

A. D. Masegosa
DeustoTech, Faculty of Engineering, University of Deusto, Bilbao, Spain

IKERBASQUE, Basque Foundation for Science, Bilbao, Spain
e-mail: ad.masegosa@deusto.es

I. Triguero
Computational Optimisation and Learning (COL) Lab, School of Computer Science,
University of Nottingham, Nottingham, UK
e-mail: Isaac.Triguero@nottingham.ac.uk

© Springer Nature Switzerland AG 2020
O. Llanes Santiago et al. (eds.), *Computational Intelligence in Emerging Technologies for Engineering Applications*, Studies in Computational Intelligence 872, https://doi.org/10.1007/978-3-030-34409-2_11

187

Keywords Traffic forecasting · Supervised learning · Machine learning · Automated machine learning · Computational intelligence · Intelligent transportation systems

11.1 Introduction

Urban development, population growth, and high motorisation rates have increased levels of congestion in cities around the world. One well-established strategy to tackle congestion is the design, development, and implementation of traffic forecasting (TF) systems. TF can be defined as the prediction of near future traffic conditions (e.g., speed, travel time) for single locations, road segments, or entire networks [33].

The recent emergence of telecommunications technologies integrated to transportation infrastructure generates vast volumes of traffic data. This unprecedented data availability and growing computational capacities have incremented the use of machine learning (ML) to address TF. From a data-driven perspective TF can be addressed using different modelling approaches, such as a supervised regression problem [1, 9], a supervised classification problem [1, 16], or a clustering-pattern recognition problem [30, 34]. Nevertheless, the supervised regression approach is typically the most widely used modelling perspective in the TF literature. During the last decades, the number of academic publications about TF approached as a supervised regression problem has increased extensively. From a ML perspective [3], a supervised TF regression problem is focused on building a predictive model using historical data to make predictions of continuous traffic measures, based on unseen data.

The transportation literature reports a great number of ML algorithms that can be used for the prediction of traffic, such as neural networks (NNs), support vector machines (SVMs), k-nearest neighbours (k-NN), and random forest (RF), among others [14]. However, given the broad wide range of ML methods, there are no clear baselines that guide the process of selecting the most appropriate algorithm and its best hyper-parameter setting given the characteristics of the TF problem at hand. In the ML area, this challenge is known as the model selection problem (MSP) [12], and automated machine learning (AutoML) [12, 35] has been one of the most successful approaches to address it so far. AutoML aims at automatically finding the ML algorithm and hyper-parameters configuration pair which maximises a performance measure on given data, using an optimisation strategy that minimises a pre-defined loss function.

Although AutoML methods have approached the MSP with high performance in other research areas [17, 37], to the best of authors' knowledge there are only two works that have tackled the MSP in TF [1, 31]. On the one hand, Vlahogianni [31] proposed an AutoML method that handles the prediction of speed in a time horizon of 5 min using a supervised regression approach. Contrary to [31], Angarita-Zapata et al. [1] focused on a TF supervised classification problem to predict the level of service through multiple time horizons and using a different AutoML method (Auto-WEKA). Notwithstanding, in spite of the progress achieved by the aforementioned works, AutoML in the transportation area is still in its infancy and there are TF supervised regression problems [2] that still need to be addressed in order to develop more reliable TF systems.

In this book chapter, our objective is to continue deepening into the benefits of AutoML for TF from a supervised regression perspective, following the research line proposed in [1]. Specifically, we use Auto-WEKA, a well-known AutoML method, on a subset of families of TF problems characterised by having loop detectors, as traffic data source, and scales of predictions focused on the point and the road segment levels within freeway and urban environments. We compare the AutoML method versus the general approach in TF, which consists of selecting the best of a set of commonly used ML algorithms. Concretely, we contrast Auto-WEKA results with four state-of-the-art ML algorithms (NN, SVM, k-NN, and RF) in the task of forecasting traffic speed, using data taken from the Caltrans Performance Measurement System (PeMS) and the Madrid (Spain) City Council. The main contributions of this work are:

- Exploring, in a more deeply way, the benefits of AutoML for TF supervised regression problems.
- Determining suitable ML algorithms to approach the prediction of traffic at scales of predictions focused on the point and the road segment levels within, freeway and urban environments.

The rest of this chapter is structured as follows. Section 11.2 presents background and related work about ML and AutoML methods in the area of TF. Section 11.3 exposes the methodology followed in this work. Then, Sect. 11.4 shows main results obtained by the AutoML method and the baseline algorithms considered. Finally, the main conclusions of the chapter are discussed in Sect. 11.5.

11.2 Background

This section reviews literature related to ML and AutoML in the context of TF. We start presenting a brief history of how TF has evolved in the last years (Sect. 11.2.1). Then, Sect. 11.2.2 summarises the ML modelling perspectives used to approach the prediction of traffic. Later, Sect. 11.2.3 discusses ML methods for TF. Finally, Sect. 11.2.4 reviews AutoML methods.

11.2.1 A Brief History of Traffic Forecasting

TF is a relevant research because of its active role in intelligent transportation systems (ITSs) to address traffic congestion. The main objective of TF is the prediction of near future traffic measures based on past traffic data [33]. Three decades ago, transportation research was focused on predicting traffic at a single location using traffic theory models [11] and classical statistical methods [10]. However, these two approaches are not able to deal, in an efficient way, with uncertainty and big volumes of traffic data.

Recently, the emergence of sensing and telecommunications technologies integrated to ITSs started to generate vast volumes of traffic data, which in turn caused a switch in the modelling paradigm towards a data-driven approach [20]. Since that time, a variety of methods have been proposed placing special emphasis on computational intelligence-based approaches, such as NNs [15, 36], fuzzy logic [4, 5], and bio-inspired algorithms [16, 24], among others [33].

Currently, although some TF literature still relies on statistical methods, ML methods have attracted the interest of the transportation community and they are present in a wide proportion of contemporary research (see the review published by Ergamun and Levinson [6]). As computational capacities have increased, more complex scenarios with different road settings can be tackled with ML (e.g., network-wide predictions) due to its ability to predict traffic without the need of knowing theoretical traffic mechanisms [14]; in this way, leaving behind traditional approaches in the prediction of traffic.

11.2.2 Machine Learning Modelling Approaches for Traffic Forecasting

TF, from a data-driven perspective, is facilitated by sensing infrastructure of ITSs. Some technologies, such as automatic vehicle identification, electronic tolls, and GPS, collect individual traffic data related to each vehicle on the road; meanwhile, others collect macroscopic traffic measures (averages of many vehicles) as vehicle detection stations (VDS). Taking as input the data provided by any of the mentioned data sources, when the objective is to predict a continuous traffic variable, the possible ML modelling approaches can be supervised regression or clustering-pattern recognition. In the first case, the focus is on using ML algorithms to learn a functional form based on the input data, without prior models or data distribution assumptions [6]. In this context, the goal is to approximate the learned mapping function in such a way that when the model faces new and unseen traffic data, it is able to make accurate predictions.

In the second case, clustering-pattern recognition, the objective is finding the relationships of different locations by characterising similar traffic measures values from one road to another, and grouping them in clusters that divide the network into correlated groups. Once the clusters have been identified, the next step is to use a supervised regression perspective to predict the traffic conditions, cluster by cluster, based on historical traffic data belonging to each group.

Finally, when the objective is forecasting a discrete traffic measure, the modelling approach is supervised classification that also learns a mapping function based on historical data. For instance, ML methods can forecast the level of service (LoS) of a specific road. LoS is a categorical variable that measures the quality of the traffic through letters from A to E in a gradual way, being category A moderate traffic and category E extended delays [26]. It is important to clarify that the forecasting of discrete variables could be also addressed as a supervised regression problem in some occasions, predicting either speed or density (continuous values), and then discretising these predictions to obtain the categorical outputs.

Regardless of the aforementioned modelling approaches, ML methods for traffic prediction are based on forecasting traffic based on historical data and their objective is to predict traffic in similar conditions in which this data was observed. In this context, traffic forecasting under severe changes, such as new road infrastructure or traffic control policies, is out of the scope of ML and simulation-based approaches become alternatives that are more suitable [23].

In this work, we centre on ML applied to VDS data from a supervised regression approach because of two reasons. First, VDS data is the most common type of data available and used in transportation literature [20]. Secondly, the supervised regression approach is, by far, the most widely used modelling perspective to predict traffic [2].

11.2.3 Machine Learning Algorithms for TF Using VDS Data

ML methods applied to TF can be categorised into single or hybrid. The first type corresponds to adaptations of existing ML algorithms that can be classified as parametric and non-parametric [32]. The parametric category assumes the relationship between the explanatory and response variables as known; meanwhile, the non-parametric ones are able to model nonlinear relationships without requiring the mentioned assumptions. Commonly non-parametric algorithms are NNs, SVMs, k-NN, and RF [14, 33].

As mentioned before, the other approach of ML algorithms is hybridisation. Within it, two or more algorithms, from ML or even other areas, are combined to find synergies that improve their isolated performance. Some recent examples are [18], where authors integrate a Boltzmann machine with recurrent NNs, and [16], where genetic algorithms are integrated with fuzzy systems.

Despite the great variety of ML methods, dealing with the MSP in TF is not a trivial task, as mentioned before. The general approach to tackle the MSP in TF consists of testing a set of algorithms with multiple hyper-parameter combinations and select the best one. This requires expert knowledge and a lot of human effort. Nowadays, AutoML has received a lot of attention in ML because of its promising results in dealing with the MSP with low human intervention.

11.2.4   AutoML in Traffic Forecasting

As stated above, AutoML deals with MSP as an optimisation problem whose objective consists of finding the ML algorithm, from a pre-defined base of algorithms, and its hyper-parameter configuration that maximises an accuracy measure on a given ML problem. In this sense, AutoML aims to improve the current way of building ML applications by automating the application of ML algorithms to data-sets, in such a way that enables human users avoiding tedious tasks (e.g., hyper-parameter optimisation). Although current AutoML methods have already produced impressive results, the field is still far from being mature.

The first AutoML method in tackling simultaneously the selection of algorithm and hyper-parameters was Auto-WEKA [29]. It uses Bayesian optimisation to search for the best pair (algorithm, hyper-parameter setting), considering a base of 39 algorithms implemented in WEKA (a well-known open-source ML software that contains algorithms for data analysis and predictive modelling). Subsequently, Komer et al. [13] and Feurer et al. [7] developed Hyperopt-sklearn and Auto-sklearn, respectively. These two frameworks automatically select ML algorithms and hyper-parameter values from scikit-learn.[1] In the case of [13], the AutoML method uses Hyperopt Python library for the optimisation process, concretely a Bayesian optimization method as Auto-WEKA. Meanwhile, Auto-sklearn stores the best combination of ML algorithm and hyper-parameters that have been found for each previous ML problem, and using meta-learning it chooses a starting point for a sequential optimisation process.

More recently, Sparks et al. [27] proposed a method that supports distributed computing for AutoML, and Sabharwal et al. [25] developed a cost-sensitive training data allocation method that assesses a pair (algorithm, hyper-parameters setting) on a small random sample of the data-set, and gradually expands it over time to re-evaluate it when one combination is promising. Then, Olson and Moore [21] designed a framework for building and tuning classification and regression ML pipelines. It uses genetic programming to construct flexible pipelines and to select an algorithm in each pipeline stage. However, TPOT does not exhaustively test all different combinations of hyper-parameters which in turn causes that some promising configuration may be ignored.

Lately, Swearingen et al. [28] built ATM, which is a collaborative service to build optimised ML pipelines. This framework has a strong emphasis on parallelisation enabling the distribution of a single combination (algorithm, hyper-parameter setting) in a cluster to process it in a more efficient way. Currently, ATM uses the same base of algorithms from scikit-learn, and it finishes the optimisation process after either a fixed number of iterations or after expending a time budget defined by the human user. One year later, Mohr et al. [19] developed ML-Plan, a framework for building ML pipelines based on hierarchical task networks. ML-Plan is initialised

---

[1] Scikit-learn is a Python library of ML algorithms: http://scikit-learn.org.

with a fixed set of pre-processing algorithms, classification algorithms, and their respective potential hyper-parameters. Nevertheless, ML-Plan only considers a supervised classification approach, ignoring the supervised regression perspective that, as it was stated before, is the most common approach in TF.

For this research, we select Auto-WEKA because of a twofold reason. First, its wider variety base of regression algorithms in comparison with the other approaches reviewed. Second, unlike the aforementioned methods that only consider a pre-defined set of hyper-parameters combinations, Auto-WEKA has no limitations in the hyper-parameter space to be explored.

Moving from general-purpose AutoML methods to the transportation area, to the best authors' knowledge, only two works have tackled the MSP in TF [1, 31]. Angarita et al. [1] used Auto-WEKA and compared it to the general approach (which consists of selecting the best of a set of algorithms) over a multi-class imbalanced classification TF problem, predicting traffic level of service at a fixed location through multiple time horizons. In the case of Vlahogianni [31], the author proposed a meta-modelling technique that, based on surrogate modelling and a genetic algorithm with an island model, optimises both the algorithm selection and the hyper-parameter setting. The AutoML task is performed from an algorithms base of three ML methods (NN, SVM, and radial base function) that forecast average speed in a time horizon of 5 min, using a regression approach.

The main differences between this research and the two aforementioned works lay on the addressed TF problems and ML modelling approach used. Regarding the problems, we predict traffic speed for TF problems characterised by having scales of predictions at the point and the road segment level, within freeway and urban environments. This means that we consider problems that take into account the temporal dimension of traffic, on the one hand, and the temporal-spatial component of traffic data on the other hand. Lastly, with respect to the modelling approach, we use an AutoML method for a supervised regression approach that considers a much broader base of algorithms that the one used by Vlahogianni [31]. In the case of the work of Angarita et al. [1], the same AutoML method is considered but for a supervised classification approach whereas in this work we are considering a regression approach.

11.3 Methodology

This research seeks to keep exploring the benefits that AutoML can bring to TF. To accomplish such purpose, we compare to what extent the results of AutoML differ from the general approach in TF, in which a set of baseline algorithms (BAs) is tested over the forecasting problem at hand, and the one with best performance metrics is chosen. We select Auto-WEKA, as AutoML method, and NN, SVM, k-NN, and RF, as the BAs that represent the general approach. The following parts of this sections are devoted to give more details about how Auto-WEKA finds, in a iterative way, the best combination of ML algorithm and hyper-parameters setting (Sect. 11.3.1); the raw data stored to represent the prediction of traffic speed in freeway and urban environments (Sect. 11.3.2); and the data-sets generated as well as the experimental set-up of this work (Sect. 11.3.3).

11.3.1 Auto-WEKA

Auto-WEKA approaches the algorithm selection problem through a Bayesian optimisation method. It considers the space of WEKA's ML algorithms $X = \{X^{(1)}, ..., X^{(k)}\}$ and their hyper-parameter spaces $A = \{A^{(1)}, ..., A^{(k)}\}$ to identify the combination of algorithm $X^{(i)} \in X$ and hyper-parameters $A^{(i)} \in A$, which minimises cross-validation loss (Eq. (11.1)), where $\gamma\left(X_A^{(i)}, D_{train}^{(i)}, D_{test}^{(i)}\right)$ denotes the loss achieved by algorithm $X^{(i)}$ with hyper-parameters $A^{(i)}$ when trained on training data-set $D_{train}^{(i)}$ and evaluated on test data-set $D_{test}^{(i)}$.

$$X_A^* * = \underset{X^{(i)} \in X, A^{(i)} \in A}{argmin} \frac{1}{K} \sum_{i=1}^{k} \gamma\left(X_A^{(i)}, D_{train}^{(i)}, D_{test}^{(i)}\right) \tag{11.1}$$

Thornton et al. [29] call this the combined algorithm selection and hyper-parameter optimisation (CASH) problem: determining $argmin_{\theta \in \Theta} f(\Theta)$ wherein each configuration $\theta \in \Theta$ contains the choice of algorithm $X^{(i)} \in X$ and its hyper-parameters setting $A^{(i)} \in A$. With this problem definition, the Bayesian optimisation strategy fits a probabilistic model to capture the relationship between different hyper-parameter configurations and their performance; it then uses this model to select the most promising hyper-parameter setting, assesses it, updates the model with the result of configuration chose, and iterates until a pre-defined time budget is reached.

One drawback of the Bayesian optimisation approach is its high computational cost at the moment of initialising the search for the most promising hyper-parameter setting. Besides, as the space of algorithms and hyperameters increases, the computational cost of evaluating them also increments. To overcome this issue, Feurer et al. [7] proposed a method to warm-start the Bayesian search. Concretely, the authors use a meta-learning approach that quickly suggests some instantiations of ML algorithms with their hyper-parameter setting that are likely to perform well. Meta-learning performs a pre-selection of promising configurations that are fed into the optimisation procedure, which ultimately is in charge of doing a fine-grained optimisation of them. Thus, it is possible to decrease in an efficient way the computational costs associated with broad spaces of algorithms and hyper-parameters.

Taking into account that Auto-WEKA does not incorporate any mechanism to deal with the aforementioned issue, in this work, we consider the full range of ML algorithms that the AutoML includes.

11.3.2 Raw Data

Freeway data used in this work is provided by the Caltrans Performance Measurement System[2] whose information is collected, in real time every 30 s, from nearly 40,000 individual detectors spanning the freeway system across the metropolitan area of California (USA). According to recent literature, this data source is highly used in the area of TF because of its high quality data, availability of various traffic measures, and its public accessibility.

The route selected for our experiments is the California Interstate I405-S. It is a heavily trafficked freeway by commuters along its entire length [22]. Particularly, we focus on the loop detectors shown in Fig. 11.1, where the detector marked with a ⋆ symbol represents the forecast target location. The traffic measure collected from the detectors is speed in an aggregation time of 5 min within the time window from March 1, 2019 to April 7, 2019 (38 days of data).

Contrarily, the urban data included in this research is the one obtained from the Madrid Open Data Portal.[3] The Madrid City Council provides through this website access to traffic data around the whole city, publishing 15 min aggregates and live

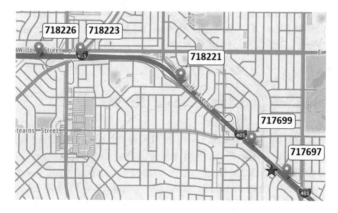

Fig. 11.1 Location of five freeway sensors in California State (USA). The detector marked with a ⋆ symbol represents the forecast target location

---

[2]http://pems.dot.ca.gov.

[3]https://datos.madrid.es/portal/site/egob/.

Fig. 11.2 Location of five urban sensors in Madrid city (Spain). The detector marked with a ⋆ symbol represents the forecast target location

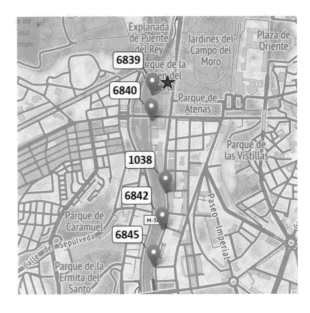

5 min aggregates of flow, occupancy, and speed data in more than 3600 measuring stations (loops).

We chose the M-30 motorway that circles the central districts of Madrid and that is considered the busiest Spanish road because of its traffic jams. On this route, we focus on the loop detectors depicted in Fig. 11.2 where again the ⋆ symbol represents the forecast target location. From them, we extract traffic speed data, in an aggregation time of 15 min, for the period lapsed between February 2, 2019 and February 28, 2019 (27 days of data).

11.3.3 Data-Sets and Experimental Set-Up

In this chapter, we approach two types of TF regression problems with different instances of them that are described below. The first type corresponds to the prediction of traffic at a target location, in a freeway environment; on the one hand, using only past traffic data of this location (temporal data, T), and then considering historical traffic data coming from the target location and from four downstream positions (temporal and spatial data, TS). Besides, in both instances, the input is enriched with calendar data (CD).

The second kind of TF problems is focused on forecasting traffic speed within an urban context. Repeatedly, the predictions are done for a single target location considering exclusively historical data of this spot; and on the other hand, taking into account past traffic data of the objective location together with other four downstream positions. Again, the input data in both instances is complemented with calendar data.

For the two families of TF problems described, we generate 18 data-sets in which speed is the traffic measure to be predicted. In the freeway case, time horizons wherein speed is predicted are 5, 15, 30, 45, and 60 min using data granularity of 5 min (granularity means how often the traffic measure is aggregated). Differently, for the urban TF problems, the forecasting time steps are 15, 30, 45, and 60 min with data granularity of 15 min. To better identify the data-sets, they are named following the next structure: *Context_InputData_Granularity_TimeHorizon*.

Attributes of freeway data-sets where the input is composed of only traffic data from the target location together with calendar data are: *Day of the week; Minute of the day; Traffic speed of the objective spot at past 5, 10, 15, 20, 25, 30, 35, 40, and 45 min; and Current traffic speed in such point*. In the case of freeway data-sets where the input consists of historical speed taken from the target location and from four downstream detectors, the attributes are: *Day of the week; Minute of the day; Traffic speed of the target position and four downstream locations at past 5, 10, 15, and 20 min; and Current speed of these five spots*.

Attributes of urban data-sets in which the input comprises traffic data of the target spot and calendar information are: *Day of the week; Minute of the day; Traffic speed of the objective spot at past 15, 30, 45, 60, 75, 90, 105, 120, and 135 min; and Current traffic speed in this point of interest*. Contrarily, urban data-sets, wherein the input is past traffic speed stored from the target location and from four downstream positions in addition to calendar, have the following attributes: *Day of the week; Minute of the day; Traffic speed of the target position and four downstream locations at 15, 30, 45, and 60 min in the past; and Current speed of these five positions*.

Lastly, for the experimentation with Auto-WEKA, three execution times (ET) were considered: 15, 150, and 300 min. These correspond to the time that the method takes to find the best ML algorithm and its hyper-parameter configuration for a given data-set. Furthermore, five repetitions with different initial seeds were carried out for each execution time. In the case of the BAs, we test them using WEKA. The process of evaluating every BA over a data-set was done with five repetitions with different initial seeds, and using the default hyper-parameter setting offered by WEKA. We have not performed any optimisation or extra-adjustment of the BAs' hyper-parameters because our aim is to compare the performance of AutoML versus BAs using the same human effort for both of them in order to make a fairer comparison.

11.4 Results

This section presents the results obtained with the experimental set-up proposed in the previous section. We evaluated the performance of the AutoML method and the BAs using the metric *Root-Mean-Square Error (RMSE)*, which is applied for regression problems to measure the average magnitude of the error between the predictions of a learning model and the actual values extracted from the raw data. Its calculation is expressed as $RMSE = \sqrt{\frac{1}{n} \sum_{i=1}^{n} (y_i - \overline{y_i})^2}$ wherein n corresponds to the number of samples in the data-set.

Table 11.1 shows the mean and standard deviation (between brackets) of the $RMSE$ values obtained by both Auto-WEKA and the BAs over all repetitions for each data-set. $RMSE$ values in bold indicate the best result in every data-set achieved from either any of the BAs or any of the Auto-WEKA's execution times.

As it can be seen in Table 11.1, the AutoML method performs better than the BAs along eight of the data-sets. In all the other cases, RF or NN obtain better results than Auto-WEKA although with small improvements ranging from 0.01 to 1.31 in the $RMSE$ values. These results are interesting because in order to get the conclusion that RF and NN are the best BAs in those cases, the human user should run all BAs over all data-sets and compare their performance among them, which is a time consuming task. However, running Auto-WEKA only once, and therefore employing less human effort, the user can achieve similar or better results than those obtained with the best BAs.

Regarding data-sets characteristics, we can see that they do influence the differences between results of Auto-WEKA and BAs. Concretely, for all urban data-sets with a granularity of 15 min (with the exception of the data-set $UbTS+CD15m15$), the AutoML method obtains the best $RMSE$ values. On the other hand, RA works especially well on freeway data-sets with the shortest and longest time horizons to be predicted, excluding both $FwT + CD5m30$ and $FwTS + CD5m30$ data-sets in which the AutoML get the best $RMSE$ performance.

Another interesting aspect is the relation between the execution time and the performance of the models provided by Auto-WEKA. Longer execution times contribute to obtaining better results, particularly, in the urban data-sets with longer time horizons. In the case of freeway data-sets where RF and NN algorithms are the best ones, the results improve when the Auto-WEKA's execution time increases from 15 to 150 min, but they are worse when we pass from 150 to 300 min. Similar to what happened in [1], we observed that this worsening is due to the over-fitting produced by the hyper-parameters selected by Auto-WEKA. This result indicates that it is necessary to introduce mechanisms in the AutoML method to deal with over-fitting, especially when execution times are high.

To assess whether the differences in performance observed in Table 11.1 are significant or not, we made use of non-parametric statistical tests. Two statistical tests have been applied, following the guidelines proposed in [8]. First, the Friedman's test for multiple comparisons has been applied to check whether there are significant differences among the three execution times of Auto-WEKA. Given that the p-value returned by this test is 0.35, the null hypothesis cannot be rejected in any of the cases. According to the Friedman's average ranking, shown in Table 11.2, 300 meT is the best execution time of the AutoML method confirming that, for this type of methods, the longer the execution time, the better.

In order to assess if the differences observed between the best AutoML method (300 mET) and the BAs are significant or not, we also used Friedman's non-

Table 11.1 Mean $RMSE$ values and their standard deviations (in brackets) obtained, for freeway (Fw) and urban (Ub) data-sets, by the AutoML method and the BAs

Data-sets	Auto-WEKA			Baseline algorithms			
	15 mET	150 mET	300 mET	k-NN	NN	RF	SVM
$Fw\_T+CD\_5m\_5$	2.87 (0.08)	2.87 (0.08)	2.91 (0.06)	4.25 (0.14)	2.93 (0.20)	**2.86 (0.06)**	2.90 (0.09)
$Fw\_T+CD\_5m\_15$	5.81 (0.33)	5.80 (0.28)	5.82 (0.34)	6.66 (0.22)	5.90 (0.45)	**5.16 (0.19)**	5.68 (0.18)
$Fw\_T+CD\_5m\_30$	7.35 (0.85)	**6.76 (0.41)**	6.99 (0.68)	8.30 (0.39)	9.05 (1.59)	7.06 (0.13)	8.19 (0.23)
$Fw\_T+CD\_5m\_45$	8.30 (1.09)	7.83 (1.12)	8.53 (0.30)	8.72 (0.20)	10.26 (1.15)	**7.70 (0.25)**	9.65 (0.17)
$Fw\_T+CD\_5m\_60$	9.12 (1.87)	9.01 (1.67)	9.61 (1.70)	9.01 (0.26)	10.90 (0.74)	**7.99 (0.08)**	10.56 (0.09)
$Fw\_TS+CD\_5m\_5$	1.19 (0.05)	1.16 (0.01)	1.17 (0.03)	1.46 (0.03)	1.44 (0.29)	**1.13 (0.05)**	1.11 (0.03)
$Fw\_TS+CD\_5m\_15$	1.92 (0.00)	2.00 (0.47)	2.01 (0.55)	1.78 (0.06)	2.16 (0.24)	**1.64 (0.03)**	1.86 (0.05)
$Fw\_TS+CD\_5m\_30$	2.12 (0.37)	2.37 (0.47)	**1.90 (0.41)**	1.95 (0.13)	2.60 (0.26)	1.91 (0.08)	2.43 (0.05)
$Fw\_TS+CD\_5m\_45$	2.50 (0.48)	2.33 (0.49)	2.14 (0.49)	**2.05 (0.09)**	2.92 (0.24)	2.06 (0.07)	2.82 (0.05)
$Fw\_TS+CD\_5m\_60$	3.17 (0.63)	2.82 (0.69)	2.26 (0.49)	**2.16 (0.09)**	2.89 (0.15)	**2.16 (0.12)**	3.10 (0.11)
$Ub\_T+CD\_15m\_15$	**5.62 (0.15)**	5.76 (0.26)	5.71 (0.36)	7.74 (0.40)	7.68 (1.27)	5.77 (0.03)	6.05 (0.11)
$Ub\_T+CD\_15m\_30$	**5.71 (0.29)**	5.97 (0.57)	5.74 (0.35)	8.20 (0.37)	8.02 (1.03)	5.80 (0.23)	6.33 (0.25)
$Ub\_T+CD\_15m\_45$	5.68 (0.14)	5.73 (0.15)	**5.65 (0.03)**	8.45 (0.20)	8.25 (1.88)	6.16 (0.26)	6.80 (0.37)
$Ub\_T+CD\_15m\_60$	5.91 (0.12)	**5.85 (0.13)**	5.88 (0.25)	8.52 (0.60)	7.25 (0.70)	5.98 (0.42)	7.05 (0.42)
$Ub\_TS+CD\_15m\_15$	8.97 (0.46)	8.84 (0.38)	8.83 (0.18)	10.42 (0.72)	14.81 (0.93)	**7.92 (0.30)**	8.45 (0.35)
$Ub\_TS+CD\_15m\_30$	7.91 (0.23)	7.80 (0.17)	**7.61 (0.23)**	12.95 (0.80)	17.18 (1.82)	9.34 (0.53)	10.75 (0.66)
$Ub\_TS+CD\_15m\_45$	9.89 (0.18)	9.56 (0.23)	**9.54 (0.24)**	13.96 (0.79)	19.02 (2.94)	9.74 (0.51)	11.53 (0.41)
$Ub\_TS+CD\_15m\_60$	9.25 (0.09)	9.07 (0.24)	**8.94 (0.11)**	13.07 (0.52)	17.09 (0.96)	9.77 (0.91)	11.94 (0.84)

Table 11.2 Friedman's average ranking for the auto-WEKA execution times

Auto-WEKA mET	Avg. ranking
300 mET	1.8333
150 mET	1.8889
15 mET	2.2778

Table 11.3 Friedman's average ranking and adjusted p-values obtained through Holm post-hoc test using RF as control algorithm

Algorithms	Avg. ranking	Adj. p-values
RF	1.6111	–
Auto-WEKA (300 mET)	2	4.60597 e−1
SVM	3.0556	**1.2264 e−2**
k-NN	3.7778	**1.18 e−4**
NN	4.5556	**0**

parametric test. Considering that the p-value returned by these tests was 0, the null hypothesis could be rejected. The mean ranking returned by the test is displayed in Table 11.3, confirming the better global results of RF against the others BAs and Auto-WEKA 300 mET. At the same time, it also shows the better global results of the AutoML method versus k-NN, NN, and SVM.

Holm post-hoc test has also been applied using RF as control algorithm (because it is the method that achieved the best overall performance) to assess the significance of the differences in performance with respect to the other algorithms. Table 11.3 presents the adjusted p-values returned by this test. In order to highlight significant differences, those p-values lower than 0.05 are shown in bold. Looking at Table 11.3, there are important differences in the test's outcomes. It can be said that RF results improve significantly the rest of BAs, but not the 300 mET of Auto-WEKA.

To finalise with this section, we analyse the ML methods selected by Auto-WEKA over all data-sets. Table 11.4 summarises how many times an algorithm is selected to forecast traffic speed along the data-sets. It is important to clarify that Auto-WEKA has a base of 39 algorithms and the ones that were not suggested for the data-sets evaluated are not included in Table 11.4. As each data-set was evaluated with three Auto-WEKA's running times along five repetitions in each of them, one algorithm can be chosen at most 15 times per data-set.

According to the results of Table 11.4, for freeway data-sets, IBk (k-NN) is the most selected method with the exception of data-sets $FwT + CD5m5$, $FwT + CD5m15$, and $FwTS + CD5m5$ wherein two tree-based algorithms (M5 and RF) and one regression algorithm are the most chosen. On the other hand, for urban data-sets, tree-based algorithms (RF and M5) are the most chosen algorithms, excluding the data-set $UbT + CD15m30$ in which the method with the highest frequency is bagging. In the cases of data-sets $UbT + CD15m45$ and $UbTS + CD15m60$, tree-based algorithms got the selection frequency with ensemble methods (random committee and bagging).

In general, the three most chosen algorithms (RF, IBk, M5) along all data-sets belong to tree-based and lazy families of methods. This is in concordance with the results obtained by Angarita et al. [1] wherein RF was the most selected ML

Table 11.4 ML methods selected by Auto-WEKA and absolute frequency in which they were suggested for freeway and urban data-sets

ML methods	Fw_T+CD_5m_5	Fw_T+CD_5m_15	Fw_T+CD_5m_30	Fw_T+CD_5m_45	Fw_T+CD_5m_60	Fw_TS+CD_5m_5	Fw_TS+CD_5m_15	Fw_TS+CD_5m_30	Fw_TS+CD_5m_45	Fw_TS+CD_5m_60	Subtotal Fw	Ub_T+CD_15m_15	Ub_T+CD_15m_30	Ub_T+CD_15m_45	Ub_T+CD_15m_60	Ub_TS+CD_15m_15	Ub_TS+CD_15m_30	Ub_TS+CD_15m_45	Ub_TS+CD_15m_60	Subtotal Ub	Total Fw + Ub
M5	8	0	0	0	0	4	0	0	0	0	**12**	1	0	0	0	1	5	12	6	**25**	**37**
Linear regression	1	0	0	0	0	7	0	0	0	0	**8**	0	0	0	0	0	0	0	0	**0**	**8**
Random committee	2	0	0	0	0	0	0	3	0	0	**5**	0	0	0	0	0	0	0	0	**0**	**5**
Bagging	3	0	0	1	0	0	0	1	1	0	**6**	2	4	4	2	3	4	2	6	**27**	**33**
Additive regression	1	0	1	0	2	0	0	0	1	1	**6**	2	1	2	1	2	2	0	0	**10**	**16**
Random forest	0	7	4	2	5	2	5	2	2	1	**30**	7	3	4	5	8	1	0	1	**29**	**59**
Random committee	0	6	2	1	3	0	0	0	0	2	**14**	2	5	3	3	1	1	0	0	**15**	**29**
Vote	0	1	0	0	0	0	0	0	0	0	**1**	0	0	0	0	0	0	0	0	**0**	**1**
LWL	0	1	0	0	0	1	0	0	2	1	**5**	1	0	1	1	0	0	0	0	**3**	**8**
IBk	0	0	6	10	5	0	6	5	7	6	**45**	0	1	1	0	0	0	0	0	**2**	**47**
KStar	0	0	1	0	0	0	3	3	1	3	**11**	0	0	0	0	0	0	0	0	**0**	**11**
SMOreg	0	0	0	1	0	1	1	0	0	1	**4**	0	0	0	0	0	0	0	0	**0**	**4**
Random subspace	0	0	0	0	0	1	0	1	1	0	**3**	0	1	0	0	0	0	0	0	**1**	**4**
Gaussian processes	0	0	0	0	0	0	0	0	0	0	**0**	0	0	0	2	0	0	0	0	**2**	**2**
J48	0	0	0	0	0	0	0	0	0	0	**0**	0	0	0	2	0	1	1	2	**6**	**6**

Values in bold indicate that one algorithm has been suggested by Auto-WEKA at least once per data-set algorithm to address the TF classification problem approached by the authors. Furthermore, the relevance of hyper-parameter tuning can be appreciated through the case of IBk algorithm. Concretely, the instance of this method within the BAs (k-NN) did not achieve competitive results; however, in the case of Auto-WEKA, its performance was improved, without the need of human effort. The better adjustment of hyper-parameters done by the AutoML method, make the IBk algorithm be among the three most selected methods.

11.5   Conclusions

In this paper, we have focused on deepening into the benefits of AutoML for supervised regression in the field of TF. To accomplish such purpose, we have compared to what extent the results of AutoML differ from the general approach in TF. We used Auto-WEKA as AutoML method and NN, SVM, k-NN, and RF as BAs. Concretely, our comparisons were made based on predicting traffic speed, over multiple time horizons ahead, for two different scenarios for the TF regression problems with different instances of them. The first type corresponds to the prediction of traffic at a target location, in a freeway environment; on the one hand, using only past traffic data of the target location, and, on the other hand, considering historical traffic data coming from the target location and from four downstream positions. The second type of TF problems is focused on forecasting traffic speed within an urban context, using the same variants as for the freeway scenario described above.

From the results we drawn interesting conclusions. From a computer science perspective, the AutoML method improves three out of the four BAs, and obtains similar results to RF (the best BA) without statistically significant differences. With a lower human effort, the user can expect similar or even better results (in the case of urban data-sets) than the best BA. Besides, another interesting conclusion is that higher execution times for Auto-WEKA not always lead to better results as we can expect. With some preliminary tests we detected that is was due to over-fitting issues.

From a transportation approach, tree-based algorithms and IBk are suitable methods to make predictions, in freeway contexts, at a target location using either traffic data obtained only from that location or data also provided by other locations surrounding the target position. On the other hand, tree-based and ensembles algorithms seem to be the best approach to forecast traffic speed for urban environments.

Further research lines that we aim to explore in the future are: (1) comparing optimisation and meta-learning strategies to find the best pair (algorithm and hyper-parameter setting); (2) integrating data pre-processing techniques to the AutoML process; and (3) exploring the benefits and recommendations of algorithms done by AutoML methods at the moment of approaching more complex families of TF regression problems.

Acknowledgements This project has received funding from the European Union's Horizon 2020 research and innovation programme under grant agreements No 636220 and No 815069, and the Marie Sklodowska-Curie grant agreement No. 665959.

References

1. Angarita-Zapata, J.S., Triguero, I., Masegosa, A.D.: A preliminary study on automatic algorithm selection for short-term traffic forecasting. In: Del Ser J., Osaba E., Bilbao M.N., Sanchez-Medina J.J., Vecchio M., Yang X.S. (eds.) Intelligent Distributed Computing XII, pp. 204–214. Springer, Berlin (2018)
2. Angarita-Zapata, J.S., Masegosa, A.D., Triguero, I.: A taxonomy of traffic forecasting regression problems from a supervised learning perspective. IEEE Access, 1–1 (2019). https://doi.org/10.1109/ACCESS.2019.2917228
3. Charte, D., Charte, F., García, S., Herrera, F.: A snapshot on nonstandard supervised learning problems: taxonomy, relationships, problem transformations and algorithm adaptations. In: Progress in Artificial Intelligence, pp. 1–14. Springer, Berlin (2018)
4. Chen, J.F., Lo, S.K., Do, Q.H.: Forecasting short-term traffic flow by fuzzy wavelet neural network with parameters optimized by biogeography-based optimization algorithm. In: Computational Intelligence and Neuroscience, pp. 1–13 (2018)
5. Chen, W., An, J., Li, R., Fu, L., Xie, G., Bhuiyan, M.Z.A., Li, K.: A novel fuzzy deep-learning approach to traffic flow prediction with uncertain spatial-temporal data features. Futur. Gener. Comput. Syst. **89**, 78–88 (2018)
6. Ermagun, A., Levinson, D.: Spatiotemporal traffic forecasting: review and proposed directions. Transp. Rev. **38**(6), 786–814 (2018)
7. Feurer, M., Klein, A., Eggensperger, K., Springenberg, J., Blum, M., Hutter, F.: Efficient and robust automated machine learning. In: Advances in Neural Information Processing Systems, vol. 28, pp. 2962–2970 (2015)
8. Garcia, S., Fernandez, A., Luengo, J., Herrera, F.: Advanced nonparametric tests for multiple comparisons in the design of experiments in computational intelligence and data mining: experimental analysis of power. Inf. Sci. **180**(10), 2044–2064 (2010)
9. Howell, S.: Meta-analysis of machine learning approaches to short-term urban traffic prediction. In: Scottish Transport Applications and Research Conference (STAR), pp. 1–15 (2018)
10. Karlaftis, M., Vlahogianni, E.: Statistical methods versus neural networks in transportation research: differences, similarities and some insights. Transp. Res. C: Emerg. Technol. **19**(3), 387–399 (2011)
11. Kerner, B.S.: The physics of traffic. Phys. World **12**(8), 25–30 (1999)
12. Kerschke, P., Hoos, H., Neumann, F., Trautmann, H.: Automated algorithm selection: survey and perspectives. Evol. Comput. **27**, 1–47 (2018)
13. Komer, B., Bergstra, J., Eliasmith, C.: Hyperopt-sklearn: automatic hyperparameter configuration for scikit-learn. In: Proceedings of SciPy, p. 33–39 (2014)
14. Laña, I., Del Ser, J., Velez, M., Vlahogianni, E.I.: Road traffic forecasting: recent advances and new challenges. IEEE Intell. Transp. Syst. Mag. **10**(2), 93–109 (2018)
15. Liu, Z., Guo, J., Cao, J., Wei, Y., Huang, W.: A hybrid short-term traffic flow forecasting method based on neural networks combined with K-nearest neighbor. J. Traffic Transp. Technol. **30**(4), 445–456 (2018)
16. Lopez-Garcia, P., Onieva, E., Osaba, E., Masegosa, A.D., Perallos, A.: A hybrid method for short-term traffic congestion forecasting using genetic algorithms and cross entropy. IEEE Trans. Intell. Transp. Syst. **17**(2), 557–569 (2016)
17. Luo, G.: A review of automatic selection methods for machine learning algorithms and hyperparameter values. Netw. Model. Anal. Health Inform. Bioinform. **5**(1), 5–18 (2016)

18. Ma, X., Yu, H., Wang, Y., Wang, Y.: Large-scale transportation network congestion evolution prediction using deep learning theory. PLoS One **10**(3), 1–17 (2015)
19. Mohr, F., Wever, M., Hüllermeier, E.: ML-Plan: automated machine learning via hierarchical planning. Mach. Learn. **107**(8), 1495–1515 (2018)
20. Oh, S., Byon, Y.J., Jang, K., Yeo, H.: Short-term travel-time prediction on highway: a review of the data-driven approach. Transp. Rev. **35**(1), 4–32 (2015)
21. Olson, R.S., Bartley, N., Urbanowicz, R.J., Moore, J.H.: Evaluation of a tree-based pipeline optimization tool for automating data science. In: Proceedings of the Genetic and Evolutionary Computation Conference 2016, pp. 485–492. ACM, New York (2016)
22. Park, J., Murphey, Y.L., McGee, R., Kristinsson, J.G., Kuang, M.L., Phillips, A.M.: Intelligent trip modeling for the prediction of an origin-destination traveling speed profile. IEEE Trans. Intel. Transp. Syst. **15**(3), 1039–1053 (2014). https://doi.org/10.1109/TITS.2013.2294934. http://ieeexplore.ieee.org/document/6728714/
23. Pell, A., Meingast, A., Schauer, O.: Trends in real-time traffic simulation. Transp. Res. Proc. **25**, 1477–1484 (2017)
24. Rahimipour, S., Moeinfar, R., Hashemi, S.M.: Traffic prediction using a self-adjusted evolutionary neural network. J. Mod. Transp., 1–11 (2018). https://doi.org/10.1007/s40534-018-0179-5
25. Sabharwal, A., Samulowitz, H., Tesauro, G.: Selecting near-optimal learners via incremental data allocation. In: Proceedings of the Thirtieth Conference on Artificial Intelligence, pp. 2007–2015 (2016)
26. Skycomp, I.B.M.: Major High- way Performance Ratings and Bottleneck Inventory. Maryland State Highway Administration, the Baltimore Metropolitan Council and Maryland Transportation Authority, State of Maryland, Maryland (2009)
27. Sparks, E.R., Talwalkar, A., Haas, D., Franklin, M.J., Jordan, M.I., Kraska, T.: Automating model search for large scale machine learning. In: Proceedings of System-on-Chip Conference, pp. 368–380 (2015)
28. Swearingen, T., Drevo, W., Cyphers, B., Cuesta-Infante, A., Ross, A., Veeramachaneni, K.: ATM: a distributed, collaborative, scalable system for automated machine learning. In: 2017 IEEE International Conference on Big Data, pp. 151–162 (2017)
29. Thornton, C., Hutter, F., Hoos, H.H., Leyton-Brown, K.: Auto-WEKA. In: Proceedings of the 19th International conference on Knowledge discovery and data mining, pp. 847–855 (2013)
30. Vlahogianni, E.I.: Enhancing predictions in signalized arterials with information on short-term traffic flow dynamics. J. Intell. Transp. Syst. **13**(2), 73–84 (2009)
31. Vlahogianni, E.I.: Optimization of traffic forecasting: intelligent surrogate modeling. Transp. Res. C: Emerg. Technol. **55**, 14–23 (2015)
32. Vlahogianni, E.I., Golias, J.C., Karlaftis, M.G.: Short-term traffic forecasting: overview of objectives and methods. Transp. Rev. **24**(5), 533–557 (2004)
33. Vlahogianni, E.I., Karlaftis, M.G., Golias, J.C.: Short-term traffic forecasting: where we are and where we're going. Transp. Res. C: Emerg. Technol. **43**, 3–19 (2014)
34. Xia, J., Huang, W., Guo, J.: A clustering approach to online freeway traffic state identification using ITS data. J. Civ. Eng. **16**(3), 426–432 (2012)
35. Yao, Q., Wang, M., Chen, Y., Dai, W., Yi-Qi, H., Yu-Feng, L., Wei-Wei, T., Qiang, Y., Yang, Y.: Taking human out of learning applications: a survey on automated machine learning. CoRR (2018)
36. Yu, H., Wu, Z., Wang, S., Wang, Y., Ma, X.: Spatiotemporal recurrent convolutional networks for traffic prediction in transportation networks. Sensors **17**(7), 1501(1–16) (2017)
37. Zöller, M.A., Huber, M.F.: Survey on automated machine learning. CoRR (2019)

Chapter 12
Multi-Robot Coalition Formation and Task Allocation Using Immigrant Based Adaptive Genetic Algorithms

Amit Rauniyar and Pranab K. Muhuri

Abstract In multi-robot systems, coordination and cooperation among group of robots are very important to execute the tasks assigning the available resources. Multi-robot coalition formation (MRCF) is a well-known problem that deals in the formation of groups of robots to handle and execute the tasks simultaneously and efficiently. The involvement of different robots form coalitions with different efficiencies and costs due to the differences in the number and the capabilities of the robots. Therefore, the problem becomes finding an optimal coalition of robots for the completion of tasks with the main objective to maximize the total efficiency. We develop and implement different variants of GA as a solution technique. Also, in order to enhance the diversity of solutions and premature convergence of SGA (Standard GA), we develop and incorporate immigrants based schemes into SGA and develop RIGA (Random Immigrants GA) and EIGA (Elitism-based Immigrants GA) to find optimal task allocation. It is important to govern the rate of changes during the evolution of solutions. Therefore, we also introduce and integrate adaptive settings of genetic operations to maintain the adaptability of environment and develop adaptive variants of GAs. We term them as aRIGA (adaptive RIGA) and aEIGA (adaptive EIGA). Extensive simulation experiments and statistical analysis on the results obtained have demonstrated that RIGA and EIGA produce better solutions than SGA with both fixed and adaptive genetic operators. Among them, aEIGA outperforms all.

Keywords Multi-robot · Coalition formation · Task allocation · Genetic algorithm · Adaptive · Immigrants · Evolutionary algorithms · Elitism

A. Rauniyar · P. K. Muhuri (✉)
Department of Computer Science, South Asian University, New Delhi, India
e-mail: pranabmuhuri@cs.sau.ac.in

© Springer Nature Switzerland AG 2020
O. Llanes Santiago et al. (eds.), *Computational Intelligence in Emerging Technologies for Engineering Applications*, Studies in Computational Intelligence 872, https://doi.org/10.1007/978-3-030-34409-2_12

12.1   Introduction

The advancement in robotics has brought the employment of multiple robots to solve complex tasks together in several domains like manufacturing process, health care, transportation, etc. A multi-robot system (MRS) offers more robustness and flexibility to the system than a single-robot system in terms of their scopes and utilities [1]. A task may consist of various sub-tasks which may require different robots bearing different tools to execute them. This brings the scope of multi-robots task assignment/allocation (MRTA) to be addressed to execute the tasks efficiently. The available robots in MRS need to coordinate together and forms a coalition to execute the assigned task. A coalition is defined as a group of robots that can combine together to form a team [2]. Different robotic coalitions may execute a particular task at different efficiencies and costs due to the differences in the numbers and capabilities of their member robots. Thus, allocation of tasks depends on these differences [3]. Therefore, to handle the multiple available tasks of a system, multiple such coalitions need to be formed which can be termed as multi-robot coalition formation (MRCF) problem [4]. Hence, the goal becomes to find a coalition structure of available robots for the completion of tasks with higher efficiency. The set of all disjoint coalitions is called the coalition structure.

The problem belongs to the category NP-hard optimization problems [4]. The main complexity is to make coalition considering the available resources with robots and resources required for the tasks to find optimal coalition structure. Several, well-designed coalition formation algorithms exist in literature that comprises of traditional approaches including game theory, distributed problem solving, auction-based, behavior-based, and market-based [2, 5–15]. A fine survey on cooperative heterogeneous multi-robot is done by Rizk et al. [16]. The complexity lies in the design of methods with the addition of real-world constraints like computation time, cost, communication costs, and fault tolerance. Genetic algorithm (GA) being a meta-heuristic and evolutionary approach has been widely used to design solution techniques to solve a broad range of such problems [17–19]. Recently, several studies have been reported for solving the MRCF problems for task allocation using Standard GA (SGA) [3, 20–22]. The main limitation with SGA is that it reduces the diversity quickly from the population as it searches the global optimum with a strong selection strategy and small mutation rate. Some modified versions of SGA might be used to find a series of optima by enhancing the diversity in a population. This can be done by introducing the concept of immigrants which replaces a fraction of the population from an earlier generation [23–25]. Also, an adaptive technique to tune the crossover and mutation rate can be applied to maintain the diversity in a population without affecting its convergence properties [26, 27].

Therefore, in this chapter, we have presented and investigated six novel variants of GA based solution approaches to handle the tasks allocation issues in the MRCF problem. We have proposed and evaluated an SGA and two dynamic variants of dynamic GA, RIGA (Random Immigrants Genetic Algorithm) and EIGA (Elitism-based Immigrants Genetic Algorithm) for MRCF. The approaches have integration of adaptive technique of parameters and mutation after generation of immigrants, namely aSGA, aRIGA, and aEIGA [28, 29]. The development of algorithms starts with the designing of components of SGA for the problem. Then we incorporate immigrants and adaptive crossover/mutation rates schemes to maintain population diversity and better convergence. In immigrants schemes, a certain number of immigrants are generated and added to the pool of population by replacing the worst individual from the current generation. In case of adaptive parameters of crossover and mutation rate, the algorithm automatically sets these rates to enhance the quality of the degraded population. For the comparison purpose, we implemented all six variants. The simulation results show that the modified immigrants based GAs have outperformed the SGA.

Further, the chapter is organized as follows: Sect. 12.2 discusses some related works. The formation of robot coalition structure and problem formulation are described in Sect. 12.3. In Sect. 12.4, details about the elements used for developing SGA are discussed along with a numerical example. The details of dynamic GA variants for the MRCF problem are discussed in Sect. 12.5. Section 12.6 provides the simulation experiments and comparative study using the obtained results. At last a summary of the chapter is drawn with some discussion on future work in Sect. 12.7.

12.2   Related Work

Several works have focused on evolutionary approaches for the MRCF problem. The main difference between them has been the chromosomal representation of the coalition structure. Particle swarm optimization (PSO) has also been used for various coalition problems [30, 31] with the main aim to maximize the utility of system and completion of maximum number of tasks. A quantum-inspired ant colony optimization [33] has been developed for coalition formation problem where each ant is regarded as a quantum individual. Their strategy depicts better diversity than basic ACO without premature convergence. Yang et al. [22] proposed GA with two-dimensional binary chromosomes encoding that helps to track only a valid chromosome during the crossover. The main complexity lies in scanning the chromosome for making unique coalition, i.e., a robot can join only one coalition. But the results of the simulation show that their algorithm has outperformed one of the traditional approaches [3]. Liu et al. [20] has proposed a technique to generate variable length binary chromosome for GA with one-point crossover to detect the highest possible fitness value. The working of the algorithm has been illustrated with simple examples. In a recent work by Li et al. [21], a skillful coding policy for quantum evolutionary algorithm for MRCF has been designed to decrease the complexity of an algorithm that speeds up the searching time for the optimal solution. Such algorithms have good performance due to great exploration ability which can be further improved [14]. Many other significant works on the study of coordination and coalition formation in such systems have been reported in recent times [34–36]. Despite these advanced algorithms, to the best of our knowledge dynamic variants of GA have not been studied so far. This work aims to demonstrate the benefits of using dynamic variants of GA for the MRCF problem.

12.3 Problem Formulation

This section explains the elements required to understand the model of multi-robot coalition formation problem [2, 28, 29] we have considered in this chapter.

12.3.1 Formation of Robot Coalition

A robot coalition C is a set of one or more robots to complete a task in the system. A robot can involve in only one coalition for a task execution at a time. That means a task is allocated to a coalition having unique members. It is known as non-additive environment. The number of possible coalition with n robots in a system is

$$\sum_{i=1}^{n} C_n^i = 2^n - 1 \qquad (12.1)$$

12.3.2 Model of Coalition Structure Formation

Coalition structure (CS) is a combination of distinct coalitions for the completion of the tasks list. Every robot has certain feature capabilities which determine its efficiency. A task requires certain resources in order to get executed. Thus the certain cost is associated with the requirement of resources by the task. When robots are grouped together for a task execution, then the efficiency of coalition becomes the sum of each member's efficiency. The total available features capability in the coalition and the capability utilized by task give the cost and profit associated with coalition. This helps to determine the quality of coalition formed. Hence, the main goal is to maximize the efficiency as well as the profit of the system after the formation of the coalition for all tasks. We assume all robots are available and the robots can execute only one task at a time.

A centralized framework is designed where the central robot has knowledge about all other robots and tasks to be allocated. The central robot evaluates the coalition for all available tasks. The following entities describe the model of the system:

1. Suppose a set of n robots $R = \{R_1, R_2, \ldots, R_n\}$, a set of m tasks $T = \{T_1, T_2, \ldots, T_m\}$, and a set of r types of features $F = \{f^1, f^2, \ldots, f^r\}$ are present in the system.
2. Each robot R_i has r-dimension features capability $F_i = \{f_i^1, f_i^2, \ldots \ldots, f_i^r\}$, where $f_i^j \geq 0$ and $1 \leq i \leq n, 1 \leq j \leq r$.
3. f_i^j describes the feature f^j property of robot R_i with some positive quantity. This parameter determines the capability of R_i to complete a task or any action.
4. To complete a task T_i, it requires a set of features of certain capabilities $F_{T_i} = \{f_{T_i}^1, f_{T_i}^2, \ldots \ldots, f_{T_i}^r\}$, where $f_{T_i}^j \geq 0$ and $1 \leq i \leq m, 1 \leq j \leq r$.
5. There exists a capability vector $F_C = \{F_C^1, F_C^2, \ldots, F_C^r\}$ in coalition C and F_C is the total capability vectors of all robots.
6. In order to execute a task T_i, a coalition of robots C must satisfy

$$F_{T_i}^j \leq F_C^j, 1 \leq j \leq r \qquad (12.2)$$

That is the sum of total features capability in a coalition should be either greater than or equals to the sum of required capability for the execution of the task T_i. For each coalition (C) a cost Cost_C is associated with it which is the sum of total cost for each member to execute T_i by C.

7. After task T_i is completed, profit is gained which can be symbolized as Profit_{T_i}.
8. A task can be completed by different coalitions, each having a unique capability amount which can be termed as efficiency value, Value_C of coalition C.
9. Value_C is determined by Cost_C and Profit_{T_i} for the completion of task T_i.
10. If the coalition C satisfies relation (12.2), Value_C obtained is positive. Otherwise it is zero as Profit_{T_i} becomes 0.
11. So, for all tasks in the list, the combination of coalitions formed is termed as coalition structure (CS). There exist Cost_{CS}, Profit_{CS}, and Value_{CS} associated with it.

$$\text{Cost}_{CS} = \sum_{i=1}^{k} \text{Cost}_{Ci}, 1 \leq k \leq m \qquad (12.3)$$

$$\text{Profit}_{CS} = \sum_{i=1}^{k} \text{Profit}_{Ci}, 1 \leq k \leq m \qquad (12.4)$$

$$\text{Value}_{CS} = \sum_{i=1}^{k} \text{Value}_{Ci}, 1 \leq k \leq m \qquad (12.5)$$

The problem can be described as, for a given set of tasks $T = \{T_1, T_2, \ldots, T_m\}$, each T_k requires a set of capability vector of set F_k, and each robot R_i has its own capability vector F_i. The problem is how to assign each task in the list $T_i, \ldots, T_j$, where $T_i, T_j \in T$ and $1 \leq i \leq m, 1 \leq j \leq m, i \neq j$ to some CS of robots with

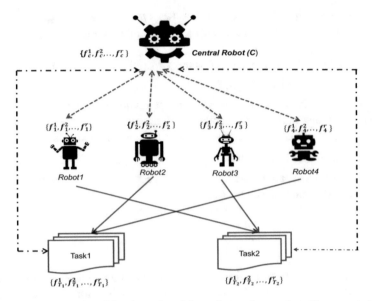

Fig. 12.1 An example of coalition formation of four robots and two tasks with one central robot

an objective to maximize the Value$_{CS}$. A typical scenario is demonstrated in Fig. 12.1 where a central robot makes two distinct coalitions among 4 different robots to complete two tasks with different resources requirements.

12.4 SGA for MRCF for Task Allocation

In this section, we are going to describe the implementation of SGA for MRCF for task allocation [28, 29]. The implementation starts with the initialization of population by mapping tasks to random robots and forming coalition. The fitness score is calculated to select the best individual for the generation of offspring. The genetic operators, crossover, and mutation are applied to get a new generation which leads to the optimal solution.

12.4.1 Genetic Encoding

Genes of a chromosome are represented using decimal number. The length of a chromosome at a time is equal to the number of available free robots. Assume n no. of robots are available in the system. Each individual g consists of n genes g_1, $g_2, \ldots, g_n$ and each gene g_i can take value of j, where $1 \leq j \leq n$. Thus, the value j assigned at ith gene can be interpreted as task j is allocated to ith robot.

Robots:

R$_1$	R$_2$	R$_3$	R$_4$	R$_5$	R$_6$	R$_7$	R$_8$	R$_9$	R$_{10}$

Chromosome

T$_1$	T$_3$	T$_4$	T$_1$	T$_1$	T$_2$	T$_4$	T$_1$	T$_2$	T$_1$

(a)

Robots:

1	2	3	4	5	6	7	8	9	10

Chromosome:

4	3	2	2	4	4	1	4	1	4

(b)

Fig. 12.2 (a) A chromosome model. (b) A chromosome generated as per code

Robot Task	R$_1$	R$_2$	R$_3$	R$_4$	R$_5$	R$_6$	R$_7$	R$_8$	R$_9$	R$_{10}$
T$_1$	1	0	0	1	1	0	0	1	0	1
T$_2$	0	0	0	0	0	1	0	0	1	0
T$_3$	0	1	0	0	0	0	0	0	0	0
T$_4$	0	0	1	0	0	0	1	0	0	0

Fig. 12.3 Binary chromosome model showing coalition of robots for task

Suppose in a system there are ten robots and four tasks to be completed. Figure 12.2a, b gives the genotype model we have considered here. Figures illustrate that task T_1 is allocated to robots' coalition of $R_1, R_4, R_5, R_8, R_{10}$, while T_2 is allocated to coalition of robots R_6 and R_9 and so on. Thus, the chromosome for each coalition can be represented in binary form too. The matrix shown in Fig. 12.3 represents whether the robot is available in coalition of task T or not. The binary matrix can be obtained from the chromosome we designed. This helps us to eliminate the invalid individuals and accept the individual presenting unique coalitions. Let us define an individual $C = (r_1, r_2, \ldots, r_n)$ as a coalition result, where a gene r_i represents whether robot R_i is present in the coalition for task T_j or not. If it is present, $r_i = 1$; otherwise, $r_i = 0$.

12.4.2 Population Initialization

The initial population of chromosomes is generated randomly for certain number of population size. The population of individuals represents the variety of solution which helps to explore the search space. Each individual is regarded as a candidate solution.

12.4.3   Fitness Function and Evaluation

The main goal is to maximize the efficiency (Value$_{CS}$) of system. For calculation of fitness value, we need to consider the total cost involved and profit gained in the system after coalition. First we calculate the profit and cost associated with coalition for a task. The profit is determined by the values of all capability vectors required by task T_i for its execution. However, the cost of coalition is the sum of all capability vectors of each member in the coalition. The profit for task and cost of coalition can be obtained from the following equations:

$$\text{Profit}_{T_i} = \left(f_{T_i}^1 + 2.f_{T_i}^2 + \cdots + r.f_{T_i}^r \right), \tag{12.6}$$

where $f_{T_i}^j$ is the value of capability of feature j required by task T_i and $1 \leq j \leq r$, r is the number of features. If relation (12.2) is satisfied, then the profit is positive; otherwise, profit is equal to 0.

$$\text{Cost}_{C T_i} = \sum_{j=1}^{k} \left(f_j^1 + 2.f_j^2 + \cdots + f_j^r \right), \tag{12.7}$$

where C_{T_i} is the coalition k of robots involved to complete task T_i. Thus, total value of the fitness of coalition structure formed for all task may be given by

$$F(\text{CS}) = \text{Value}_{CS} = \sum_{i=1}^{m} \text{Profit}_{T_i} / \sum_{i=1}^{m} \text{Cost}_{T_i}, \ 1 \leq i \leq m, \tag{12.8}$$

where Value$_{CS}$ is efficiency of total system for m number of tasks.

12.4.3.1   Computation Experiment

Let us suppose there are ten robots available in the system with four different features capabilities mentioned in Table 12.1.

Table 12.1 Capability value of robots

Feature capability	Robots									
	R_1	R_2	R_3	R_4	R_5	R_6	R_7	R_8	R_9	R_{10}
F^1	85	45	75	43	100	3	56	46	100	22
F^2	35	39	10	71	68	1	26	31	37	96
F^3	36	53	8	67	27	36	19	97	40	69
F^4	62	73	18	29	35	86	55	86	1	94

Also given that three tasks $T = \{T_1, T_2, T_3\}$ enter the system and require four different robots capabilities $F_{T_i} = \left\{ f_{T_i}^1, f_{T_i}^2, f_{T_i}^3, f_{T_i}^4 \right\}$ for completion. Assume a coalition structure, CS $= \{2\ 2\ 1\ 3\ 1\ 3\ 3\ 1\ 3\ 2\}$ is generated as one of the candidate solutions for the tasks in list T. That means, T_1, T_2, and T_3 can be executed by coalition of robots $\{R_3, R_5, R_8\}$, $\{R_1, R_2, R_{10}\}$, and $\{R_4, R_6, R_7, R_9\}$, respectively. Therefore, the coalition Cost$_{CT_1}$ for completing task T_1 will be

$$\text{Cost}_{CT_1} = \sum_{j=R_3, R_5, R_8} \left(f_j^1 + 2.f_j^2 + 3.f_j^3 + 4.f_j^4 \right),$$

$$\text{i.e., Cost}_{CT_1} = F_C^1 + 2.F_C^2 + 3.F_C^3 + 4.F_C^4,$$

where $F_C^1 = f_3^1 + f_5^1 + f_8^1$; $F_C^2 = f_3^2 + f_5^2 + f_8^2$; $F_C^3 = f_3^3 + f_5^3 + f_8^3$ and $F_C^4 = f_3^4 + f_5^4 + f_8^4$.

$F_C^1 = 75 + 100 + 46 = 221$; $F_C^2 = 10 + 68 + 31 = 109$; $F_C^3 = 8 + 27 + 97 = 132$; and $F_C^4 = 18 + 35 + 86 = 139$.

Thus the numerical value of Cost$_{CT_1} = 1*221 + 2*109 + 3*132 + 4*139 = 1391$. Similarly, we calculate the cost for task T_2 and T_3. Table 12.2 shows the total cost involved in completing these tasks.

The profit for T_1 can be calculated as Profit$_{T_1} = \left\{ 1.f_{T_1}^1 + 2.f_{T_1}^2 + 3.f_{T_1}^3 + 4.f_{T_1}^4 \right\}$. If the coalition cannot perform the task T_1 then Profit$_{T_1} = 0$. For our example, available capabilities are greater than required capabilities of all three tasks, i.e., Eq. (12.2) is satisfied. So, the profit value will be positive. Table 12.3 consists of details for required capabilities to complete the tasks and profit value obtained after its completion.

Table 12.2 Cost for completing tasks

T_i	F_C^r				Cost$_{CT_i}$
	F_C^1	F_C^2	F_C^3	F_C^4	
T_1	221	109	132	139	1391
T_2	152	170	158	229	1882
T_3	202	135	162	171	1642
				Total	4915

Table 12.3 Profit for tasks

T_i	$f_{T_i}^r$				Profit$_{T_i}$
	$f_{T_i}^1$	$f_{T_i}^2$	$f_{T_i}^3$	$f_{T_i}^4$	
T_1	110	105	130	112	1158
T_2	144	135	147	105	1275
T_3	141	121	104	126	1199
				Total	3632

Finally, the fitness value of the candidate solution considered is $F(CS) = \frac{3632}{4915} = 0.7389$. This concludes that the coalition structure (CS) for tasks completion utilizes almost 74% of resources available.

12.4.4 Selection Method

Selection plays a vital role in enhancing the average value of the population by passing high quality chromosomes to the next generation. We have implemented binary tournament selection without replacement method to choose parents for generating offspring. In tournament selection, t numbers of chromosomes are randomly selected from the population. Here, t is known as a tournament size that represents the number of individuals selected randomly. This technique selects the best individual with higher fitness as a parent through the tournament. The winner of the tournament competition is inserted into the mating pool and the completion is repeated until the pool for generating new offspring is filled up [37]. This selection strategy provides selective pressure through a tournament completion among selected individuals. As the mating pool comprises of winners, it has higher average population fitness value. This fitness difference provides the selection pressure that helps GA to improve the fitness of succeeding generations. Binary tournament selection is a variant of the tournament where two chromosomes with higher fitness values are selected randomly. The same procedure is repeated for all the chromosomes [38].

12.4.5 Genetic Operators

There are two important genetic operators, which are crossover and mutations. We now explain each of them below:

12.4.5.1 Crossover

One-point crossover is implemented. A random number is generated between $R \in [0, 1]$ if $R < p_c$, the crossover is done. The crossover point is decided by generating a random integer between 1 and n. The part of individual on the right of crossover point is exchanged to form two new offspring. Figure 12.4 below shows a one-point crossover operation.

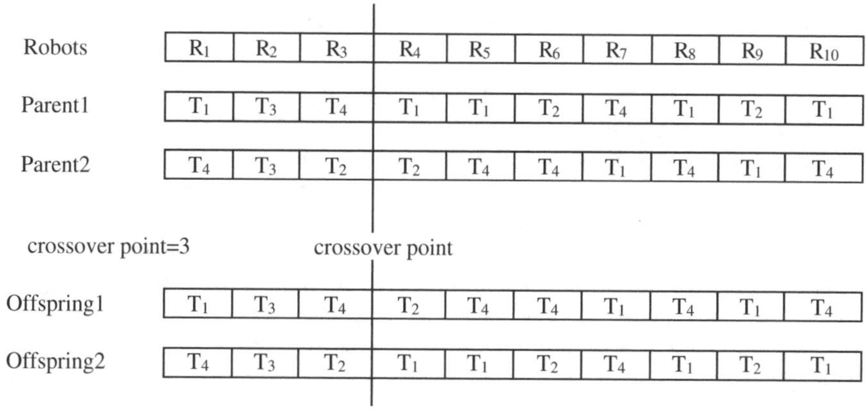

Robots	R_1	R_2	R_3	R_4	R_5	R_6	R_7	R_8	R_9	R_{10}
Parent1	T_1	T_3	T_4	T_1	T_1	T_2	T_4	T_1	T_2	T_1
Parent2	T_4	T_3	T_2	T_2	T_4	T_4	T_1	T_4	T_1	T_4

crossover point=3   crossover point

Offspring1	T_1	T_3	T_4	T_2	T_4	T_4	T_1	T_4	T_1	T_4
Offspring2	T_4	T_3	T_2	T_1	T_1	T_2	T_4	T_1	T_2	T_1

Fig. 12.4 One-point crossover

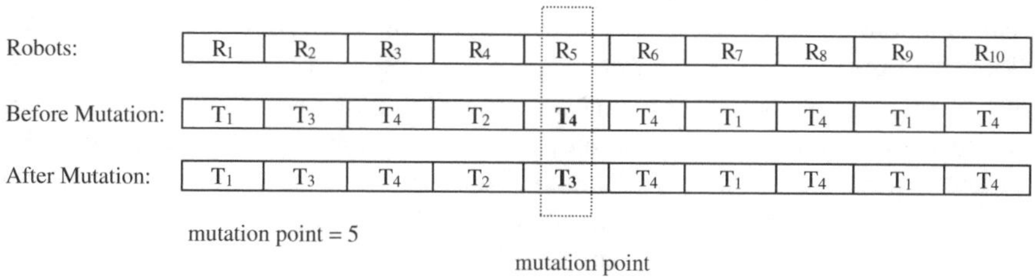

Robots:	R_1	R_2	R_3	R_4	R_5	R_6	R_7	R_8	R_9	R_{10}
Before Mutation:	T_1	T_3	T_4	T_2	T_4	T_4	T_1	T_4	T_1	T_4
After Mutation:	T_1	T_3	T_4	T_2	T_3	T_4	T_1	T_4	T_1	T_4

mutation point = 5

mutation point

Fig. 12.5 Mutation

12.4.5.2   Mutation

The mutation changes the allocation of tasks from one robot to others. Each gene in a chromosome is liable for mutation with probability p_m. A mutation point is fixed by generating a random integer number between 1 and n. After applying the mutation operator, the allocation of task T_4 at R_5 has changed to T_3. Now, R_5 joins to the coalition with R_2 for completion of task T_3. Figure 12.5 demonstrates an example for mutation operation.

12.5   Dynamic Genetic Algorithm(s) for the MRCF

The population evolution is no longer better as it adopts a set of individuals with fixed rate for the genetic operators [21, 28, 29, 39, 40] which somehow limits the possibility of adding diversified individuals. Dynamic GA versions are applied to maintain the level of diversity in population.

12.5.1 Genetic Algorithms with Immigrants Scheme

The population's survival is ensured by the diversity in nature. As the members in population represent potential solutions, they are useful in tracking the change in environment. SGA with its typical parameter is often unable to explore different part of the search space as its adaptiveness is limited to find a single solution [25]. Since SGA runs quickly to search an optimum location, some modified versions of it might be used to find a series of optima by increasing the diversity in the population. This can be done by introducing the concept of immigrants which replaces a fraction of the population from an earlier generation [23–25]. The immigrants schemes have been successfully adapted by several researchers to find a solution for dynamic optimization problems [24]. In dynamic environments, the convergence becomes a big issue for SGA. Since early convergence restricts the evolution of better individuals, SGA requires keeping a certain diversity level in population to maintain the adaptability of the environment. In order to resolve this convergence issue, random immigrants schemes such as RIGA and EIGA are quite useful. RIGA (Random Immigrant GA) was first proposed by [23] with the motivation from the flux of immigrants that stroll in and out of population between two generations in nature. In every generation, certain individuals of the current generation are replaced by newly generated random individuals. This maintains the diversity level within the population [32, 40]. The individuals in the population can be substituted either by replacing random individuals or replacing the worst [39]. In our case, we have used the strategy to replace the worst individuals of current population. The random immigrants population size is normally set to a small value (r_{ri}, ratio of the number of immigrants to the population size of N) in order to avoid the disruption during search. Generally, the ratio is set to 0.2. The difference between RIGA and SGA is that after fitness evaluation of population, $r_{ri} * N$ worst individuals in the populations are substituted by random immigrants. This scheme may degrade the performance of algorithm if the newly replaced random immigrants disturb the searching. There may arise a situation where random immigrants do not affect the searching scope as the individuals in earlier generation may be quite fit. Thus, a new approach called elitism-based immigrants was introduced by Yang et al. [42] to overcome such problems.

In case of SGA, the best individual generated at a generation is called as elite and it is updated every generation. As we have discussed earlier, due to strong selection strategy, the individuals in the population will ultimately converge to optimal solution. The EIGA scheme combines the idea of both elitism and random immigrants methods. This may adapt to sustain the diversity in the population. In EIGA, after the genetic operations, at each generation t, the elite $E(t - 1)$ from previous generation is used as base to generate immigrants. Within EIGA $r_{ei} * N$ number of elitism-based immigrants are generated from $E(t - 1)$ generations iteratively. Here, r_{ei} is the ratio of elitism-based immigrants to the population size N. A standard mutation is done on generated individuals with a probability p_m^i to make the individuals more diversified. These generated immigrants substitute the worst individuals in the current population. The use of elite from previous population helps to improve the performance in searching the solution.

12.5.2 Adaptive Rate of Genetic Operators

The choice of the value of p_c and p_m critically affects the behavior and performance of the GA. During crossover, randomized exchange of genetic information between solutions takes place. It helps to explore the search space and achieve better solution by the combination of information from good solutions. Whereas mutation plays a vital role to restore the lost or unexplored genetic material with the modification in solutions obtained. It helps to come out of local optima as well. In case of fixed rate of genetic operators, the algorithm may converge quickly without exploring the possible search scopes for optimal solutions. Adaptive technique of crossover and mutation is used to sustain the diversity and without affecting faster convergence [27]. The comparison of SGA with fixed rate of operators and dynamic (adaptive) rate has proved dynamic one to be much better [26]. The adaptive technique finds the gap between the best and worst fit individuals after which the probability of crossover (μ_c) and mutation (μ_m) is tuned for efficient results without affecting other properties of convergence.

$$\mu_c = \frac{k_c \left(f_{\max} - f_c \right)}{f_{\max} - f_{\mathrm{avg}}} \tag{12.9}$$

$$\mu_m = \frac{k_m \left(f_{\max} - f_m \right)}{f_{\max} - f_{\mathrm{avg}}}, \tag{12.10}$$

where $f_{\max}$ is the maximum fitness value, f_{avg} is the average fitness value, f_c is the fitness value of the best chromosome for the crossover, f_m is the fitness value of the chromosome to be mutated and, k_c and k_m are positive real constants less than 1.

12.6 Experimental Study

We investigated the MRCF problem with a traditional SGA and two dynamic GAs, RIGA and EIGA. Further, we integrated the adaptive methods of genetic operations to modify SGA, RIGA, and EIGA. We named them as aSGA, aRIGA, and aEIGA. The flowchart shown in Fig. 12.6 accumulated view of the working of all designed GAs together. Further, this section discusses about the parameter settings for the experiment and provides comparative analysis using obtained results.

Fig. 12.6 A flowchart for a combined view of investigated algorithms

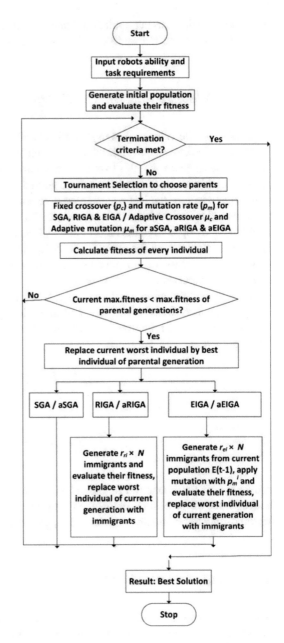

12.6.1 Experimental Settings

We have conducted simulation experiments using a popular dataset [20, 21] consisting of ten robots. Each robot has four different features capabilities. The quantitative values for each feature capability of the robots are given in Table 12.1.

Table 12.4 Required capability values by task

| | Tasks | | | |
Required capability	T_1	T_2	T_3	T_4
F^1	50	40	10	80
F^2	60	30	20	30
F^3	30	40	50	20
F^4	40	45	60	10

Here, we have considered four tasks. Each task may require a varying number of available robotic features for their executions. The specific requirements for all the four tasks are mentioned in Table 12.4 below. Experiments have been done using a population of 100 individuals for 500 generations. The probability of crossover p_c and mutation p_m is set to 0.9 and 0.2, respectively. For immigrants schemes, the number of immigrants to the population size, i.e., r_{ri} and r_{ei} are set to 0.2. In order to generate new immigrants, mutation probability was set to 0.8 in EIGA and aEIGA. For adaptive schemes, the constants k_c and k_m used in (12.9) and (12.10) are set to 0.8 and 0.2, respectively. Tournament selection with a tournament size of two has been used to select the best individual from the current population. We have performed ten independent runs of each algorithm for the given parameters.

12.6.2 Efficiency and Average Efficiency

It is necessary to measure the performance of an evolutionary algorithm to analyze its competence in terms of solution generation. Best-of-generation (BOG) fitness is an optimality-based performance measure to evaluate the ability of an algorithm in finding the solutions with the best objective values [32, 41]. This measure helps to compare the performance of two or more algorithms quantitatively. For each algorithm, at every generation, the best individual is selected and the best efficiency value at the current generation is noted. The overall performance based on the optimality of a GA can be calculated by

$$\overline{F}_{\text{BOG}_i} = \frac{1}{G} \times \sum_{i=1}^{G} \left(\frac{1}{R} \times \sum_{j=1}^{j=R} F_{\text{BOG}_{ij}} \right), \quad (12.11)$$

where $\overline{F}_{\text{BOG}_i}$ is the overall performance of the algorithm for given problem, i.e., the average of best-of-generation efficiency value (in our case) over the runs, G is the number of generation, R is the number of runs of algorithm, and $F_{\text{BOG}_{ij}}$ is the best efficiency value of generation i of run j. This measure provides the method to compare the performance of algorithm quantitatively [32, 41].

It would also be reasonable to compare EAs according to the average efficiency of the population since in dynamic environments the diversity of population fluctuates with the evolution. The average efficiency is the mean of the efficiency values of the entire population. The arrival of immigrants into the current pool of individuals changes the population and we get new average population efficiency at every generation. It is useful in depicting the convergence of individual near to the best efficiency value.

We have executed ten independent runs for each algorithm and best-of-generation efficiency value of each generation is recorded for performance analysis. The result of simulation shows that with an increase in generations, better results are produced. The experiment shows significant improvement at the end of evolution for all the techniques.

12.6.3 Comparative Results for GA Variants with Fixed Rate Genetic Operators

Table 12.5 provides the comparative results for SGA and two dynamic variants RIGA and EIGA with a fixed rate of mutation and genetic operations. For SGA, the average efficiency value is converged to the best efficiency values, which means the evolution stuck in local minima due to lack of diversity in the population. This brings the scope of immigrants scheme to be incorporated into GA to enhance the diversity of solutions. It can be observed that RIGA and EIGA have outperformed SGA with good convergence rate and best fitness value. As discussed earlier, in the case of RIGA, we can observe that the approach has given a better solution than SGA methods. But on the other hand, it has disrupted on an average fitness of the population. This has occurred because of immigrants with lower quality. In the case of EIGA, as we are choosing the elite one thus, both the best fit value and average fitness of population improve during evolution. The feasibility of generated coalition structure is examined with the best efficiency value that is obtained after 500 generations. Figure 12.7 shows the comparison of $\overline{F}_{BOG_i}$ obtained using a fixed rate of operators, respectively, with varying generations. Similarly, the plot of average efficiency values of population for 500 generations over 10 runs is shown in Fig. 12.8.

Table 12.5 Comparative results for algorithms with fixed rate of genetic operators

Algorithms	Best efficiency (10 runs)	Average value of efficiency of population (10 runs)	Average execution time
SGA	0.6009	0.6009	1.0779
RIGA	0.8120	0.4620	2.8516
EIGA	0.8467	0.6634	2.6020

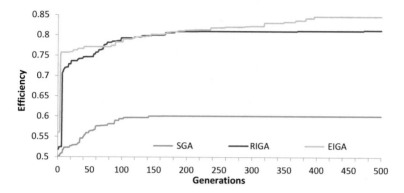

Fig. 12.7 Best efficiency values with varying generations for a fixed rate of operators

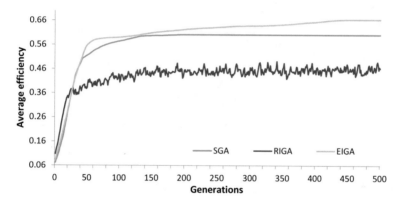

Fig. 12.8 Average value of efficiency with varying generations for a fixed rate of operators

12.6.4   Comparative Results for GA Variants with the Adaptive Rate of Genetic Operators

Though RIGA and EIGA have been able to generate better solutions, the exploration of search space cannot be done exhaustively with a fixed rate of crossover and mutation. Therefore, it is necessary to tune up these parameters relatively to scan the search spaces and obtain possible best solutions. The comparative results for all three adaptive variants aSGA, aRIGA, and aEIGA are summarized in Table 12.6. From Tables 12.5 and 12.6, it can be observed that all three adaptive variants generate solutions with better efficiency compared to respective fixed rate variants. Among all, aEIGA outperforms in terms of the best efficiency and on an average generates better solutions. Figure 12.9 illustrates the evaluation of adaptive variants of algorithms using $\overline{F}_{\text{BOG}_i}$. There is a considerable improvement in efficiency and better convergence. The average efficiency values of the population for 500 generations over 10 runs are depicted in Fig. 12.10.

Table 12.6 Comparative results for algorithms with the adaptive rate of genetic operators

Algorithms	Best efficiency (10 runs)	Average value of efficiency of population (10 runs)	Average execution time
aSGA	0.6542	0.6518	1.4352
aRIGA	0.9046	0.5239	3.0014
aEIGA	0.9573	0.6761	3.1403

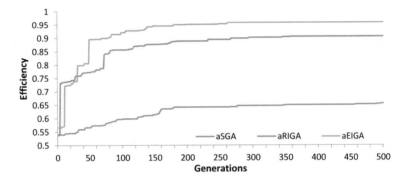

Fig. 12.9 Best efficiency values with varying generations for the adaptive rate of operators

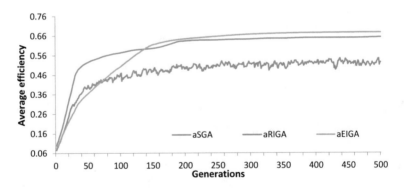

Fig. 12.10 Average value of efficiency with varying generations for the adaptive rate of operators

A combined comparison of optimal efficiency and average efficiency for all methods is shown in Figs. 12.11 and 12.12, respectively. The average running time of algorithms is shown in Fig. 12.13. Among adaptive and fixed rate genetic operations, the adaptive techniques have shown significant improvements. Among the designed algorithms for MRCF problem, EIGA has outperformed others in both cases.

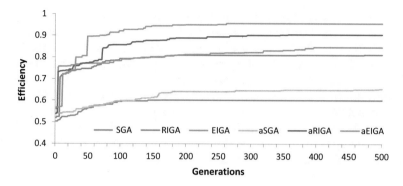

Fig. 12.11 Best efficiency with varying generations

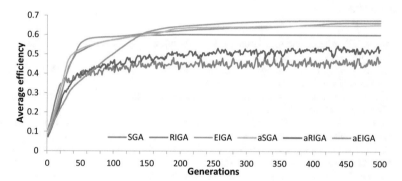

Fig. 12.12 Average value of efficiency with varying generations

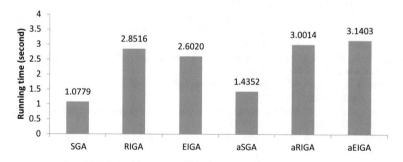

Fig. 12.13 Average execution time of algorithms

12.7 Conclusion and Future Work

In this chapter, the novelty of immigrant-based genetic algorithms is presented to find efficient coalitions of robots to execute the tasks in multi-robot systems. We have implemented six different versions of genetic algorithms viz., SGA, random immigrants GA and elitism-based immigrant GA and their adaptive versions to solve the addressed problem. Simulation experiments have demonstrated all the six variants were able to produce a feasible solution, but aEIGA outperforms among all. The experiment is done on a small instance, so our future work shall focus on accommodating a larger dataset and simulating using a simulated and real environment. Further, we will identify other critical objectives of this problem and formulate it as a multi-objective problem addressing several other constraints. The formulation can be applied to check the applicability of our approaches to a real-world application. The domain can be explored using other computational intelligence tools to the application as well.

References

1. Gerkey, B.P., Mataric, M.J.: A Framework for Studying Multi-Robot Task Allocation. Kluwer Academic, Amsterdam (2003)
2. Vig, L., Adams, J.A.: Coalition formation: from software agents to robots. J. Intell. Robot. Syst. **50**(1), 85–118 (2007)
3. Shehory, O., Kraus, S.: Task allocation via coalition formation among autonomous agents. In: IJCAI, vol. 1, pp. 655–661 (1995)
4. Sandholm, T., Larson, K., Andersson, M., Shehory, O., Tohmé, F.: Coalition structure generation with worst case guarantees. Artif. Intell. **111**(1), 209–238 (1999)
5. Adams, J.A.: Coalition formation for task allocation: theory and algorithms. Auton. Agent. Multi-Agent Syst. **22**(2), 225–248 (2011)
6. Agarwal, M., Kumar, N., Vig, L.: Non-additive multi-objective robot coalition formation. Expert Syst. Appl. **41**(8), 3736–3747 (2014)
7. Balch, T., Arkin, R.C.: Behavior-based formation control for multi-robot teams. IEEE Trans. Robot. Autom. **14**(6), 926–939 (1998)
8. Dias, M.B., Goldberg, D., Stentz, A.: Market-based multirobot coordination for complex space applications. In: The 7th International Symposium on Artificial Intelligence, Robotics and Automation in Space (i-SAIRAS) (2003)
9. Dias, M.B., Zlot, R., Kalra, N., Stentz, A.: Market-based multirobot coordination: a survey and analysis. Proc. IEEE. **94**(7), 1257–1270 (2006)
10. Gerkey, B.P., Mataric, M.J.: Sold!: auction methods for multirobot coordination. IEEE Trans. Robot. Autom. **18**(5), 758–768 (2002)
11. Parker, L.E., Tang, F.: Building multirobot coalitions through automated task solution synthesis. Proc. IEEE. **94**(7), 1289–1305 (2006)
12. Shehory, O., Kraus, S.: Methods for task allocation via agent coalition formation. Artif. Intell. **101**(1), 165–200 (1998)
13. Tang, F., Parker, L.E.: A complete methodology for generating multi-robot task solutions using ASyMTRe-D and market-based task allocation. In: Proceedings 2007 IEEE International Conference on Robotics and Automation, pp. 3351–3358. IEEE, Roma (2007)
14. Vig, L., Adams, J.A.: Multi-robot coalition formation. IEEE Trans. Robot. **22**(4), 637–649 (2006)
15. Zhang, Y., Parker, L.E.: IQ-ASyMTRe: forming executable coalitions for tightly coupled multirobot tasks. IEEE Trans. Robot. **29**(2), 400–416 (2013)
16. Rizk, Y., Awad, M., Tunstel, E.W.: Cooperative heterogeneous multi-robot systems: a survey. ACM Comput. Surv. **52**(2), 29 (2019)
17. Goldberg, D.E.: Genetic Algorithms. Pearson Education, New Delhi (2006)
18. Mishra, B.S.P., Mishra, S., Singh, S.S.: Parallel multi-criterion genetic algorithms: review and comprehensive study. Int. J. Appl. Evol. Comput. **7**(1), 50–62 (2016)

19. Sivanandam, S.N., Deepa, S.N.: Introduction to Genetic Algorithms. Springer, Berlin (2007)
20. Liu, H.Y., Chen, J.F.: Multi-robot cooperation coalition formation based on genetic algorithm. In: 2006 International Conference on Machine Learning and Cybernetics, pp. 85–88. IEEE, Dalian (2006)
21. Li, Z., Xu, B., Yang, L., Chen, J., Li, K.: Quantum evolutionary algorithm for multi-robot coalition formation. In: Proceedings of the First ACM/SIGEVO Summit on Genetic and Evolutionary Computation, pp. 295–302. ACM, New York (2009)
22. Yang, J., Luo, Z.: Coalition formation mechanism in multi-agent systems based on genetic algorithms. Appl. Soft Comput. **7**(2), 561–568 (2007)
23. Cobb, H.G., Grefenstette, J.J.: Genetic Algorithms for Tracking Changing Environments. Naval Research Lab, Washington DC (1993)
24. Cheng, H., Yang, S., Cao, J.: Dynamic genetic algorithms for the dynamic load balanced clustering problem in mobile ad hoc networks. Expert Syst. Appl. **40**(4), 1381–1392 (2013)
25. Grefenstette, J.J.: Genetic algorithms for changing environments. In: PPSN, vol. 2, pp. 137–144 (1992)
26. Omara, F.A., Arafa, M.M.: Genetic algorithms for task scheduling problem. J. Parallel Distr. Comput. **70**(1), 13–22 (2010)
27. Srinivas, M., Patnaik, L.M.: Adaptive probabilities of crossover and mutation in genetic algorithms. IEEE Trans. Syst. Man Cybern. **24**(4), 656–667 (1994)
28. Muhuri, P.K., Rauniyar, A.: Immigrants based adaptive genetic algorithms for task allocation in multi-robot systems. Int. J. Comput. Intell. Appl. **16**(04), 1750025 (2017)
29. Rauniyar, A., Muhuri, P.K.: Multi-robot coalition formation problem: task allocation with adaptive immigrants based genetic algorithms. In: 2016 IEEE International Conference on Systems, Man, and Cybernetics (SMC), pp. 000137–000142. IEEE, Budapest (2016)
30. Liu, S.H., Zhang, Y., Wu, H.Y., Liu, J.: Multi-robot task allocation based on swarm intelligence. J. Jilin Univ. **40**(1), 123–129 (2010)
31. Manathara, J.G., Sujit, P.B., Beard, R.W.: Multiple UAV coalitions for a search and prosecute mission. J. Intell. Robot. Syst. **62**(1), 125–158 (2011)
32. Yu, X., Tang, K., Chen, T., Yao, X.: Empirical analysis of evolutionary algorithms with immigrants schemes for dynamic optimization. Memet. Comput. **1**(1), 3–24 (2009)
33. Yu, Z., Shuhua, L., Shuai, F., Di, W.: A quantum-inspired ant colony optimization for robot coalition formation. In: 2009 Chinese Control and Decision Conference, pp. 626–631. IEEE, Shanghai (2009)
34. Agarwal, M., Agrawal, N., Sharma, S., Vig, L., Kumar, N.: Parallel multi-objective multi-robot coalition formation. Expert Syst. Appl. **42**(21), 7797–7811 (2015)
35. Guerrero, J., Oliver, G., Valero, O.: Multi-robot coalitions formation with deadlines: complexity analysis and solutions. PLoS One. **12**(1), e0170659 (2017)
36. Mouradian, C., Sahoo, J., Glitho, R.H., Morrow, M.J., Polakos, P.A.: A coalition formation algorithm for Multi-Robot Task Allocation in large-scale natural disasters. In: 2017 13th International Wireless Communications and Mobile Computing Conference (IWCMC), pp. 1909–1914. IEEE, Valencia (2017)
37. Blickle, T., Thiele, L.: A mathematical analysis of tournament selection. In: ICGA, pp. 9–16 (1995)
38. Lee, S., Soak, S., Kim, K., Park, H., Jeon, M.: Statistical properties analysis of real world tournament selection in genetic algorithms. Appl. Intell. **28**(2), 195–205 (2008)
39. Vavak, F., Fogarty, T.C.: Comparison of steady state and generational genetic algorithms for use in non-stationary environments. In: Proceedings of IEEE International Conference on Evolutionary Computation, 1996, pp. 192–195. IEEE, Piscataway (1996)
40. Yang, S.: Genetic algorithms with memory-and elitism-based immigrants in dynamic environments. Evol. Comput. **16**(3), 385–416 (2008)
41. Nguyen, T.T., Yang, S., Branke, J.: Evolutionary dynamic optimization: a survey of the state of the art. Swarm Evol. Comput. **6**, 1–24 (2012)
42. Yang, S.: Genetic algorithms with elitism-based immigrants for changing optimization problems. In: Workshops on Applications of Evolutionary Computation, pp. 627–636. Springer, Berlin, Heidelberg (2007)

Chapter 13
Lidar and Non-extensive Particle Filter for UAV Autonomous Navigation

José Renato G. Braga, Haroldo F. de Campos Velho, and Elcio H. Shiguemori

Abstract Unmanned aerial vehicle (UAV) is a technology employed for several applications nowadays. One important UAV research topic is the autonomous navigation (AN). The standard procedure for AN is to fuse the signals from an inertial navigation system (INS) and from a global navigation satellite system (GNSS). Our approach to perform the autonomous navigation uses a computer vision system, instead of GNSS signal, associated to the visual odometry. The two techniques applied to estimate the UAV position are combined by a non-extensive particle filter. However, the development of a computer vision system for estimating the UAV position in a situation of flight over water-covered areas and flight in low light conditions is a challenge. Our approach uses images from an active sensor called light detection and ranging (LiDAR) to allow the flight in such conditions.

Keywords Unmanned aerial vehicle (UAV) · Autonomous navigation · LiDAR data · Non-extensive particle filter · Data fusion

13.1 Introduction

The unmanned aerial vehicles (UAVs) have been applied to several areas, such as land-use monitoring, search and rescue operations, monitoring of environmental impacts, engineering projects, and others. There is an expectation that the use

J. R. G. Braga · H. F. de Campos Velho (✉)
Instituto Nacional de Pesquisas Espaciais (INPE), São José dos Campos, SP, Brazil
e-mail: haroldo.camposvelho@inpe.br

E. H. Shiguemori
Instituto de Estudos Avançados (IEAv), Departamento de Ciência e Tecnologia Aeroespacial (DCTA), Trevo Coronel Aviador José Alberto Albano do Amarante 01 - Putim, São José dos Campos, SP, Brazil
e-mail: elcio@ieav.cta.br

© Springer Nature Switzerland AG 2020
O. Llanes Santiago et al. (eds.), *Computational Intelligence in Emerging Technologies for Engineering Applications*, Studies in Computational Intelligence 872, https://doi.org/10.1007/978-3-030-34409-2_13

Fig. 13.1 A DJI phantom
quadcopter UAV—photo from
the Wikipedia

of UAVs will increase in near future, due to the low costs of development and operation. In fact, the main advantage of the UAV is the absence of onboard crew, reducing the risks to human life. An example of the UAV—a quadcopter—is shown in Fig. 13.1. With the increase of UAV employment, the development of autonomous navigation systems has been one of the hot research topics. The main task for the autonomous navigation systems is to estimate the UAV position. A strategy applied for estimating the UAV position is the use of information from inertial navigation system (INS) combined with information from a global navigation satellite system (GNSS). The GNSS signal can suffer an outage due to natural phenomena or from a malicious attack. Therefore, the development of image processing strategies, using images captured and processed during the flight time, can be applied to determine the UAV location [8].

However, the development of an image processing system for estimating the UAV position during the flight over water-covered areas (for example: over the ocean) and under low or no light conditions is a challenge. Our approach uses images from an active sensor called light detection and ranging (LiDAR) to allow the flight under such conditions. The proposed scheme estimates the aircraft position by applying data fusion from two positioning techniques, i.e., computer vision and visual odometry. The Kalman filter (KF) could be applied to carry out data fusion. However, we cannot guarantee the Gaussian statistics for the data, as it is assumed by the KF approach. In such cases, particle filters can be applied.

Here, the data fusion is performed by a non-extensive particle filter (NExt-PF) [6, 7]. For the validation of the proposed method, a simulation using images from a LiDAR sensor is employed. The approach results are promising for the UAV position estimation in regions covered by water or in regions with low light conditions.

The text is organized with five sections. In the second section, two image processing approaches applied to UAV positioning are explained. Section 13.3 describes the method for data fusion based on a new version of particle filter. The experiment for autonomous navigation with LiDAR images by using data fusion from two image processing schemes is commented in Sect. 13.4. Finally, Sect. 13.5 addresses some remarks and conclusion.

13.2   Autonomous Navigation by Image Processing

The INS signal can be used to estimate the UAV position. However, the INS has a drift error, which needs to be corrected. Indeed, the drift error could become obsolete the INS information (large estimation error) within a few seconds. Our goal here is to avoid the use of GNSS signal, for example the global positioning system (GPS), for correcting the INS information. An alternative is to embed in the aircraft a sensor to capture images to estimate the UAV position. Here, visual odometry and computer vision system are two techniques applied to estimate the UAV position by image processing.

Visual odometry determines the position of the aircraft by processing two subsequent images of the same scene. Image correlation determines the best overlap between two images. A brief description on the VO is presented in Sect. 13.2.1. Another approach is based on a computer vision system (CVS), correlating a geo-referenced image, commonly named as reference image, and an image without geographic coordinate information, to incorporate geographic location to each pixel of the second image. A CVS scheme can be established by extracting edges between images, and evaluating the correlation between the segmented images. Our approach employs a self-configuring neural network for edge extraction. Section 13.2.2 describes the CVS technique used here.

A data fusion method combining both schemes for the UAV positioning mentioned before, visual odometry and computer vision, is explained in Sect. 13.3. The data fusion method is called *non-extensive particle filter*.

13.2.1   Visual odometry for UAV Positioning

Visual odometry (VO) is a technique to estimate the vehicle position and orientation by processing the changes in the images caused by its movement [13] using one (monocular VO) or multiple sensors (stereo VO). Here, the monocular VO is applied to estimate the movement of outdoor UAVs. The VO basic principle is detecting interest points in the image and extracts a data structure which describe them. From the interest points matching, it is possible to estimate the vehicle motion.

The VO is carried out by four steps:

a) Image sequence,
b) Detecting points,
c) Matching between points,
d) Motion.

The operation *Image Sequence* uses two images from the sensor: the first image captured at instant $t - \Delta t$, and the second one at instant t. The value of Δt must be appropriate to display the most of the scene for both images. *Detecting Points* operation finds the interest points in both images. For each point, an attribute vector (data structure) is computed: the descriptor process. One of the most successful descriptors is the speeded up robust features (SURF) [3]. The *Matching between points* performs the identification of interest points using the attribute vector. The matching is done through a similarity metric, for example, Euclidean distance. The last step of the monocular VO algorithm is the *Motion* estimation, which determines the movement of the UAV using pairs of corresponding interest points. The eight point algorithm is often applied for motion estimation. This algorithm can be described by calculating the fundamental matrix **F**:

$$\left(\mathbf{x'}\,\mathbf{F}\,\mathbf{x}\right) = 0\,. \tag{13.1}$$

The main property of the fundamental matrix **F** is the condition: for a set of eight corresponding points $\mathbf{x} \leftrightarrow \mathbf{x'}$, there is just one matrix **F** satisfying Eq. (13.1). The eight point algorithm finds the fundamental matrix, and using the singular value decomposition (SVD) is possible to obtain the vehicle motion, represented by the rotation matrix and the translation vector:

$$\mathrm{SVD}(\mathbf{F}) = \mathbf{K}^T\,\mathbf{R}[\mathbf{t}]_\mathbf{x}\,\mathbf{K}^{-1}, \tag{13.2}$$

where **K** is the matrix of the intrinsic sensor parameters linked to the vehicle, **R** is the rotation matrix, and $[\mathbf{t}]_\mathbf{x}$ is the representation of the cross product of the translation vector.

13.2.2 Computer Vision Based System

The computer vision technique employed here is multi-step process—see Fig. 13.2a and 13.2b for the aerial and reference images, respectively—and its main objective is to perform the correlation between two images—see Fig. 13.2c. For the first step, the UAV and reference images are changed to gray-scale. The reference image covers the area overflown by UAV. After that, a median filter [9] is applied to the reference and UAV images to obtain noiseless images. Image edge extraction is performed by a multi-layer perceptron (MLP) neural network [11], using back-propagation training algorithm. Finding an optimal topology for a neural network is not an easy task. The best neural network topology can be determined by minimizing a functional—see Eq. (13.4). The optimization problem is solved by using a new metaheuristic called multi-particle collision algorithm (MPCA) [12]. The last step of the computer vision approach is the correlation process, which determines the best overlap of the reference image and the image caught by UAV during the flight.

Fig. 13.2 Procedures of the computer vision approach for a UAV positioning system (**a**) aerial (UAV) image, (**b**) reference (georeferenced) image, (**c**) UAV positioning by correlation of segmented images

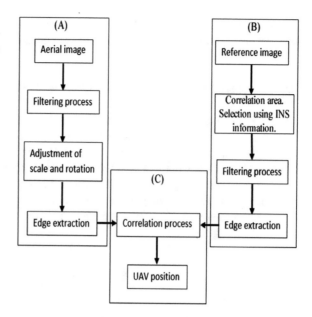

13.2.2.1   Multilayer Perceptron Neural Network (MLP-NN)

Artificial neural network is a mathematical model, inspired by human brain processes. The goal is to emulate the intelligent procedure for solving problems. The MLP is a feedforward neural network widely used. The architecture of this neural network has three main features: one input layer—composed of non-computational neurons, one or more hidden layers composed of computational units (neurons), and one output layer—also composed of computational neurons. The MLP-NN is a supervised neural network, where the neurons are fully connected. The connection weights need to be computed by a training algorithm. There are two steps for the training process: the propagation phase and the back-propagation phase. During the propagation phase, the input data is presented to the MLP by the input layer, and propagated up to the output layer, which finds its response $y(n)$. In this phase, the neural connection values do not change. In the back-propagation phase, the error $e(n)$ is computed by the square difference between the neural network response $y(n)$ and the target values, and the error $e(n)$ is back-propagated through the MLP to update the weight values. The algorithm used to update the weight connection values is written as [11]

$$w_{ij}(n+1) = w_{ij}(n) + \alpha \left[w_{ij}(n) - w_{ij}(n-1) \right] + \eta\, \delta(n)\, y(n), \qquad (13.3)$$

where w_{ij} is the entry of the weight matrix $\mathbf{W}$, linking the i-th input to the j-th neuron, α is the momentum value, η is the learning rate value, $y(n)$ is the output produced by the MLP, and $\delta(n) \equiv \partial e(n)/\partial \mathbf{W}$ is the error gradient. The number of hidden layers, the number of neurons for each layer, type of activation function, momentum, and learning rate are parameters defined by the MLP user through a process of trial and error or automatically discovered by techniques such as MPCA.

13.2.2.2 Multiple Particle Collision Algorithm (MPCA)

MLP has been used in several applications. MPCA is a metaheuristic inspired considering a neutron traveling inside of a nuclear reactor, where two phenomena are noted: absorption and scattering [12]. MPCA automatically finds the parameters associated to an ANN optimal architecture [2] for a specific application. The MPCA is an extension of the particle collision algorithm (PCA) [15]. MPCA is executed with a cooperative searching process to find the parameter values to optimize an objective function. The objective function which determines a MLP optimal architecture is given by

$$f_{\text{obj}} = \text{penality} \left[\frac{\rho_1 \, E_{\text{train}} + \rho_2 \, E_{\text{gen}}}{\rho_1 + \rho_2} \right] \tag{13.4}$$

$$\text{penality} = c_1 \, e^{\{\#\text{neurons}\}} + c_2 \, \{\#\text{epoch}\} + 1 \tag{13.5}$$

here, $\rho_1 = \rho_2 = 1$ are adjustment parameters [2, 5] that modify the relevance attributed to E_{train} (training error) and E_{gen} (generalization error), respectively. The penality is metric of complexity of a neural network, and it is used to find the simplest architectures (lowest numbers of neurons, with fastest convergence for weight computation) for the MLP, with $c_1 = 5 \times 10^8$ and $c_2 = 5 \times 10^5$ [2].

13.3 Non-extensive Particle Filter and Data Fusion

The fundamental idea underlying the sequential Monte Carlo method is to represent the probability density function (PDF) by a set of samples with their associated weights. This set of samples is also referred to particles [10, 14, 16]. In the particle filter, an estimation of a posteriori distribution is obtained by resampling with replacement from a priori ensemble. Since the particle filter does not require linearity and Gaussian assumptions, it is applicable to general nonlinear problems. In particular, particle filter can be applied to cases in which the relationship between a state and observed data is nonlinear.

Two properties are relevant issues for the particle filter: the Bayes' theorem and the Markov property. The Bayes' theorem can express by the expression to calculate the conditional probability:

$$P(A|B) = \frac{P(B|A) \, P(A)}{P(A)}. \tag{13.6}$$

The probability $P(B)$ is understood as a normalization factor. The Markov process is characterized by the property:

$$p(u_n|u_{n-1}, \ldots, u_2, u_1) = p(u_n|u_{n-1}) . \tag{13.7}$$

For the current application, the distribution $p(u_n|u_{n-1})$ represents the vector with the entries being the UAV position estimations computed by visual odometry and computer vision system for the data fusion.

The algorithm for the particle filter implementation can be written as:

a) Compute the initial particle ensemble:

$$\left\{ u_{0|n-1}^{(i)} \right\}_{i=0}^{M} \sim \rho_{u_0}(u_0)$$

the initial distribution $(\rho_{u_0}(u_0))$ is a free choice—our choice: $\rho_{u_0}(u_0) \sim N(0, 5)$—the Gaussian distribution with zero mean and variance $\sigma^2 = 5$.

b) Compute:

$$r_n^{(i)} = p(z_n|u_{n|n-1}) = p_{et}(z_n - h(u_n, t_n))$$

where z_n denotes observations, and $p_{et}(z_n - h(u_n, t_n))$ is the likelihood function.

c) Normalize:

$$\hat{r}_n^{(i)} \frac{r_n^{(i)}}{\sum_{j=1}^{M} r_n^{(j)}} .$$

d) Resampling: extract particles with substitution, according to (a standard notation is used here—see [7, 16]):

$$Pr\left\{ u_{n|n}^{(i)} = u_{n|n-1}^{(j)} \right\} = \hat{q}_n^{(j)} , \qquad i = 1, \ldots, M .$$

e) Time up-dating: compute the new particles:

$$u_{n+1|n}^{(i)} = f\left(u_{n+1|n}^{(i)}, t_n \right) + \mu_n , \quad \text{with: } \mu_n \in N(0, 1)$$

where: $u_{n+1|n}^{(i)} \sim p\left(u_{n+1|n}^{(i)} | u_{n|n}^{(i)} \right)$, and $i = 1, \ldots, M$.

f) Set $t_{n+1} = t_n + \Delta t$, and go to the step (b).

The kernel coming from the application of the Bayes' theorem and from the Markov property, suggesting the following choice:

$$\underbrace{p(u_n|z_n)}_{\text{posterior}} \propto \underbrace{p(z_n|u_n)}_{\text{likelihood}} \underbrace{p(u_n|z_{n-1})}_{\text{prior}} . \tag{13.8}$$

13.3.1 Non-extensive Particle Filter

A non-extensive form of entropy has been proposed by Tsallis [17, 18]:

$$S_q(p) = \frac{k}{q-1}\left[1 - \sum_{i=1}^{N} p_i^q\right], \tag{13.9}$$

where p_i is a probability, and q is a free parameter—it is called the non-extensivity parameter. In thermodynamics, the parameter k is known as the Boltzmann's constant. The Tsallis' entropy reduces to the usual Boltzmann–Gibbs–Shannon formula when $q \to 1$.

The equiprobability condition produces the maximum for the extensive and non-extensive forms to the entropy function, leading to the distributions [18]:

$$q > 1 : \quad p_q(x) = \alpha_q^+\left[1 - \frac{1-q}{3-q}\left(\frac{x}{\sigma}\right)^2\right]^{-1/(q-1)} \tag{13.10}$$

$$q = 1 : \quad \frac{1}{\sigma}\left[\frac{1}{2\pi}\right]e^{-(x/\sigma)^2/2} \tag{13.11}$$

$$q < 1 : \quad p_q(x) = \alpha_q^-\left[1 - \frac{1-q}{3-q}\left(\frac{x}{\sigma}\right)^2\right]^{1/(q-1)} , \tag{13.12}$$

where—below: $\Gamma(.)$ is the gamma function [1]:

$$\sigma^2 = \frac{\int_{-\infty}^{+\infty} x^2 [p_q(x)]^q \, dx}{\int_{-\infty}^{+\infty} [p_q(x)]^q \, dx} ,$$

$$\alpha_q^+ = \frac{1}{\sigma}\left[\frac{q-1}{\pi(3-q)}\right]^{1/2} \frac{\Gamma(1/(q-1))}{\Gamma((3-q)/2(q-1))} ,$$

$$\alpha_q^- = \frac{1}{\sigma}\left[\frac{1-q}{\pi(3-q)}\right]^{1/2} \frac{\Gamma((5-3q)/2(q-1))}{\Gamma((2-q)/(1-q))} .$$

The distributions above apply if $|x| < \sigma[(3-q)/(1-q)]^{1/2}$, and $p_q(x) = 0$ otherwise. For distributions with $q < 5/3$, the standard central limit theorem applies, implying that if p_q is written as a sum of M random independent variables, in the limit case $M \to \infty$, the probability density function for $p_q(x)$ in the distribution space is the normal (Gaussian) distribution. However, for $5/3 < q < 3$ the Levy-Gnedenko's central limit theorem applies, resulting for $M \to \infty$ the Lévy's distribution as the probability density function for the random variable p_q. The index in such Lévy's distribution is $\gamma = (3-q)/(q-1)$ [18].

The purpose is to use the Tsallis' thermostatistics (13.10) or (13.12) for substituting the Gaussian function to represent the likelihood operator in step-(b) of the particle filter [7]. The idea is to explore the property of this thermostatistics to access different attractors in the distribution space.

13.4   Autonomous Navigation with Non-extensive Particle Filter

Visual odometry and computer vision approaches have been applied as techniques to estimate UAV position for autonomous navigation. Braga [4] studied considering these techniques separately and as combined methods (data fusion). He showed that better results are provided when both methodologies are used together. A particle filter can be used to combine the cited methods. The ensemble Kalman filter is one type of particle filter, but, as already mentioned, we cannot assume the Gaussian statistics. From this consideration, a more general formulation for the particle filter is employed [10]. In addition, different likelihood operators will produce better results to the particle filter estimation. Indeed, Braga has used a non-extensive particle filter (NExt-PF) to UAV positioning with visible band camera [4] on board of the helicopter RMAX (Yamaha Motor Company), used for testing in the Linköping University (Sweden). Braga did a parametric investigation to determine the best value for the non-extensive parameter q, and he has obtained the value $q = 2.57$ [4, 6].

For the visual odometry, the algorithm SURF was used to identify the points of interest, and the RANSAC (RANdom SAmple Consensus) filter was applied to remove false corresponding points. For the computer vision system, a MLP neural network was trained to carry out the edges extraction for the objects in the images. The edge patterns used for training the MLP to perform the edge detection are shown in Fig. 13.3, and Table 13.1 shows the MLP neural network topology determined by the MPCA optimizer.

Our study is focused on the autonomous navigation over water-covered surface or under low or no-light scenario by using LiDAR data. Our experiment is performed with NExt-PF to do the data fusion with visual odometry and computer vision techniques. Just for comparison, the results are shown for two values of the non-extensive parameter: $q = 1.00$ and $q = 2.57$.

Numerical results were obtained using LiDAR data from the INFOMAR project (INtegrated Mapping FOr The Sustainable of Ireland's MArine Resources) support by the Geological Survey of Ireland (GSI). The project is producing a bathymetric mapping of the Ireland coast for helping the marine exploration of minerals. The LiDAR sensor used is the LADS Mk II Ground System. This sensor is able to obtain data up to 60 m depth, depending on the level of water turbidity.

From the LiDAR data, it is possible to obtain the surface digital model (SDM)—see Fig. 13.4. The resolution of the SDM in Fig. 13.4 is equal to 5 m/pixel. The

Fig. 13.3 Patterns for MLP-NN for the learning phase

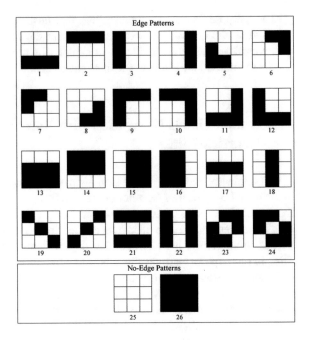

Table 13.1 Neural network architecture and parameters computed from the MPCA metaheuristics

MLP-NN characteristics	Parameters/type
Neurons for the input layer	9
Neurons for the output layer	1
Number of hidden layers	1
Hidden layer neurons	18
Activation function	tanh
Learning rate	0.73
Momentum	0.85

Fig. 13.4 Estimated UAV trajectory for NExt-PF (blue color)—red color is the true UAV trajectory

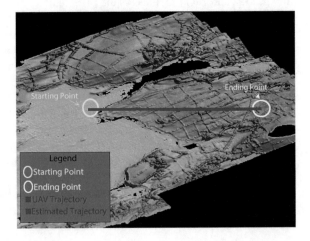

Table 13.2 UAV trajectory correction under different methods—DF: data fusion

Method	Average error
VO	2.7572 m
CVS	1.8315 m
DF: NExt-PF: $q = 1.00$	1.6885 m
DF: NExt-PF: $q = 2.57$	0.9857 m

maximum depth for the image from the LiDAR sensor is 35 m. There is no data by LiDAR for the black zones in the image—the water layer is deeper than 35 m. On the SDM, a UAV trajectory was simulated, where 92 sub-images were defined for representing UAV images from the LiDAR sensor during the flight. The simulated trajectory has about of 1 km of extension. For each sub-image, some rotation and scale noise were introduced to emulate images caught during the flight. The synthetic images were used for visual odometry (VO) and computer vision system (CVS) to estimate the UAV position. The results from VO and CVS were fused by NExt-PF to obtain the final UAV position estimation. The UAV simulated trajectory (red line) is shown in Fig. 13.4, with yellow and white circles representing the start and final points—the blue color is the estimated trajectory by NExt-PF.

Table 13.2 shows the average error for the UAV positioning using visual odometry (VO) alone, computer vision system (CVS) alone, data fusion (VO+CVS) by NExt-PF with $q = 1$, and by NExt-PF with $q = 2.57$. From the table, we can verify that the worst result was obtained with visual odometry, and the best result for the UAV trajectory was computed by data fusion with non-extensive particle filter with $q = 2.57$—the smallest error among the employed techniques.

13.5 Conclusions

The UAV positioning by image processing was effective using visual odometry and computer vision system. The VO has a disadvantage to present a cumulative error, and CVS needs reference marks or images to estimate the UAV position. The data fusion approach can overcome the mentioned disadvantages. A novel approach for data fusion applied to the UAV positioning was employed by using the non-extensive particle filter. The particle filter formulation also allows to determine the confidence interval encapsulating the uncertainties linked to the present estimation problem—not shown here, but Braga and co-authors have applied the same formulation with visible band camera for the UAV autonomous navigation, and they computed the confidence interval for the flight trajectory.

In this study, the marks were obtained from bathymetric map generated by LiDAR data. From the comparison between bathymetric maps generated by the Geological Survey of Ireland (GSI) and another one obtained by the LiDAR on-board the drone, the cited image processing strategies were used to estimate the UAV position. The data fusion process by the non-extensive particle filter presented better results than standard particle filter for the UAV positioning.

Acknowledgements Authors want to thank to CNPq, Capes, and Fapesp, Brazilian agencies for research support.

References

1. Andrews, G.E., Askey, R., Roy, R.: Gamma and beta functions. In: Special Functions. Cambridge University Press, Cambridge (1999)
2. Anochi, J.A., Campos Velho, H.F., Furtado, H.C.M., Luz, E.F.P.: Self-configuring two types neural networks by MPCA. In 2nd International Symposium on Uncertainty Quantification and Stochastic Modeling (Uncertainties 2014), Rouen, France, pp. 429–436 (2014)
3. Bay, H., Ess, A., Tuytelaars, T, Gool, L.V.: Speeded-up robust features (SURF). Comput. Vis. Image Underst. **110**, 346–359 (2008)
4. Braga, J.R.G.: UAV autonomous navigation by LiDAR images, PhD thesis on Applied Computing, National Institute for Space Research (INPE), São José dos Campos (SP), Brazil (2018, in Portuguese)
5. Braga, J.R.G., Campos Velho, H.F., Conte, G., Doherty, P., Shiguemori, E.H.: An image matching system for autonomous UAV navigation based on neural network. In: 14th International Conference on Control, Automation, Robotics and Vision (ICARCV), Phuket, Thailand, pp. 1–6 (2016)
6. Braga, J.R.G., Shiguemori, H.E., Campos Velho, H.F.: Determining the trajectory of unmanned aerial vehicles by a novel approach for the particle filter. In: Etse, J.G., Luccioni, B.M., Pucheta, M.A., Storti, M.A. (eds.) Mecánica Computacional – special issue, vol. XXXVI, pp. 683–692 (2018)
7. Campos Velho, H.F., Furtado, H.C.M.: Adaptive particle filter for stable distribution. In: Constanda, C., Harris, P.J. (eds.) Integral Methods in Science and Engineering, pp. 47–57. Birkhäuser, Basel (2011)
8. Conte, G., Doherty, P.: An integrated UAV navigation system based on aerial image matching. In: IEEE Aerospace Conference, Big Sky, MT, USA, pp. 1–10 (2008)
9. Gonzalez, R.C., Woods, R.E.: Digital Image Processing, 4th edn. Pearson, London (2017)
10. Gordon, N.J., Salmond, D.J., Smith, A.F.M.: Novel approach to nonlinear/non-Gaussian Bayesian state estimation. IEE Proc. F Radar Signal Process. **140**, 107–113 (1993)
11. Haykin, S.: Neural Networks: A Comprehensive Foundation. Prentice Hall, Upper Saddle River (1998)
12. Luz, E.F.P., Becceneri, J.C., Campos Velho, H.F.: A new multi-particle collision algorithm for optimization in a high performance environment. J. Comput. Interdiscip. Sci. **1**, 3–10 (2008)
13. Nishar, A., Richrds, S., Breen, D., Robertson, J., Breen, B.: Thermal infrared imaging of geothermal environments and by an unmanned aerial vehicle (UAV): a case study of the Wairakei–Tauhara geothermal field, Taupo, New Zealand. Renew. Energy **4**, 136–145 (2016)
14. Ristic, B., Arulampalam, S., Neil Gordon, N.: Beyond the Kalman Filter: Particle Filters for Tracking Applications. Artech House Radar Library, Boston (2004)
15. Sacco, W.F., Oliveira, C.R.E.: A new stochastic optimization algorithm based on a particle collision metaheuristic. In: 6th World Congress of Structural and Multidisciplinary Optimization (WCSMO), Rio de Janeiro, 30 May–03 June 2005. http://citeseerx.ist.psu.edu/viewdoc/versions?doi=10.1.1.80.6308
16. Shon, T., Gustafsson, F., Nordlund, P.-J.: Marginalized particle filters for mixed linear/nonlinear state-space models. IEEE Trans. Signal Process. **53**, 2279–2289 (2005)
17. Tsallis, C.: Possible generalization of Boltzmann-Gibbs statistics. J. Stat. Phys. **52**, 479–487 (1988)
18. Tsallis, C.: Non additive entropy and non-extensive statistical mechanics–an overview after 20 years. Braz. J. Phys. **39**(2A), 337–356 (2009)

Chapter 14
Quasi-Optimization of the Time Dependent Traveling Salesman Problem by Intuitionistic Fuzzy Model and Memetic Algorithm

Ruba Almahasneh, Boldizsar Tuu-Szabo, Peter Foldesi, and Laszlo T. Koczy

Abstract The Traveling Salesman Problem (TSP) is an NP-hard graph search problem. Despite having numerous modifications of the original abstract problem, Time Dependent Traveling Salesman Problem (TD TSP) was one of the most realistic extensions under real traffic conditions. In TD TSP the edges between nodes are assigned higher costs (weights), if they were traveled during the rush hour periods, or crossed the traffic jam regions, such as the city center(s). In this paper we introduce an even more real-life motivated approach, the Intuitionistic Fuzzy Time Dependent Traveling Salesman Problem (IFTD TSP), which is a further extension of the TSP, and also of the classic TD TSP, with the additional notion of using intuitionistic fuzzy sets for the definition of uncertain costs, time, and space of the rush hour—traffic jam region affecting graph sections. In IFTD TSP we use fuzzy memberships and non-memberships sets for estimating the vague costs between nodes in order to quantify the behavior of traffic jam regions, and the rush hour periods. Since intuitionistic fuzzy sets are generalizations of classic fuzzy sets, our

R. Almahasneh (✉)
Telecommunications and Media Informatics, Budapest University of Technology and Economics, Budapest, Hungary
e-mail: mahasnehr@tmit.bme.hu

B. Tuu-Szabo
Department of Information Technology, Széchenyi István University, Gyor, Hungary
e-mail: tuu.szabo.boldizsar@sze.hu

P. Foldesi
Department of Logistics Technology, Széchenyi István University, Gyor, Hungary
e-mail: foldesi@sze.hu

L. T. Koczy
Telecommunications and Media Informatics, Budapest University of Technology and Economics, Budapest, Hungary

Department of Information Technology, Széchenyi István University, Gyor, Hungary
e-mail: koczy@tmit.bme.hu

© Springer Nature Switzerland AG 2020
O. Llanes Santiago et al. (eds.), *Computational Intelligence in Emerging Technologies for Engineering Applications*, Studies in Computational Intelligence 872, https://doi.org/10.1007/978-3-030-34409-2_14

239

approach may be considered an extension and substitution of the original abstract TD TSP problem, even, of the (classic) Fuzzy TD TSP. Lastly, DBMEA (Discrete Bacterial Memetic Evolutionary Algorithm) was applied on the IFTD TSP model, the results of the simulation runs based on some extensions of the benchmarks generated from the original TD TSP data set showed quite good and promising preliminary results.

Keywords Intuitionistic fuzzy sets · Traveling salesman problem · Time dependent traveling salesman problem · Fuzzy cost · Fuzzy jam region · Fuzzy rush hour period · Discrete bacterial memetic algorithm

14.1 Introduction

Logistics businesses are competing highly to apply new technologies on their supply chain models to increase parcels demand, expedite deliveries, and ultimately, to delight customers. However, one of their biggest expenses and challenges comes from the delivery process, as it is the most expensive and time-consuming part of SCM (supply chain management). If we look at the delivery problem in a wider perspective, it fits in the typical framework of the abstract Traveling Salesman Problem (TSP) model. The TSP originally attempts to find the optimal (shortest) route starting from the company headquarters so that all locations on the agenda are visited exactly once and then the salesman returns to the starting point [1]. In the language of graph theory, it means that the shortest Hamiltonian cycle in the graph including all nodes (cities or stops) has to be visited. Indeed, this abstract problem models many real-life applications, such as optimal delivery in logistics, VLSI design, manufacturing optimization, and many more. The TSP being an NP-hard problem, it is known that no algorithm or method may deliver the optimal solution in the general case. Thus, *quasi-optimal* solutions, or methods with no guaranteed exact results are sought for, in order to find the practical optimum, a close enough approximation of the exact goal. In the literature, there have been numerous attempts published, which apply various approaches in order to possibly find the optimum (least cost) route, or a solution close to it. These methods may be classified into three categories. First, the *exact solver, however, this type has* strong limitations concerning the speed and maximal size of the problem. Second, *approximate solvers* with guaranteed result of some quality. This type, however, often does not produce very good results; particularly it is possible that at the end, there is 50% error in the optimum. Third, *heuristic methods*, which often determine the optimal route, and for larger instances, often give good approximations, yet there is no guarantee for a good solution. The efficiency of such methods is always based on a large amount of successful search on *benchmark instances* found in the large existing repositories of the TSP and related problems. However, including the effects of rush hours and traffic jam regions will considerably increase the model complexity, particularly when simulating actual real-life cases.

There have been many algorithms to quantify such exactly intangible factors; TD TSP may be considered as one of the very hard combinatorial optimization problems with real-life motivation [2, 3]. The original abstract problem assumes that a fixed part of the graph (the traffic jam region) has time dependent costs, namely each graph edge is assigned a cost (weight) in the non-rush hour period, and another (higher) one in the rush hour time. The aim is to find the tour with the lowest cost in the same way as in the case of the TSP. Despite its ability to achieve good results in determining the overall cost, a major pitfall in the TD TSP is that a single concrete number is used to represent the proportional jam factor. This is rather unrealistic representation, as usually, the traffic jams build up gradually, and their respective effects are not uniformly affecting the area—in some border regions there may be a much smaller increase of the costs, compared to the real core. That concrete jam factor value is used to multiply the edge lengths (costs) between the nodes in the indicated region, to quantify the jam factor's effect. Realizing the limitation and impreciseness of jam factors representation in TD TSP was our driving motivation in this research paper. In particular, the impracticality of actualizing such model on real-life scenarios poses a huge challenge. We claim that the unrealistic simplification of the rush hours and the jam regions calculation was overcome in the IFTD TSP model.

There are several generalizations of fuzzy set theory satisfying various objectives; among them, the notion of *intuitionistic fuzzy sets* (IFSs) introduced by Atanassov [4] is an interesting and practically useful one—even though, identification of the intuitionistic membership functions still bears serious difficulties at present. Classic fuzzy sets are special IFSs [5, 6], with identical membership and complemented non-membership function. In fact, there are situations where IFS theory is more convenient to deal with [6]; we propose that such is the case in the TD TSP with jam factors. The literature shows several successful IFSs-based models where uncertainty factors were effectively represented [7], which fact supports the expectations that this particular application may be also successful. In fact, IFS theory has been utilized in various different areas that have to do with decision-making under vagueness and with unstructured problems and indeed proved being successful [8].

14.2   Classic TSP Cost Calculation

The original TSP was first formulated in 1930, and till present, is one of the most studied optimization problems [9]. The ultimate goal is to find the route that allows the salesman to visit all destination points with the minimum overall traveled cost. TSP can be defined as a graph search problem with edge weights as per (1). To formulate the symmetric case with n nodes (cities) $C_{ij} = C_{ji}$, so a graph can be considered where there is only one arc between every two nodes. Let $X_{ij} = \{0,1\}$ be the decision variable ($i = 1, 2, \ldots, n$ and $j = 1, 2, \ldots, n$), and $X_{ij} = 1$ means that the arc connecting node i to node j is an element of the tour. Let

$$x_{ij} = 0 \ (i = 1, 2, \ldots, n)$$

$$G_{\text{TSP}} = \left(V_{\text{cities}}, E_{\text{conn}} \right)$$

(14.1)

$$V_{\text{cities}} = \{v1, v2, \ldots, vn\}, E_{\text{conn}} \subseteq \{(vi, vj) \,|\, i \neq j\}$$

$$C : V_{\text{cities}} \times V_{\text{cities}} \rightarrow R, C = \left(C_{ij} \right)_{n \times n}$$

C is called cost matrix, where (C_{ij}) represents the cost of going from city i to city j. The goal is to find the directed Hamiltonian cycle with minimal total length. Formulated in another way, the goal is to find a sequence of vertices that produces the least total cost.

$$\left(\sum_{i=1}^{n-1} C_{pi, Pi+1} \right) + C_{pn, p1}$$

(14.2)

Depending on the properties of the cost matrix TSPs are divided into two classes: symmetric and asymmetric ones. If the cost matrix is symmetric ($C_{ij} = C_{ji}$) for all i and j, then the TSP is called symmetric, else it is asymmetric.

The calculation of time required to cover the distance between cities is vital. Since the cost elements are time dependent, the actual cost between two cities can be accurately calculated only if the total time elapsed is precisely known. If a cost element is growing then the required time is growing as well, since the velocity of the travel salesman is supposed to be constant.

14.2.1 Fuzzification of the TSP

In 1965, fuzzy sets as an extension of the classical notion of set were introduced by Zadeh [10]. The fuzzy sets are generalized from classical sets, since the indicator functions in classical sets can be seen as special cases of the membership functions if the latter only take values from {0,1}, while elements in fuzzy sets may assume any value in [0,1]. In classical set theory, according to a bivalent condition, the elements membership in a set is assessed in binary terms. A membership function valued in the real unit interval [0, 1] can help to describe any phenomenon more efficiently, wherever the notion of "belonging" is vague, gradual, nondeterministic (uncertain) [5].

Solutions of the TSP presented in the literature quantify costs of travel between nodes (cities) based on Euclidean distances. The problem in such approaches from the real-life point of view is that trip costs are here constant crisp values. However, in practical applications distances (edge costs) are often fuzzy, due to the external circumstances that affect the overall cost of any trip or any section of it. Analyzing actual salesman tours, particularly in city centers, the topography and rush hours in various locations are factors that must be looked at as uncertain values, easily interpretable as fuzzy numbers. Hence, the cost data for estimated tour distance between two nodes is not constant anymore. On the contrary, it can be more appropriate to represent this imprecision by using fuzzy numbers; even, by intuitionistic fuzzy numbers, and afterwards these fuzzy numbers may be summed up in a tour in order to calculate the total distance, which is again an (intuitionistic) fuzzy value. Fuzzification of TSP was presented in a previous research paper [11], where we introduced a fuzzy model for the Time Dependent Traveling Salesman Problem (3FTD TSP). The model redefined the TSP entirely and transformed it to the Triple Fuzzy extension (3FTD TSP) with fuzzy costs, fuzzy jam regions, and fuzzy rush hours with fuzzy factors. The model was able to express the costs influenced by the traffic jam factors in jam regions and rush hours and eventually quantify the overall tour length of the salesman trip more accurately corresponding to the real-life situation [11]. Afterwards, we applied the heuristic DBMEA simulation on the 3FTD TSP model, and the first results obtained unambiguously confirm the effectiveness and the predictability of the proposed technique [11]. This supports the generality or universality of the method that has been already demonstrated on a series of different abstract graph search problems. As mentioned before, afterwards, we extended the model 3FTD TSP using intuitionistic fuzzy theory as will be explained in details in the coming sections.

14.3 The Discrete Bacterial Memetic Evolutionary Algorithm

The Bacterial Evolutionary Algorithm (BEA) was first introduced by Nawa and Furuhashi in order to discover the optimal parameters of a fuzzy rule based system [12]. At the beginning, BEA had been inspired by the evolution of bacteria [13], in contrary to the general genetic evolution. Afterwards, we introduced memetic versions, one using discrete local search for discrete problems, and it has been used in solving several different NP-hard problems. The DBMEA is a memetic version of the algorithm, a combination of the bacterial evolutionary algorithm as global search, and 2-opt and 3-opt (with some enhanced techniques) as local search [3]. Memetic algorithms in general combine the global search evolutionary algorithms with local search methods, which leads to eliminating the disadvantages of both methods. By this novel combined approach significant improvement in the performance of the classical evolutionary algorithms was achieved [14]. Since its run time of DBMEA is more predictable than that of the Concorde algorithm [15], or even, of the more efficient Helsgaun–Lin–Kernighan method [16], DBMEA may considered efficacious, especially for large-sized problems with a complicated structure. In the case of the TD TSP, it gave a better solution than the best-known value in the literature, and produced smaller average runtime than the state-of-the-art methods for the problem [2]. Thus, it was a rather obvious choice (without a real alternative) to apply this heuristics for the IFTD TSP model.

14.4 The Intuitionistic Fuzzy Time Dependent Traveling Salesman Problem (IFTD TSP)

14.4.1 Preliminary Definitions

Let us start with a short review of basic concepts and definitions related to intuitionistic fuzzy sets which are used in the upcoming sections in a more analytical formal model

Definition 1 Let a universal set E be fixed. An intuitionistic fuzzy set or IFS A in E is an object having the form

$$A = \{\langle x, \mu_A(x), \nu_A(x) \rangle \mid x \epsilon\ E\} \tag{14.3}$$

where $0 \le \mu_A(x) + \nu_A(x) \le 1$.

The difference $\pi_A(x) = 1 - (\mu_A(x) + \nu_A(x))$ is called the *hesitation part*, which may cater to either the membership value or to the non-membership value, or to both

Definition 2 If A and B are two IFSs of the universal set E, Then
$A \subset B$ iff

$$\forall x \in E, \begin{bmatrix} \mu_A(x) \le \mu_B(x) \text{ and} \\ \nu_A(x) \ge \nu_B(x) \end{bmatrix} \tag{14.4}$$

$A \supset B$ iff $B \subset A$

$$A = B \text{ iff } \forall x \in E, \begin{bmatrix} \mu_A(x) = \mu_B(x) \text{ and} \\ \nu_A(x) = \nu_B(x) \end{bmatrix} \tag{14.5}$$

$$A \cap B = \{\langle x, \min(\mu_A(x), \mu_B(x)), \max(\nu_A(x), \nu_B(x))\rangle \mid x \in E\} \tag{14.6}$$

$$A \cup B = \{\langle x, \max(\mu_A(x), \mu_B(x)), \min(\nu_A(x), \nu_B(x))\rangle \mid x \in E\} \tag{14.7}$$

It is clear that classic fuzzy set has the form

$$\{\langle x, \mu_A(x), \mu_{A^c}(x)\rangle \mid x \in E\}$$

Definition 3 Let X and Y be two fuzzy sets. An intuitionistic fuzzy relation (IFR) R from X to Y is an IFS of $X \times Y$, characterized by the membership function μ_R and the non-membership function v_R over $X \times Y$. An IFR R from X to Y will be denoted by $R (X \rightarrow Y)$.

Definition 4 If A is an IFS of X, the max–min–max composition of the IFR R $(X \rightarrow Y)$ with A is an IFS B of Y denoted by $(B = R \circ A)$ and is defined by the membership function

$$\mu_{R \circ A}(y) = \vee_x [\mu_A(x) \wedge \mu_R (x, y)] \tag{14.8}$$

and the non-membership function

$$v_{R \circ A}(y) = \wedge_x [v_A(x) \vee v_R (x, y)] \tag{14.9}$$

Previous formulas hold for all Y.

Definition 5 Let $Q (X \rightarrow Y)$ and $R (Y \rightarrow Z)$ be two IFRs. The max–min–max composition $(R \circ Q)$ is the intuitionistic fuzzy relation from X to Z, defined by the membership function

$$\mu_{R \circ Q} (x, z) = \vee_y \left[\mu_Q (x, y) \wedge \mu_R (y, z) \right] \tag{14.10}$$

and the non-membership function given by

$$v_{R \circ Q} (x, z) = \wedge_y \left[v_Q (x, y) \vee v_R (y, z) \right]$$
$$\forall (x, z) \in X \times Z \text{ and } \forall y \in Y \tag{14.11}$$

Let A be an IFS of the set J, and R be an IFR from J to C. Then the max–min–max composition (14.10) B of IFS A with the IFR R $(J \rightarrow C)$ denoted by $B = A \circ R$ denotes the cost of the edges as an IFS B of C with the membership function given by

$$\mu_B(c) = \vee_{j \in J} [\mu_A(j) \wedge \mu_R (j, c)] \tag{14.12}$$

and the non-membership function given by

$$v_B(c) = \wedge_{j \in J} [v_A(j) \vee v_R (j, c)]$$
$$\forall c \in C. \text{ (Here } \wedge = \text{Min and } \vee \text{ Max)} \tag{14.13}$$

If the state of the edge E is described in terms of an IFS A of J; then E is assumed to be the assigned cost in terms of IFSs B of C, through an IFR R from J to C. At present it is assumed to be given by a knowledge base directory (or by experts who are able to translate the jam degrees of association and non-association according to geographical areas) on the destination cities and the extent (membership) to which each one is included in the jam region. Later, we intend to develop automated ways for the identification of the functions by using input–output data populations representing the behavior of the traffic jam areas. This will be translated to the degrees of association and non-association, respectively, between jam and cost.

Now, let us expand this concept to a finite number of edges E that form a whole tour for a salesman. Let there be n edges E_i; $i = 1; 2; \ldots n$, in a trip (from starting point to final destination). Thus $e_i \in E$. Let R be an IFR ($J \to C$) and construct an IFR Q from the set of edges E to the set of jam factors J. Clearly, the composition T of IFRs R and ($T = R \circ Q$) give the cost for each edge from E to C given by the membership function

$$\mu_T (e_i, c) = \vee_{j \in J} \left[\mu_Q (e_i, j) \wedge \mu_R (j, c) \right] \tag{14.14}$$

and the non-membership function given by

$$\nu_T (e_i, c) = \wedge_{j \in J} \left[\nu_Q (e_i, j) \vee \nu_R (j, c) \right] \tag{14.15}$$

$$\forall e_i \in E \text{ and } c \in C$$

For given R and Q, the relation ($T = R \circ Q$) can be computed. From the knowledge of Q and T, an improved version of the IFR R can be computed, for which the following holds valid:

1. $J_R = \mu_R - \nu_R \bullet \pi_R$ is greatest
2. The equality $T = R \circ Q$ is retained

Obviously, the proposed improved version of R will be a more significant IFR translating the higher degrees of association and lower degrees of non-association of J as well as lower degrees of hesitation to the cost evaluation C. If almost equal values for different C in T are obtained, then we consider the case for which hesitation is least. From a refined version of R one may infer the cost from jam factors in the sense of a paired value, one being the degree of association and other the degree of non-association. Hence, we are one step closer to make a more realistic description of traffic jam factors as part of investigating the extended TD TSP problem.

14.4.2 Formulation of an IFTD TSP by Extending a TD TSP Case

Let there be a simple tour that has only 12 edges ($E1 \ldots E12$) as shown in Fig. 14.1. Each edge connects two nodes. Thus, each edge eventually will have a jam cost (maybe, close to zero), depending on the jam area(s) (and the membership degrees with which) it crosses. Table 14.1 shows each edge and the jam factors

Fig. 14.1 Simple example for a tour

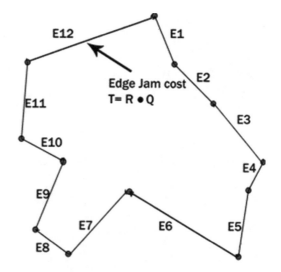

Table 14.1 Route1 = (edge 1 ... edge12)

(Q)	Jam region1	Jam region2	Jam region3	Jam region4
$E1$	(0.8, 0.1)	(0.6, 0.1)	(0, 1)	(0, 1)
$E2$	(0, 1)	(0, 1)	(0.2, 0.8)	(0.6, 0.1)
$E3$	(0.8, 0.1)	(0.8, 0.1)	(0, 1)	(0, 1)
$E4$	(0, 1)	(0, 1)	(0, 0.6)	(0.2, 0.7)
$E5$	(0.8, 0.1)	(0.8, 0.1)	(0, 0.6)	(0.2, 0.7)
$E6$	(0, 0.8)	(0.4, 0.4)	(0, 1)	(0, 1)
$E7$	(0, 1)	(0, 1)	(0.6, 0.1)	(0.1, 0.7)
$E8$	(0, 0.8)	(0.4, 0.4)	(0.6, 0.1)	(0.1, 0.7)
$E9$	(0.6, 0.1)	(0.5, 0.4)	(0, 1)	(0, 1)
$E10$	(0, 1)	(0, 1)	(0.3, 0.4)	(0.7, 0.2)
$E11$	(0, 0.8)	(0.4, 0.4)	(0.6, 0.1)	(0.1, 0.7)
$E12$	(0.4, 0.4)	(0.6, 0.1)	(0, 1)	(0.2, 0.8)

associated. The main goal is to quantify the jam factors, which will be multiplied by the physical distance (Euclidean cost) between two nodes. Eventually, this way we are quantifying the jam cost for each edge that is part of that tour path. The intuitionistic fuzzy relation $Q(E \rightarrow J)$ is given as shown in Table 14.1. Let the set of jam costs be C as given in Table 14.2. The intuitionistic fuzzy relation $R(J \rightarrow C)$ is as given in Table 14.2. The composition $(T = R \circ Q)$ will be as given in Table 14.3. We calculated jam region cost factors (c_{jam}) as given in Table 14.4 where the four cost factors are ($c_1 = 1.2$, $c_2 = 1.5$, $c_3 = 2$, $c_4 = 5$) with weighted average calculations:

$$c_{jam_e} = \frac{\sum_i j_i \times c_i}{\sum_i j_i} \qquad (14.16)$$

Table 14.2 Jam Costs

Jam area (R)	Cost factor 1 ($c1$)	Cost factor 2 ($c2$)	Cost factor 3 ($c3$)	Cost factor4 ($c4$)
Jam region1	(0.4, 0)	(0.7, 0)	(0.3, 0.3)	(0.1, 0.7)
Jam region2	(0.3, 0.5)	(0.2, 0.6)	(0.6, 0.1)	(0.2, 0.4)
Jam region3	(0.1, 0.7)	(0, 0.9)	(0.2, 0.7)	(0.8, 0)
Jam region4	(0.4, 0.3)	(0.4, 0.3)	(0.2, 0.6)	(0.2, 0.7)

Table 14.3 $T = R \circ Q$

Jam cost (T)	Cost factor1	Cost factor2	Cost factor3	Cost factor4
$E1$	(0.4, 0.1)	(0.7, 0.1)	(0.6, 0.1)	(0.2, 0.4)
$E2$	(0.4, 0.3)	(0.4, 0.3)	(0.2, 0.6)	(0.2, 0.2)
$E3$	(0.4, 0.1)	(0.7, 0.1)	(0.6, 0.1)	(0.2, 0.4)
$E4$	(0.2, 0.7)	(0.2, 0.7)	(0.2, 0.7)	(0.2, 0.6)
$E5$	(0.3, 0.1)	(0.7, 0.1)	(0.6, 0.1)	(0.2, 0)
$E6$	(0.3, 0.5)	(0.2, 0.6)	(0.4, 0.4)	(0.2, 0.4)
$E7$	(0.1, 0.7)	(0.1, 0.7)	(0.2, 0.7)	(0.6, 0.1)
$E8$	(0.3, 0.5)	(0.2, 0.6)	(0.4, 0.4)	(0.6, 0.1)
$E9$	(0.4, 0.1)	(0.6, 0.1)	(0.5, 0.3)	(0.2, 0.4)
$E10$	(0.4, 0.3)	(0.2, 0.6)	(0.2, 0.6)	(0.3, 0.4)
$E11$	(0.3, 0.5)	(0.2, 0.6)	(0.4, 0.4)	(0.6, 0.1)
$E12$	(0.4, 0.4)	(0.4, 0.4)	(0.6, 0.1)	(0.2, 0.4)

Table 14.4 Intuitionistic jam region cost factors for edges

J_R	(J_R1)	($C1$)	(J_R2)	($C2$)	(J_R3)	($C3$)	(J_R4)	($C4$)	Total Jam region cost factors
$E1$	0.35	1.2	0.68	1.5	0.57	2	0.04	5	1.695
$E2$	0.31	1.2	0.31	1.5	0.08	2	0.08	5	1.791
$E3$	0.35	1.2	0.68	1.5	0.57	2	0.04	5	1.695
$E4$	0.13	1.2	0.13	1.5	0.13	2	0.08	5	2.151
$E5$	0.24	1.2	0.68	1.5	0.57	2	0.2	5	2.04
$E6$	0.2	1.2	0.08	1.5	0.32	2	0.36	5	2.917
$E7$	0	1.2	0	1.5	0.13	2	0	5	2
$E8$	0.2	1.2	0.08	1.5	0.32	2	0.57	5	3.291
$E9$	0.35	1.2	0.57	1.5	0.44	2	0.04	5	1.682
$E10$	0.31	1.2	0.08	1.5	0.08	2	0.18	5	2.388
$E11$	0.2	1.2	0.08	1.5	0.32	2	0.57	5	3.291
$E12$	0.32	1.2	0.32	1.5	0.57	2	0.04	5	1.763

The rush hour cost factors of each tour edge (c_{rush}) are determined in a similar intuitionistic model. The relations between time and the rush hour periods ($\grave{Q}$) are described with intuitionistic fuzzy functions, Fig. 14.2. An intuitionistic fuzzy relation ($\grave{R}$) is given between the rush hour periods and the cost factors similarly, as

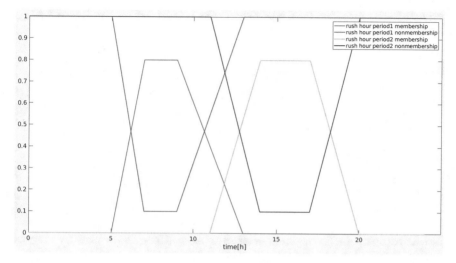

Fig. 14.2 Fuzzy membership and non-membership functions of the rush hour periods

was done for the jam regions in Table 14.2. Then the composition $\left(\dot{T} = \dot{Q} \circ \dot{R}\right)$ is calculated. Finally rush hour cost factors were calculated with weighted averaging.

The cost of the edges is calculated taking into account the two cost factors (jam region and rush hour cost factors):

- if $c_{jam_e} > 0$ (the edge belongs to at least one of the jam regions)

$$\text{AND } c_{rush_e} > 0 \text{ (in rush hours)}$$

$$C_e = c_{jam_e} \times c_{rush_e} \times \text{dist}_e \tag{14.17}$$

- else

$$C_e = \text{dist}_e$$

where dist_e is the Euclidean distance.

14.5   Computational Results

The DBMEA algorithm successfully used for a series of NP-hard graph search problems (including the TSP and TD TSP as well) was modified and tested for the new IFTD TSP problem. In the sense of the first TD TSP related publication [12] the same benchmark is set, namely the bier127 problem was selected for the test, because results for the "classic" TD TSP with this instance were available already, and so it was easier to validate the obtained results [2, 12].

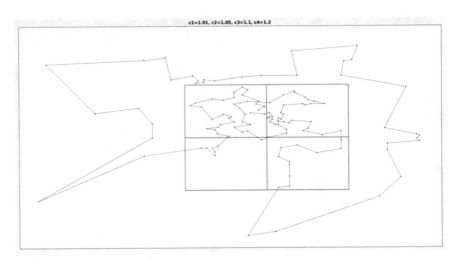

Fig. 14.3 Best tour $c_1 = 1.01$ $c_2 = 1.05$ $c_3 = 1.1$ $c_4 = 1.2$

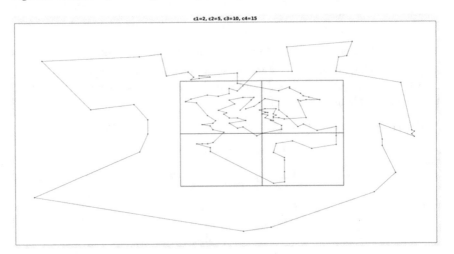

Fig. 14.4 Best tour $c_1{=}2$ $c_2{=}5$ $c_3 = 10$ $c_4 = 15$

There was a single traffic jam region defined (with coordinates of the left corner point (7080, 7200), and the width of 6920, and the height of 9490). In our test, this jam region was divided into four equal-sized subdivisions; see red rectangles in Figs. 14.3 and 14.4. Velocity v is 6000 m/h in each test. Two rush hour periods were defined as in Fig. 14.2.

Our algorithm was tested on an Intel Core i7-7500U 2.7 GHz, 8GB of RAM memory workstation under Linux Mint 18. The results were calculated by averaging five test runs as shown in Table 14.5. This Table contains the total time in hours

Table 14.5 Computational results for IFTD TSP using DBMEA

Cost factors	DBMEA		
	Best elapsed time	Average elapsed time	Average runtime [s]
$c_1 = 1.01$ $c_2 = 1.05$ $c_3 = 1.1$ $c_4 = 1.2$	20.605	20.712	170.535
$c_1 = 1.05$ $c_2 = 1.1$ $c_3 = 1.2$ $c_4 = 1.5$	21.166	21.349	176.691
$c_1 = 1.1$ $c_2 = 1.3$ $c_3 = 1.5$ $c_4 = 2$	22.061	22.186	192.313
$c_1 = 1.2$ $c_2 = 1.5$ $c_3 = 2$ $c_4 = 5$	22.885	23.124	207.919
$c_1 = 1.5$ $c_2 = 2$ $c_3 = 5$ $c_4 = 10$	23.867	24.185	195.417
$c_1 = 2$ $c_2 = 5$ $c_3 = 10$ $c_4 = 15$	23.909	24.302	207.109
$c_1 = 5$ $c_2 = 10$ $c_3 = 20$ $c_4 = 50$	24.058	24.629	124.910

required to visit each location with different cost factors because the distances between the locations are constant values, the traveling time from one location to the other changes depending on the current traffic conditions. This new problem is a generalization of the original TSP (with $c_1 = 1.2$, $c_2 = 1.5$, $c_3 = 2$, $c_4 = 5$ the edge costs are the Euclidean distances, so this parameters result in the TSP problem). As it can be seen in Table 14.5, DBMEA also found high quality solutions with big cost factors. Figure 14.3 shows the best tours for cost factor values ($c_1 = 1.01$ $c_2 = 1.05$ $c_3 = 1.1$ $c_4 = 1.2$), while Fig. 14.4 shows the best tours for cost factor values ($c_1=2$ $c_2=5$ $c_3 = 10$ $c_4 = 15$). Also, in Fig. 14.4 the locations within the jam regions are visited before or after the rush hours to reduce the total elapsed time.

14.6 Conclusion and Future Work

In this paper, we proposed a novel intuitionistic fuzzy set based model for the realistic extension of the TD TSP, namely the IFTD TSP model, to calculate more accurately jam factors and rush hours in the real-life extension of the TDT SP problem by applying the IFS theory. Clearly, this enhanced model will be a more promising approach in translating the higher degrees of association and lower degrees of non-association of the jam factor and rush hours as well as lower degrees of hesitation to any edge cost and, ideally, a more realistic calculation for the traveled routes. Our future work will focus on applying the DBMEA meta-heuristics for determining the (quasi) optimal tours on our model and test its efficiency and effectiveness on a larger number of benchmarks, with more complicated cases. It is also foreseen that the intuitionistic membership and non-membership functions will be determined by some objective identification technique, applying machine learning, clustering, evolutionary approached, or the like.

Acknowledgments This work was supported by National Research, Development and Innovation Office (NKFIH) K124055. Supported by the ÚNKP-18-3 New National Excellence Program of the Ministry of Human Capacities.

References

1. Gutin, G., Punnen, A.P.: The Traveling Salesman Problem and Its Variations, pp. 1–28. Springer, New York (2007)
2. Schneider, J.: The time-dependent traveling salesman problem. Physica A. **314**, 151–155 (2002)
3. Tüű-Szabó, B., Földesi, P., Kóczy, T.L.: Discrete bacterial memetic evolutionary algorithm for the time dependent traveling salesman problem. In: International Conference on Information Processing and Management of Uncertainty in Knowledge-Based Systems (IPMU 2018), Cadíz, Spain, pp. 523–533. Springer, Cham (2018)
4. Atanassov, R.K.: Intuitionistic fuzzy sets. Fuzzy Sets Syst. **20**, 87–96 (1986)
5. Gau, W.L., Buehrer, D.J.: Vague Sets. IEEE Trans. Syst. Man Cybern. **23**(2), 610–614 (1993)
6. Biswas, R.: On fuzzy sets and intuitionistic fuzzy sets. Notes Intuit. Fuzzy Sets. **3**, 3–11 (1997)
7. Boran, F., Genç, S., Ku, M., Akay, D.: A multi-criteria intuitionistic fuzzy group decision making for supplier selection with TOPSIS method. Expert Syst. Appl. **36**, 11363–11368 (2009)
8. Szmidt, E., Kacprzyk, J.: Intuitionistic fuzzy sets in group decision making. Notes Intuit. Fuzzy Sets. **2**(1), 11–14 (1996)
9. Applegate, D.L., Bixby, R.E., Chvátal, V., Cook, W.J.: The Traveling Salesman Problem: A Computational Study, pp. 1–81. Princeton University Press, Princeton (2006)
10. Zadeh, L.A.: Fuzzy sets. Inf. Control. **8**(3), 338–353 (1965)
11. Kóczy, L.T., Földesi, P., Tüű-Szabó, B.: Ruba Almahasneh: modeling of fuzzy rule-base algorithm for the time dependent traveling salesman problem. In: Proceedings of the IEEE International Conference on Fuzzy Systems (FUZZ-IEEE), (Under review) (2019)
12. Nawa, N.E., Furuhashi, T.: Fuzzy system parameters discovery by bacterial evolutionary algorithm. IEEE Trans. Fuzzy Syst. **7**, 608–616 (1999)

13. Moscato, P.: On Evolution, Search, Optimization, Genetic Algorithms and Martial Arts —
    Towards Memetic Algorithms, Technical Report Caltech Concurrent Computation Program,
    Report. 826. California Institute of Technology, Pasadena (1989)
14. Tüű-Szabó, B., Földesi, P., Kóczy, L.T.: The Discrete Bacterial Memetic Evolutionary
    Algorithm for Solving the One-Commodity Pickup-and-Delivery Traveling Salesman Problem.
    Springer, Cham (2018)
15. Applegate, D.L., Bixby, R.E., Chvátal, V., Cook, W.J., Espinoza, D., Goycoolea, M., Helsgaun,
    K.: Certification of an optimal tour through 85,900 cities. Oper. Res. Lett. **37**(1), 11–15 (2009)
16. Helsgaun, K.: An effective implementation of the Lin-Kernighan traveling salesman heuristic.
    Eur. J. Oper. Res. **126**, 106–130 (2000)

Chapter 15
Analyzing Information and Communications Technology National Indices by Using Fuzzy Data Mining Techniques

Taymi Ceruto, Orenia Lapeira, and Alejandro Rosete

Abstract Several indices are available to characterize information and communications technology (ICT) aspects in different countries and regions, such as ICT Development Index, the number of Internet users, and the number of supercomputers in the TOP500 list. In this chapter we show how the flexibility and expressiveness of fuzzy logic can be used to understand these indices and their relationships (by applying fuzzy data mining). Consequently, we obtain several interesting fuzzy patterns that generalize and describe this information (for all the set and for each region) in form of graphs, fuzzy clusters, and fuzzy predicates. In addition, the similarity of different patterns is studied, showing that different types of patterns are more similar than what it is normally assumed.

Keywords Fuzzy logic · Data mining · ICT indices · Fuzzy clusters · Fuzzy rules · Fuzzy predicates

T. Ceruto (✉)
Grupo de Investigación SCoDA (SoftComputing and Data Analysis), Facultad de Informática, Universidad Tecnológica de La Habana José Antonio Echeverría, Cujae, Marianao, Cuba

O. Lapeira
Departamento de Ingeniería de Software, Facultad de Informática, Universidad Tecnológica de La Habana José Antonio Echeverría, Cujae, Marianao, Cuba
e-mail: olapeira@ceis.cujae.edu.cu

A. Rosete
Departamento de Inteligencia Artificial e Infraestructura de Sistemas Informáticos, Universidad Tecnológica de La Habana José Antonio Echeverría, Cujae, Marianao, Cuba
e-mail: rosete@ceis.cujae.edu.cu

© Springer Nature Switzerland AG 2020
O. Llanes Santiago et al. (eds.), *Computational Intelligence in Emerging Technologies for Engineering Applications*, Studies in Computational Intelligence 872, https://doi.org/10.1007/978-3-030-34409-2_15

15.1 Introduction

Data from different fields grows day by day. Data mining is playing a vital role to intelligently convert available data into useful information and knowledge. Over the years, many different approaches have been developed for improving the classical techniques with fuzzy logic such as fuzzy clustering [27], fuzzy classifier systems [7, 33], fuzzy association rule mining [20, 25], fuzzy predicates [8], etc. All these efforts have enabled fuzzy data mining to be successfully applied in several areas [40] and this is due to its good properties for representing uncertain knowledge [17]. Each data mining algorithm has its own strengths and weaknesses and novel problems and challenges are raised every day [11, 12]. The objective of any data mining process is to build a predictive or descriptive model of a large amount of data that explains it and it is also able to generalize new data. For that reason one of the intentions of this chapter is to encourage new researchers to analyze the similarity of different types of fuzzy patterns.

In order to track the growth and impact of emerging information and communications technology (ICT) trends, many global indicators are defined (e.g., ICT Development Index [18, 26], Global Innovation Index [13], Internet penetration [19], and the TOP500 list [37]). ICT represents an important structural part of modern society and the countries are paying special attention to this topic and are continuously looking for ways to progress. The importance and necessity of ICT in the current age are undeniable, because they have the potential to transform the lives of people and nations. Indeed, in several papers (see Sect. 15.2.3 for details) data mining techniques have been used for analyzing ICT indices.

This chapter presents an experimental study (using 11 fuzzy ICT indices associated with 207 countries) that shows how this data can be analyzed from a fuzzy point of view in order to discover similarities among apparently not related original (crisp) ICT indices and also to find several fuzzy patterns based on data mining techniques (visualization, clusters, predicates, correlations, similarity measures) that characterize the influence of ICT over different regions on the world.

The remainder of this chapter is organized as follows. Section 15.2 introduces some basic definitions of fuzzy data mining algorithms (Sect. 15.2.1), several indices that characterize information and communications technologies (Sect. 15.2.2), and some related works (Sect. 15.2.3). The experimental results of applying fuzzy data mining algorithms over several ICT indices are given in Sect. 15.3. Finally, Section 15.4 summarizes the contributions and achievements of the paper as well as some further developments.

15.2 Fuzzy Data Mining and ICT Indices

15.2.1 Fuzzy Sets and Fuzzy Data Mining

There is a growing indisputable role of fuzzy set technology in the realm of data mining [11, 12] and those models stand out for many applications [1]. One of the main limitations of traditional (crisp) sets is that they consider that the elements in the domain only have two possible values of membership, i.e., 0 or 1. However, in fuzzy sets each element $x \in X$ has a certain degree of membership $\mu(x)$, i.e., $\mu :$ $X \rightarrow [0, 1]$. Several operations such as complement/negation, union/disjunction, and intersection/conjunction are available, and they are usually defined by using the subtraction from 1, maximum and minimum, respectively, according to Zadeh classical operators [39, 40].

Let us clarify the approach with a simple example. A simple database could represent a fuzzy characterization of the use of ICT in three dimensions or columns (mobile services, fixed services, government policy) and several rows (one for each country). For example, each column could represent the fuzzy sets M: "wide use of Mobile services", F: "good Fixed services", G: "well oriented Government policy" defined by the membership functions $\mu_M(x)$, $\mu_F(x)$, and $\mu_G(x)$. It is possible to obtain the degree of membership of each country to the classes or sets M, F, and G by using fuzzy membership functions [39, 40] (like trapezoidal, L-function, singleton or triangular functions) applied over numerical data, e.g., the ratio of mobile devices per person could be used as the raw data to obtain the degree of membership to M. For example, three countries may be described as follows [26]:

- Country C: 70% of Internet users stated they used mobile devices to access the Internet, fiber-based technology remained at 5% and there are numerous governmental programmes.
- Country E: 50% of persons with mobile phones, diverse access to several international undersea cables through three border crossings and the government has devoted substantial resources to boost infrastructure.
- Country S: Mobile-broadband penetration is in the first places of the world, the fixed market is very well developed and the government has provided the framework for this development and continues to prioritize the roll-out of high-speed networks.

The previous information may be represented in a database with tuples (M, F, G) with fuzzy values: {C:(0.7,0.1,0.85); E:(0.5,0.8,0.9); S:(1,1,1)}. It is worth mentioning that these classes are subjective, thus the functions used to obtain the fuzzy degree of membership to each class may depend on the context and the interest.

More complex fuzzy sets may be obtained by combining the sets M, F, and G. For example, the set K of countries with "wide use of Mobile services or good Fixed services, and not well oriented Government policy", i.e., $K = (M \cup F) \cap -G$, may be defined by the membership function $\mu_K(x) = min(max(\mu_M(x), \mu_F(x)), 1 - \mu_G(x))$. In the previous database, the degrees of membership of each country to the set K are {C:0.15,E:0.1,S:0}. On the other hand, the degrees of membership of each country to the set $P = M \cup -F$ are {C:0.9,E:0.5,S:1}

For an extended explanation about fuzzy sets we recommend [39, 40]. Applications of fuzzy technology can be found in artificial intelligence, computer science, control engineering, decision theory, expert systems, logic, management science, operations research, robotics, and others [40] with a recognized interpretability based on the linguistic variables [39, 40].

Fuzzy data mining includes many techniques like clustering [27], classification [27], association rule mining [15, 20, 25], predicates elicitation [8], etc. This section only introduces the basis of the descriptive models and algorithms that are used specifically in this chapter: fuzzy clustering and fuzzy predicates. A comprehensive description of many data mining techniques can be found in [38] and an interesting discussion on the synergy between data mining and fuzzy logic is available in [17].

15.2.1.1 Fuzzy Clustering

In clustering, samples without class labels are grouped into meaningful clusters according to their similarity. These clusters can be utilized to describe the underlying structure in data, which is helpful for better understanding and exploring the data distribution [38]. Clustering can be either hard or fuzzy. In the first category, the patterns are distinguished in a well defined cluster boundary region. But due to the overlapping nature of the cluster boundaries, some class of patterns may be specified in a single cluster group or dissimilar group. This property limits the use of hard clustering in real life applications. Fuzzy type clustering provides more information about the memberships of the patterns [27].

The highest effort that has been carried out for fuzzy clustering algorithms is related to a fuzzy C-means (FCM) approach [14]. In FCM a sample belongs to all the clusters with a certain fuzzy membership degree. That is why each cluster corresponds to a fuzzy set. FCM is a clustering algorithm based on partitioning, i.e., searching for the biggest similarity between objects on the same cluster and the minimum similarity among different clusters. It has the advantage of giving good modeling results in many cases [14].

15.2.1.2 Fuzzy Predicates in Normal Forms

A fuzzy predicate [8] may be interpreted as a complex fuzzy concept or class with a certain degree of membership for each object, e.g., the previously presented sets $P = M \cup -F$ and $K = (M \cup F) \cap -G$. Each fuzzy predicate is expressed as a combination of fuzzy concepts (corresponding to variables or columns in a fuzzy database) and operators (such as intersection/conjunction, union/disjunction, and complement/negation).

Several measures are available to evaluate the goodness of a fuzzy predicate in a fuzzy database [8]: truth value (TV), support (S), high pruning average (HPA), central pruning average (CPA), low pruning average (LPA), and binary support (BS). Each measure is focused on different aspects of a fuzzy set (or a fuzzy predicate). TV indicates the minimum degree of membership, while S indicates the average degree of membership. HPA, CPA, and LPA indicate the average degree of membership of different subsets of the original set: HPA corresponds to the 50% of the elements with the lowest degrees of membership, LPA the 50% with the highest degrees, and CPA the 50% with the intermediate degrees. Finally, BS indicates the percent of the whole set that have a degree of membership equal to or greater than a given value. In the previous examples, the support S of the predicate K is 0.083 (i.e., average of 0.15, 0.1, and 0) and its truth value TV is 0 (minimum of 0.15, 0.1, and 0); while the support S of P is 0.8 (i.e., average of 0.9, 0.5, and 1) and its truth value TV is 0.5 (i.e., minimum of 0.15, 0.1, and 0). If a predicate has high TV or S, it represents a good characterization of the examples in the database, e.g., based on the previous values of TV and S the countries C, E, and S are better described by the set P than by the set K.

Fuzzy predicates elicitation [8] is a novel unsupervised (descriptive) data mining task that searches for good fuzzy predicates (often expressed in conjunctive or disjunctive normal forms) to describe a fuzzy database. This task is faced as an optimization problem that tries to maximize the quality of the obtained predicates.

15.2.2   *Information and Communications Technologies (ICT)*

Information and communications technologies (ICT) include computers, networks, phones, television, radio, audio-visual equipment, etc. Taking into account the importance of ICT for social development, several initiatives exist for measuring and monitoring ICT level. In this paper we will focus on Internet penetration (IP), ICT Development Index (IDI), TOP500 list, QS rankings, SCImago rankings, and the Global Innovation Index (GII). It is worth noting that QS and SCImago rankings are not really ICT indices, i.e., they have the focus on other aspects (universities and journals, respectively) but these aspects are evidently connected with ICT development. All these indices are very well known, they are based on reliable information and they are available for the public domain [23].

15.2.2.1   Internet Penetration (IP)

The global network has become something most people find necessary and irreplaceable. Without the Internet in the twenty-first century we would not be able to conduct business, communication with friends and family, and to entertain ourselves as we can now. Over half of the world's population is online, over half of those browsed websites using mobile devices and on average, mobile Internet users spend nearly 3 h online every day. Internet penetration (IP) is only one of the extensive lists of available Internet statistics [19]. The Miniwatts Marketing Group is the organization behind the Internet World Stats [19]. Internet World Stats is an International website that features up to date world Internet usage, population statistics, social media stats, and Internet Market Research Data, for over 243 individual countries and world regions. The first overview provided is the Internet usage statistics in the World by Region, that gives a big picture of the world Internet users and population statistic. One of the most interesting indexes is the penetration rate that corresponds to the percent of the population with access.

15.2.2.2 ICT Development Index (IDI)

ITU is the United Nations' specialized agency for information and communications technologies. In 2006, the ITU was asked to develop a single index, called the ICT Development Index (IDI), to measure the extent of a country's ICT level. The overall national ICT level has the inherent power of promoting the modernization of society. IDI is designed to be global and reflect changes taking place in countries at different levels of ICT development [18]. It is a composite index that combines 11 indicators into one benchmark measure. The main objectives of the IDI are to measure [18]:

- the level and evolution over time of ICT developments within countries and the experience of those countries relative to others;
- progress in ICT development in both developed and developing countries;
- the digital divide, i.e., differences between countries in terms of their levels of ICT development; and
- the development potential of ICTs and the extent to which countries can make use of them to enhance growth and development in the context of available capabilities and skills.

Besides the IDI, ITU publishes an annual report (Measuring the Information Society Report) since 2009. The report presents a quantitative analysis of the information society and highlights emerging trends and measurement issues.

15.2.2.3 TOP500 List

The TOP500 List is compiled by Erich Strohmaier of NERSC/Lawrence Berkeley National Laboratory, Jack Dongarra of the University of Tennessee, Knoxville, Horst Simon of NERSC/Lawrence Berkeley National Laboratory, and Martin Meuer of Prometeus (and, from 1993 until his death in 2014, Hans Meuer of the University of Mannheim, Germany) [36, 37]. The TOP500 table shows the 500 most powerful commercially available computer systems known, using directly information from the suppliers and cross checking different sources of information, including position within the TOP500 ranking, manufacturer, year of installation (or last major update), number of processors (Cores), Rmax (maximal LINPACK performance achieved), and Rpeak (theoretical peak performance). In the TOP500 List table, the computers are ordered (sequentially) by Rmax, Rpeak, and memory size.

15.2.2.4 QS Rankings

Quacquarelli Symonds (QS) rankings, published annually, are designed to help prospective students find the leading schools in their field of interest [30]. QS rankings are viewed as one of the three most-widely read university rankings in the world recognized by the International Ranking Expert Group (IREG). Some of the rankings are QS World University Rankings (global overall) and the QS World

University Rankings by Subject (which name the world's top universities for the study of 48 different subjects and five composite faculty areas). The QS World University Rankings has a remarkably consistent methodological framework, compiled using six simple metrics that effectively capture university performance [30]: academic reputation (40%), employer reputation (10%), faculty/student ratio (20%), citations per faculty (20%), and faculty ratio/international student ratio (5% each). Examples of the subjects included in the ranking are computer science, sustainable energies, robotics, aeronautics, medical technology, etc. Taking into account that graduates of engineering degrees are in high demand across the globe (with developing and developed nations alike calling out for highly qualified specialists to keep their economies growing) here we focus in the ranking with more technological subjects.

15.2.2.5 SCImago Rankings

SCImago is a research group from the Consejo Superior de Investigaciones Científicas (CSIC), the universities of Granada, Extremadura, Carlos III (Madrid), and Alcalá de Henares, dedicated to information analysis, representation and retrieval by means of visualization techniques. The SCImago Journal & Country Rank (SJCR) is a publicly available portal that includes the journals and country scientific indicators developed from the information contained in the Scopus database (Elsevier) [31].

The Journal Rankings provide various listings of journals ordered by several indicators [31]: SJR (SCImago Journal Rank) indicator, H-Index, total documents (current year and 3 years), total references, total cites (3 years), citable documents (3 years), cites per document (2 years), and references per document. The country ranking parameters have a very similar set of categories [31]: documents, citable documents, citations, self-citations, cited documents, and H-Index.

As well as the SJR Portal, SCImago has developed the Shape of Science and the SCImago Institution Rankings (SIR) [32]. The Shape of Science is an information visualization project whose aim is to reveal the structure of science. The SIR is a classification of academic and research-related institutions ranked by a composite indicator that combines three sets of indicators: research performance (50%), innovation (30%), and societal (20%).

15.2.2.6 Global Innovation Index (GII)

The Global Innovation Index is the result of a collaboration between Cornell University, INSEAD, and the World Intellectual Property Organization (WIPO) as co-publishers, and their knowledge partners. The GII recognizes the key role of innovation as a driver of economic growth and well-being. It aims to capture the multi-dimensional facets of innovation and to be applicable to developed and emerging economies alike [13]. Its 80 indicators explore a broad vision of innovation, including political environment, education, infrastructure, and business sophistication [13]. It is worth mentioning that GII contains more than half of indicators which are not related to ICT directly. This fact can be justified by the assumption that the success of ICT also depends on indirect characteristics (e.g., education) more than on the technological parameters.

The GII relies on two subindices (the Innovation Input Sub-Index and the Innovation Output Sub-Index) each built around key pillars. Five input pillars capture elements of the national economy that enable innovative activities: institutions, human capital and research, infrastructure, market sophistication, and business sophistication. Two output pillars capture actual evidence of innovation outputs: knowledge and technology outputs and creative outputs. Each pillar is divided into sub-pillars and each sub-pillar is composed of individual indicators (80 in total in 2018). Sub-pillar scores are calculated as the weighted average of individual indicators; pillar scores are calculated as the weighted average of sub-pillar scores. Four measures are then calculated [13]: Innovation Input Sub-Index (average of the first five pillar scores that capture elements of the national economy that enable innovation), Innovation Output Sub-Index (average of the last two pillar scores that provides information about outputs that are the results of innovation within the economy), overall GII score (average of the Input and Output Sub-Indices), and innovation efficiency ratio (ratio of the output over the Input Sub-Index).

15.2.3 Related Works

It is claimed that ICT constitutes an important factor contributing to different fields of application and has received great attention from the data mining research community. Without the intention to be exhaustive, here are mentioned some recent papers that illustrate the different approaches and interests:

- The impact of ICT investments (hardware, software, telecommunications, and internal services spending) on human development (standard of living and health) in three contexts (high, medium, and low income countries) using regression analysis [4].
- How the ICT development level and usage (access, use, and skills) influence student achievement in reading, mathematics, and science by using a hierarchical linear modeling [35].
- What drives ICT clustering in European Cities? [5]: ICT clustering has been used to support regional economic growth and technology-based development in European regions.
- A comparative study to classify ICT developments by economies [16] based on a clustering technique to classify the economies with only two essential variables, namely the number of Internet subscribers and gross domestic product.
- Identifying core technologies based on technological cross-impacts based on an association rule mining and analytic network process approach [21].

- Inferring comprehensible business/ICT alignment rules [9] by applying a rule induction algorithm on a data set containing information from 641 organizations in 7 European countries.
- Convergence in information and communications technology (ICT) using patent analysis [22] through association rule mining.
- Data mining on ICT usage in an academic campus: by using fuzzy-rough feature selection and several classification techniques such as J48, fuzzy-rough rule induction, fuzzy nearest neighbor [3].
- Analysis of eight data mining algorithms for smarter Internet of Things (IoT) [2].

In spite of the contributions of several authors in this trending topic (other examples may be found in [10, 24, 28, 29, 34]), there is still a need to investigate the relation and the coherence of different well known ICT indicators. There are many indices which are used to measure ICT all over the world and it might be a problem to choose one to analyze the ICT development in different regions and to take important decisions. For that reason, this paper intends to contribute to the present state of art by providing evidence on relationships among several ICT indices with special emphasis on the use of fuzzy patterns. Particularly, we think that the definition of fuzzy indices based on ICT indices is convenient to homogenize the information and to allow inter-indices comparisons and visualization as it is introduced in the next sections. In this sense, we believe in the usefulness of using fuzzy concepts as *"a methodological basis in many application domains"* [17] because of its simplicity and expressiveness.

15.3 Analyzing ICT Indices by Fuzzy Data Mining

In this paper we focus on several national indices very related to the information and communications technology (ICT), specifically those presented in Table 15.1.

More precisely, the analysis is conducted in the following form:

- Section 15.3.1.1 explains how the original (crisp) ICT indices (discussed in Sect. 15.2.2 and summarized in Table 15.1) were preprocessed. Selection and exclusion criteria were explained and the way used to compute derived indices.
- Section 15.3.1.2 defines several fuzzy ICT indices which are fuzzy sets associated to the original crisp indices.
- Fuzzy indices are analyzed in Sect. 15.3.2 in terms of their values for different regions and their similarities and differences based on several fuzzy measures.
- Fuzzy patterns (particularly, clusters and predicates) obtained from the fuzzy indices by using fuzzy data mining techniques are presented and analyzed in Sect. 15.3.3. Their similarities and differences are also highlighted.

Table 15.1 Technological indices used

i	Name	Max	Source
1	IP: Internet penetration (2018)[a]	100	[19]
2	IDI: ICT development index (2017)[a]	8.98	[18, 26]
3	TOP500: TOP500 (November 2018)	227	[37]
4	QS: QS Ranking (2018)	112	[30]
5	QSTech: QS technological subjects (2018)	942	[30]
6	SciJR: Scimago Journal Ranking JR (2017)	13,949	[31]
7	SciIR: Scimago Institutions Ranking (2018)	759	[32]
8	SciCRC: Scimago Country Ranking CR-Citations (2017)	426,316	[32]
9	SciCRH: Scimago Country Ranking CR-H index (2017)	2077	[32]
10	GII: GII Score (2018)[a]	68.4	[13]
11	GIIE: GII Efficiency Ratio (2018)[a]	0.96	[13]

[a]0 is assumed when the value is not available

15.3.1 Fuzzy ICT Indices Used in Analysis

15.3.1.1 Original ICT Indices

Each original index used in this chapter is described in Table 15.1. The minimal value (worst) is 0 in all indices while the maximal value that is found in the analyzed data for each index is presented in Table 15.1. In each indicated reference (last column of Table 15.1) it is possible to get the raw data used to derive the fuzzy indices used in the further analysis. There are territories that were absent from most indices and we prefer to exclude them from the analysis. In particular, only those territories with more than two available values were included. We prefer to use the generic term *"territory"* because they differ in the associated political status. The 207 territories included are shown in Table 15.2.

According to the United Nations Classification they are distributed in seven regions: CSA (Central and Southern Asia), EUR (Europe), LCN (Latin America and the Caribbean), NAC (Northern America), NAWA (Northern Africa and Western Asia), SEAO (South East Asia, East Asia, and Oceania), and SSF (Sub-Saharan Africa). The raw data of the indices 1, 2, 8, 9, 10, and 11 were taken directly from the sources indicated in Table 15.1.

The raw (not fuzzy) data associated to the indices 3, 4, 5, 6, and 7 were computed by counting the number of entities (universities, institutions, or journals) from each territory that are included in the corresponding rankings. Index 5 corresponds to the number of universities in the QS World University Rankings By Subjects, only considering the most technological aspects (we consider the following subjects: architecture and built environment; engineering and technology; computer science; chemical engineering; civil and structural engineering; electrical and electronic

Table 15.2 Territories included in the analysis

Region	Number of territories: list
CSA	14: Afghanistan, Bangladesh, Bhutan, India, Iran, Kazakhstan, Kyrgyzstan, Maldives, Nepal, Pakistan, Sri Lanka, Tajikistan, Turkmenistan, Uzbekistan
EUR	45: Albania, Andorra, Austria, Belarus, Belgium, Bosnia and Herzegovina, Bulgaria, Croatia, Czech Republic, Denmark, Estonia, Finland, France, Germany, Greece, Hungary, Iceland, Ireland, Italy, Kosovo, Latvia, Liechtenstein, Lithuania, Luxembourg, Macedonia, Malta, Moldova, Monaco, Montenegro, Netherlands, Norway, Poland, Portugal, Romania, Russia, San Marino, Serbia, Slovakia, Slovenia, Spain, Sweden, Switzerland, Ukraine, United Kingdom, Vatican
LCN	36: Antigua and Barbuda, Argentina, Aruba, Bahamas, Barbados, Belize, Bolivia, Brazil, Chile, Colombia, Costa Rica, Cuba, Curaçao, Dominica, Dominican Republic, Ecuador, El Salvador, Grenada, Guatemala, Guyana, Haiti, Honduras, Jamaica, Mexico, Nicaragua, Panama, Paraguay, Peru, Puerto Rico, Saint Kitts and Nevis, Saint Lucia, Saint Vincent and the Grenadines, Suriname, Trinidad and Tobago, Uruguay, Venezuela
NAC	2: Canada, United States
NAWA	25: Algeria, Armenia, Azerbaijan, Bahrain, Cyprus, Egypt, Georgia, Iraq, Israel, Jordan, Kuwait, Lebanon, Libya, Morocco, Oman, Palestine, Qatar, Sahara, Western, Saudi Arabia, Sudan, Syria, Tunisia, Turkey, United Arab Emirates, Yemen
SEAO	37: Australia, Brunei Darussalam, Cambodia, China, Christmas Island, Cook Islands, Fiji, Hong Kong, Indonesia, Japan, Kiribati, Korea, Republic of, Laos, Macao, Malaysia, Marshall Islands, Micronesia, Mongolia, Myanmar, Nauru, New Zealand, Niue, North Korea, Northern Mariana Islands, Palau, Papua New Guinea, Philippines, Samoa, Singapore, Solomon Islands, Taiwan, Thailand, Timor-Leste, Tonga, Tuvalu, Vanuatu, Viet Nam
SSF	48: Angola, Benin, Botswana, Burkina Faso, Burundi, Cameroon, Cape Verde, Central African Republic, Chad, Comoros, Congo, Congo (Dem. Rep.), Ivory Coast, Djibouti, Equatorial Guinea, Eritrea, Ethiopia, Gabon, Gambia, Ghana, Guinea, Guinea-Bissau, Kenya, Lesotho, Liberia, Madagascar, Malawi, Mali, Mauritania, Mauritius, Mozambique, Namibia, Niger, Nigeria, Rwanda, Sao Tome and Principe, Senegal, Seychelles, Sierra Leone, Somalia, South Africa, South Sudan, Swaziland, Tanzania, Togo, Uganda, Zambia, Zimbabwe

engineering; mechanical, aeronautical and manufacturing engineering; mineral and mining engineering; chemistry; environmental sciences; materials science; mathematics; physics and astronomy; operational research). A university may appear in the list of several subjects and they are counted every time they appear. In spite of the fact that SCImago and QS rankings are not directly focused on ICT aspects, the clear interdependencies among universities, scientific production and ICT justify their consideration in our study.

By following the described method, it was possible to associate a numerical value (a positive indicator of quality) in each index for each territory. However, as they are defined in different intervals it is hard to compare them. In the next section we associate fuzzy values to each of them.

15.3.1.2   Fuzzy Sets Associated to ICT Indices (Fuzzy ICT Indices)

We define fuzzy sets "high values" in the domain of territories with a degree of membership related to its value in each index. More precisely, based on the raw indices described in Sect. 15.3.1.1, the analysis is conducted by using the following definitions:

- $T = \{1, 2, 3, ...t, ...\}$: Set of territories, $|T| = 207$ (see Table 15.2)
- $I = \{1, 2, 3, ...i, ...\}$: Set of technological indices, $|I| = 11$ (see Table 15.1)
- H_i: Fuzzy set defined as *"high value in index"*
- $H_i(t)$: Degree of membership of each territory to the fuzzy set H_i, i.e., *"high value in index i of territory t"*. Each degree of membership is obtained as $H_i(t) = \frac{R^i(t)}{M^i}$ where $R^i(t)$ is the raw data of the index i for the territory t and M^i is the maximal value of the index i (see Table 15.1).
- H_a: Fuzzy set defined as *"high value for all indices"*, $H_a = \cap_{\forall i} H_i$.
- H_o: Fuzzy set defined as *"high value in some index"*, $H_o = \cup_{\forall i} H_i$.
- H_*: Fuzzy set defined as *"high average value of the indices"*, $H_* = \frac{\sum_{\forall i} H_i}{|I|}$.

It is interesting to note that the form used to obtain the fuzzy values is absolute with respect to all countries, without taking into account the values of the nearby countries. We believe that other forms of obtaining the fuzzy indices must be studied in the future. Also, it is worth noting that H_a, H_o, and H_* may be interpreted as integral indices that correspond to pessimistic, optimistic, and average visions of each territory, respectively.

Figure 15.1 shows the values of the fuzzy sets H_i for each territory. The series corresponding to H_a, H_o, and H_* is also included. Each series corresponds to a set

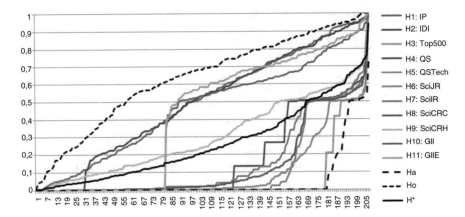

Fig. 15.1 Values of H_1–H_{11}, H_o, H_a and H_* for all territories (unpaired)

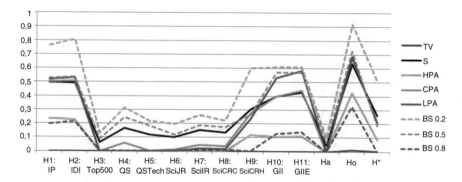

Fig. 15.2 General view of the indices H_1-H_{11}, H_o, H_a and H_*

and they are sorted in ascending order. It is worth noting that values are not paired. It may be observed that the values of the indices for most territories are smaller than 0.5. Only indices H_1 (i.e., IP), H_2 (i.e., IDI), H_{10} (i.e., GII), and H_{11} (i.e., GIIE) are over 0.5 for the territory with the central (median) position in each series. It is curious that the indices obtained by counting the number of entities (universities, institutions, or journals) are more pessimistic than the others. Indeed, for most of the indices the central median value is smaller than 0.1.

Figure 15.2 gives a general view of the indices. As they are fuzzy sets they are described in terms of measures introduced in [8] to evaluate fuzzy predicates: truth value (TV), support (S), high pruning average (HPA), central pruning average (CPA), low pruning average (LPA), and binary support (BS). In the case of BS we use 0.2, 0.5, and 0.8 to indicate three different cuts.

It may be observed in Fig. 15.2 that H_3, H_4, H_5, H_6, H_7, H_8, and H_a have small values for most of the territories (S is smaller than 0.2 and less than 30% of the territories are below 0.2 as BS 0.2 indicates). A small percent of the territories have good values in all indices because the values of BS with 0.8 (and 0.5) associated to H_a are smaller than 0.05 (0.2). Even in the optimistic view associated to H_o (*"high value in some index"*) there are a 10% of territories that are not over 0.2 (i.e., BS with 0.2 associated to H_o). Again, it may be observed that H_1, H_2, H_{10}, H_{11}, and H_o have the greatest values.

The most demanding indices are H_3-H_8 (i.e., most territories are excluded from the corresponding lists). The most inclusive indices are H_1, H_2, H_{10}, and H_{11}. A tiny part of the territories has good values in all indices and around 10% of the them have bad values in all indices.

15.3.2 Contrasting and Comparing the Fuzzy ICT Indices

15.3.2.1 Differences Among Fuzzy Indices Based on the Regions

Figure 15.3 presents the support (S) of each fuzzy index for each region. We prefer to use S because it offers a more robust vision than the other measures [8].

It is remarkable the high values of NAC, followed by EUR, and the bad values that correspond to SSF. NAC is the only region that is over 0.6 in all aspects and the only with individual values over 0.8. It is extremely separated from the rest. This may be also caused by the existence of only two territories in NAC, and this granularity may be an advantage. The second place corresponds to EUR with most values over 0.2 and some of them over 0.6. There is not a clear winner among the other regions, e.g., NAWA and LCN are better in terms of H_1 and H_2, while NAWA and CSA are better in terms of H_{10} and H_{11}. In the negative extreme, SSF has small values in all indices which are near to 0 in the indices H_3–H_8. The most remarkable differences are in the indices H_3–H_9 and in the global view given by H_a and H_*.

> In general, the best indices correspond to NAC, followed by EUR, while SSF is in the last position (worst indices).

15.3.2.2 Relations Among Fuzzy Indices

Figure 15.4 shows the values of the sets (fuzzy indices) for each territory (sorted by H_* for all series), i.e., they are paired in the sense that all values of the same territory are in the same position in the x-axis.

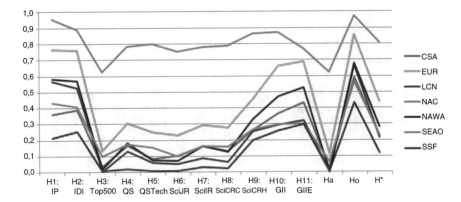

Fig. 15.3 Values of the indices H_1–H_{11}, H_o, H_a and H_* for all regions

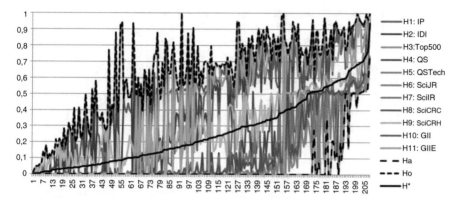

Fig. 15.4 Values of the indices H_1–H_{11}, H_o, H_a and H_* for all territories (paired)

As it may be observed there are several indices with a tendency to increase their values as H_* grows. However, in spite of the this general tendency to grow along with H_*, there are noticeable oscillations. Indeed, in terms of the Pearson correlation coefficients with respect to H_*, only two indices are greater than 0.9 (H_9 with 0.93 and H_8 with 0.91). Some correlation coefficients with respect to H_* are equal to or smaller than 0.7 (H_a with 0.68, H_3 with 0.69, H_1 with 0.70). This implies that high values in the average of the indices (H_*) are not all very strongly correlated with high values of some particular indices (e.g., H_1, H_3) or even for high values in some indices H_a.

In order to evaluate the degree of similarity among the previous fuzzy sets, we define the fuzzy set Ψ^{PQ} as follows:

- Ψ^{PQ}: Fuzzy set defined as *"Fuzzy set P is similar to fuzzy set Q"*.
- $\Psi^{PQ}(t) = 1 - |P(t) - Q(t)|$.

This implies that $\Psi^{PQ}(t)$ (the degree of membership of each territory t to the set Ψ^{PQ}) is defined as the complement of the difference between the degrees of membership to the set P and to the set Q for the territory t. Table 15.3 shows a general view of the similarity Ψ^{PQ} between each pair of the previous fuzzy ICT indices in terms of truth value (TV) and support (S).

There is a remarkable similarity among H_3, H_4, H_5, H_6, H_7, and H_8, and also between H_{10} and H_{11}, in terms of support (most of them over 0.9) and truth value (in the interval [0.35,0.65]). This implies that the values of these fuzzy indices are very similar in most countries. However, there are some remarkable differences between some indices. For example:

- the support of the similarity between H_1 and H_3 is only 0.56 with extreme cases (e.g., Andorra, Brunei, Curaçao, Denmark, and Iceland) with very high values of H_1 (greater than 0.94) and values of H_3 equal to 0;

Table 15.3 Similarity Ψ^{PQ} among predicates in terms of S[a] and TV[b]

TV/S	H_1	H_2	H_3	H_4	H_5	H_6	H_7	H_8	H_9	H_{10}	H_{11}	H_o	H_a	H_*
H_1:IP	0	0.86	0.56	0.66	0.62	0.6	0.65	0.64	0.76	0.77	0.74	0.55	0.86	0.75
H_2:IDI	0.06	0	0.56	0.66	0.62	0.61	0.65	0.63	0.76	0.82	0.79	0.56	0.86	0.75
H_3:Top500	0	0	0	0.89	**0.94**	**0.94**	**0.91**	**0.92**	0.76	0.66	0.63	**0.99**	0.43	0.8
H_4:QS	0	0	0.45	0	**0.94**	**0.93**	**0.95**	**0.94**	0.86	0.75	0.72	0.88	0.53	0.89
H_5:QSTech	0	0	0.46	0.49	0	**0.95**	**0.95**	**0.97**	0.82	0.71	0.68	**0.93**	0.49	0.86
H_6:SciJR	0	0.02	0.5	0.5	0.52	0	**0.94**	**0.95**	0.8	0.7	0.67	**0.95**	0.47	0.84
H_7:SciIR	0	0.06	0.43	0.53	0.53	0.53	0	**0.97**	0.85	0.74	0.71	0.89	0.52	0.88
H_8:SciCRC	0	0.07	0.48	0.52	0.72	0.57	0.65	0	0.83	0.73	0.7	**0.91**	0.51	0.88
H_9:SciCRH	0.06	0.2	0.37	0.48	0.48	0.46	0.52	0.53	0	0.81	0.77	0.75	0.67	**0.93**
H_{10}:GII	0	0.1	0.11	0.15	0.17	0.15	0.17	0.22	0.44	0	**0.95**	0.65	0.76	0.8
H_{11}:GIIE	0	0.1	0.02	0.02	0.04	0.02	0.04	0.08	0.3	0.73	0	0.62	0.8	0.77
H_o	0	0	0.52	0.45	0.45	0.5	0.43	0.48	0.37	0.11	0.02	0	0.42	0.79
H_a	0.28	0.06	0	0	0	0	0	0	0.06	0	0	0	0	0.63
H_*	0.14	0.29	0.41	0.55	0.61	0.5	0.64	0.68	0.63	0.52	0.34	0.41	0.14	0

[a] Values above the main diagonal correspond to support S
[b] Values below the main diagonal correspond to truth value TV

- the support of the similarity between H_1 and H_6 is only 0.6 with extreme cases (e.g., Andorra, Brunei, Curaçao, Iceland, Niue, Liechtenstein, Luxembourg) with very high values of H_1 (greater than 0.92) and values of H_6 smaller than 0.02.

In some territories, low values in the most demanding indices (e.g., H_3, H_6) are expected due to their dimensions, but the previous contrast is to be remarked in the good sense: they are good in some indices (e.g., H_1).

It is worth noting that the possibility to present all the fuzzy indices together in the previous figures is due to the common (fuzzy) scale and this is hard to do with the original crisp indices expressed in different domains. In addition, the assessment of the similarities of the indices in terms of the distribution of the values (Table 15.3) is hard to be done with the crisp indices. This analysis has allowed us to detect that the distribution of the values related to TOP500 list, QS, and SCImago are very similar.

In spite of a strong similarity among the values of some indices, the territories follow different patterns of ICT development according to their values in the fuzzy indices.

15.3.3 Fuzzy Patterns Obtained Automatically

In this section we use two fuzzy data mining techniques in order to obtain new knowledge about the presented fuzzy indices. First, fuzzy C-means (FCM) algorithm [27] is used to produce patterns in form of clusters of territories with similar values in the different indices. Second, FuzzyPred [8] is used to produce patterns in form of fuzzy predicates in normal forms (based on combination of negations, conjunctions, and disjunctions) that describe the territories. Finally, the obtained patterns are compared.

15.3.3.1 Fuzzy Clusters Based on the Indices

FCM produces clusters based on their similarities. As the obtained clusters are fuzzy, all territories have a certain degree of membership to each cluster (see Table 15.4). Here, we use the following notation:

- C^j: Prototype that represents the cluster j.
- C_j: Fuzzy set describing the membership of each territory to the cluster j.
- C: Membership of each territory to the closest cluster, $C = \cup_{\forall j} C_j$.

Table 15.4 shows the value in each fuzzy index of the prototypes of three clusters (C^b, C^m, C^w) obtained and the support (S) of the indices for the set of territories assigned to each cluster (H_i^j).

We use the implementation of FCM available in the free data mining tool Knime [6] with the standard settings (100 iterations, 3 clusters). We prefer to focus on 3 clusters because this number facilitates the analysis. In spite of the fact that all territories belong to each cluster j with a certain degree C_j, they are assigned to the cluster to which they belong with the greatest degree. FCM produce a division of the territories into three clusters: C_b (with 46 territories, 22%), C_m (77 territories, 37%), and C_w (84 territories, 41%). It is clear from Table 15.4 that cluster C_b includes the territories with the best values in all the indices, while cluster C_w includes the territories with the worst values. Cluster C_m includes territories with medium values. This is also shown in Fig. 15.5.

Table 15.4 Prototypes of the three clusters obtained by using fuzzy C-means

Cluster	H_1	H_2	H_3	H_4	H_5	H_6	H_7	H_8	H_9	H_{10}	H_{11}	H_a	H_o	H_*
C^b	0.77	0.81	0.27	0.52	0.49	0.38	0.51	0.5	0.6	0.77	0.77	0.27	0.81	0.58
H_i^b	0.75	0.78	0.28	0.54	0.51	0.39	0.54	0.52	0.6	0.75	0.76	0.21	0.86	0.58
C^m	0.54	0.56	0	0.1	0.02	0.03	0.08	0.06	0.31	0.57	0.64	0	0.64	0.27
H_i^m	0.52	0.56	0	0.11	0.01	0.04	0.08	0.06	0.32	0.6	0.68	0	0.73	0.27
C^m	0.27	0.25	0	0.01	0	0	0.01	0.01	0.12	0.03	0.03	0	0.27	0.07
H_i^w	0.33	0.27	0	0.02	0	0	0.01	0.01	0.12	0.02	0.02	0	0.42	0.07

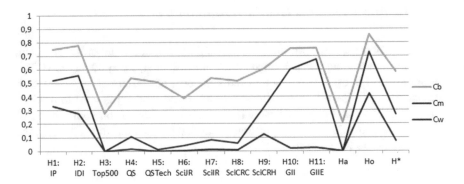

Fig. 15.5 Support of the indices for each cluster

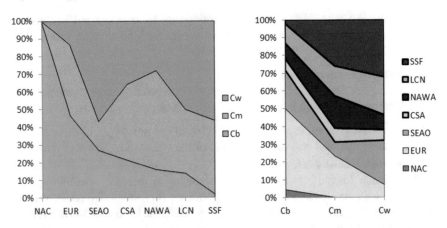

Fig. 15.6 Percent of the regions that belong to each cluster (left) and percent of the clusters that belong to each region (right)

In terms of the regions, the best cluster C_b is composed by the territories in NAC and the around the 50% of EUR which determine the 50% of this cluster, while the rest of it is composed by selected territories from almost all regions as it may be observed in Fig. 15.6 (right).

On the other hand, based on Fig. 15.6 (left) it is noticeable that NAC and EUR are the only regions with more than 80% territories in clusters C_b and C_m (followed by NAWA, around 70%). It is worth noting that in spite of the fact that the percent of SEAO that belongs to cluster C_b (best) is the third (greater than the percent of CSA, NAWA, and LCN) its percent of territories that belongs to cluster C_w (worst) is the largest in general (even worse than the percent of SSF).

A cluster with 22% of the territories (composed by NAC, a great percent of EUR and some outstanding representatives of the other regions) has the best ICT indices. Another cluster (41%) has the worst ICT indices and it is composed (mainly) by territories of SSF, LCN, and SEAO.

CSA (followed by LCN) are the regions with a percent of territories in each cluster that are more similar to the global distribution. NAC, EUR, and SEAO (slightly) are over the global tendency in the percent of territories in best cluster, while SEAO, SSF, and LCN (slightly) are over the global tendency in the percent of territories in the worst cluster.

15.3.3.2 Fuzzy Predicates Describing the Indices

In this section we obtain predicates in logical normal forms with a high degree of membership in the set of territories. This implies that the fuzzy sets defined by the obtained fuzzy predicates express conditions that are highly supported by the set of territories. Two examples of predicates would be $H_1 \cap H_2$ and $H_1 \cup H_2$. The first one has a support of 0.42, while the second has a support of 0.56. The truth value (TV) is 0 for both predicates because there are territories where both are 0. Because of the several possible combinations of logical expressions (and, or, not) with the fuzzy indices, the search for predicates with high support implies a combinatorial problem. By using FuzzyPred algorithm [8] (based on local search, with 1000 iterations and one-point mutation) the search space is explored in order to discover predicates with high support. Table 15.5 presents some of the best predicates found.

We use P_k to identify each predicate, and also the fuzzy set defined by this predicate. Consequently, $P_k(t)$ is the degree of membership of each territory to the fuzzy set defined by the predicate k. The obtained predicates with support over 0.9 and less than 5 components (in order to facilitate the interpretability) are included in Table 15.5. For example, the predicate P_5 states that the territories have low values of H_3 or low values of H_9. On the other hand the predicate P_1 states that the territories have high values of H_2, H_7, or H_{10} or low values of H_5. It may be observed that they are mostly based on combinations of not H_3, not H_5 and H_9. It is

Table 15.5 Fuzzy predicates with high support (S) and truth value (TV)

Id	Predicate	TV	S
P_1	H_2 or Not H_5 or H_7 or H_{10}	0.55	0.96
P_2	H_1 or Not H_3 or Not H_5 or H_7	0.58	0.98
P_3	H_1 or Not H_3	0.50	0.98
P_4	Not H_3 or H_9 or Not H_{10} or Not H_{11}	0.52	0.96
P_5	Not H_3 or H_9	0.52	0.95
P_6	Not H_3 or H_5 or H_9	0.52	0.96

worth noting that the predicates in Table 15.5 have values of S and TV greater than those of H_o (0.63 and 0.01, respectively). Even, the disjunction of the negation of all the fuzzy indices (considering that the support of all fuzzy indices H_i is smaller than 0.5) reach a support of 0.95 and a truth value of 0.26, i.e., the union/disjunction of the negation of all indices have worst values of TV and S than the values of P_5 with 9 more components. This implies that it is not simple to obtain combinations of conditions with high degree of membership because it implies a combinatorial search.

According to the obtained predicates, in general the territories have low values of H_3 and have high values of H_1, H_7, H_9, and H_{10}.

15.3.3.3 Relation Between Clusters, Fuzzy Predicates, and Indices

Figure 15.7 shows the set of values of the clusters and the predicates for all territories where each series is sorted independently, i.e., the series are unpaired. It is observed that the degree of membership to all clusters is around 0.1 or greater for most territories, i.e., all territories belong to all clusters with a certain observable degree.

It is also remarkable that the degree of membership to the predicates is greater than the degree of membership to the set of clusters (series C) for almost all territories. This implies that the obtained predicates are better descriptions of the territories than the description associated to the clustering.

Figure 15.8 shows the values of several measures describing the clusters and predicates. It may be observed that there are several territories that are clearly out of the clusters (truth value is 0, i.e., the minimum value of membership). However,

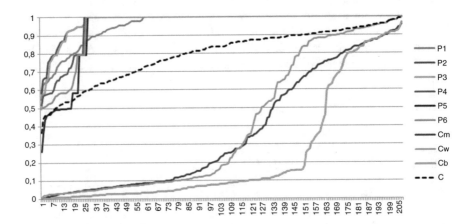

Fig. 15.7 Values of the clusters C^i and the predicates P_i for all territories (unpaired)

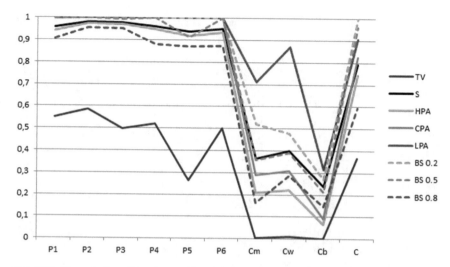

Fig. 15.8 General view of the clusters C_i and the predicates P_i

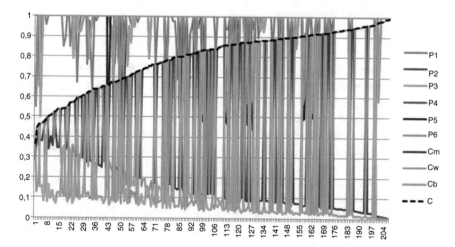

Fig. 15.9 Values of the clusters C_i and the predicates P_i for all territories (paired)

most territories belong to all the clusters with a certain degree (the binary support with 0.2 is around 0.5 for C_m and C_w and around 0.3 for C_b). Again, it is also interesting to observe that all measures of the predicates are greater than those of C, i.e., the territories are more clearly described by the predicates than by the clustering.

Figure 15.9 presents the values of membership to all the clusters and predicates sorted by C. This implies than each value of x-axis correspond to the same territory, i.e., the series are paired.

Table 15.6 Similarity Ψ^{PQ} among clusters, predicates and indices in terms of support S[a] and truth value TV[b]

TV/S	H_1	H_2	H_3	H_{10}	P_1	P_2	P_3	P_4	P_5	P_6	C_M	C_W	C_B	C
H_1	0	0.86	0.56	0.77	0.53	0.52	0.49	0.5	0.48	0.49	0.65	0.44	0.7	0.62
H_2	0.06	0	0.56	0.82	0.53	0.51	0.48	0.49	0.47	0.48	0.65	0.42	0.7	0.63
H_3	0	0	0	0.66	0.1	0.08	0.11	0.1	0.11	0.11	0.6	0.56	0.82	0.26
H_{10}	0	0.1	0.11	0	0.44	0.41	0.4	0.4	0.39	0.4	0.74	0.34	0.77	0.57
P_1	0	0	0	0	0	**0.97**	**0.94**	**0.95**	**0.93**	**0.94**	0.41	0.44	0.26	0.82
P_2	0	0	0	0	0.63	0	**0.97**	**0.98**	**0.96**	**0.97**	0.39	0.42	0.24	0.8
P_3	0	0	0	0	0.55	0.58	0	**0.99**	**0.99**	**1**	0.41	0.45	0.22	0.78
P_4	0	0	0	0	0.55	0.68	0.58	0	**0.98**	**0.99**	0.4	0.44	0.23	0.79
P_5	0	0	0	0	0.26	0.31	0.31	0.26	0	**0.99**	0.42	0.46	0.22	0.77
P_6	0	0	0	0	0.58	0.58	**0.9**	0.58	0.26	0	0.41	0.45	0.23	0.78
C_M	0.12	0.09	0.04	0.09	0	0	0	0	0	0	0	0.47	0.58	0.57
C_W	0.09	0.05	0.01	0.01	0.03	0.03	0.03	0.03	0.03	0.03	0.01	0	0.48	0.6
C_B	0.16	0.25	0.09	0.34	0	0	0	0	0	0	0.06	0.01	0	0.44
C	0.09	0.07	0.01	0.01	0.46	0.44	0.44	0.37	0.44	0.44	0.01	0.04	0.01	0

[a]Values above the main diagonal correspond to support S
[b]Values below the main diagonal correspond to truth value TV

It is remarkable the great oscillations in all series imply that each series defines different conditions. Indeed, the only Pearson correlation coefficients that are over 0.7 are among P_3, P_4, P_5, and P_6.

Table 15.6 shows the similarity Ψ^{PQ} among clusters, predicates, and some representative indices. We include in the following analysis only four indices (H_1, H_2, H_3, and H_{10}) because the other seven indices are very similar to some of the indices in this selection (see Table 15.3).

It is remarkable the similarity among the predicates P_1, P_2, P_3, P_4, and P_5 (the support of the similarity among them is greater than 0.93, and their truth value is greater than 0.55 with the exception of them with respect to P_5). This is natural because all they have high values in general. In this sense, it worth remarking that the truth value (TV) of the similarity of P_5 with respect to the others is not very high (less than 0.32) which means that P_5 reflects a different concept than the others predicates for some territories.

It is worth noting the low similarity between H_3 and the predicates which is caused by the fact that most predicates include the negation of H_3. It is also important to remark the relatively high similarity (greater than 0.7) between H_{10} with respect to C_M and C_B, i.e., it is similar the degree of accomplishment of H_{10} to the degree of membership to the clusters with medium and best values. C_M is also very similar to H_1, H_2, and H_3. It is also interesting the relatively high similarity (greater than 0.77) between C and the predicates, i.e., the conditions expressed in the predicates define similar membership than those defined by the clustering.

Figure 15.10 shows the support of the predicates and clusters for each region. For most regions all predicates are very near to 1 (they are good descriptions of these

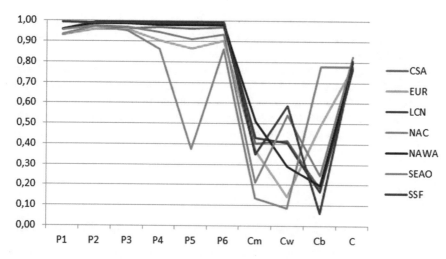

Fig. 15.10 Support of clusters C_i and predicates P_i in the regions

regions) but they have low support in NAC. It is also noticeable that EUR and SEAO are apart from the other regions. It is worth noting the low degree of membership of NAC and EUR to the worst cluster C_W, of NAC and SEAO to C_M, and of SSF (followed by LCN, CSA, and NAWA) to the best cluster C_B. In general, it is also interesting to observe that there is no direct correspondence between geographical regions and clusters.

NAC (and partially EUR and SEAO) do not follow the same status of the whole set. Indeed, predicates with high support and truth value in general, are not clearly supported by the most territories in these regions. They are a step forward with respect to the whole set.

15.4   Conclusions

This chapter has shown how the fuzzy sets can be used to analyze ICT indices that describe different technological aspects. Based on different data mining techniques (fuzzy clusters, fuzzy predicates, graphs, correlations) 11 fuzzy ICT indices are analyzed in general, and according to the different regions. Several interesting findings were commented through the paper, including the similarity among different fuzzy sets introduced and elicited by using fuzzy data mining techniques. The patterns presented along the chapter may have different implications according to the position of the different actors. For example, from the point of view of a country it may be useful to revise how it differs from the regional or world tendency. From the point of view of the producers of the ICT indices it may be useful to identify their singularity.

An interesting aspect to be explored in future studies is to consider other forms of obtaining fuzzy indices that take into account regional relative status or alternative membership functions. In addition, other data mining techniques (such as feature selection) of other parameters (e.g., more than 3 clusters) would be useful in order to discover other relevant knowledge.

References

1. Ahmad, M., Rana, A.: Fuzzy sets in data mining-a review. Int. J. Comput. Technol. Appl. **4**(2), 273–278 (2013)
2. Alam, F., Mehmood, R., Katib, I., Albeshri, A.: Analysis of eight data mining algorithms for smarter Internet of Things (IoT). Proc. Comput. Sci. **98**, 437–442 (2016)
3. Auddy, A., Mukhopadhyay, S.: Data mining on ICT usage in an academic campus: a case study. In: International Conference on Distributed Computing and Internet Technology (ICDCIT). Lecture Notes in Computer Science, vol. 8956, pp. 443–447. Springer, Berlin (2015)
4. Bankole, F., Osei-Bryson, K., Brown, I.: The impact of ICT investments on human development: a regression splines analysis. J. Global Inf. Technol. Manag. **16**(2), 59–85 (2013)
5. Belitski, M., Desai, S.: What drives ICT clustering in European cities? J. Technol. Transf. **41**(3), 430–450 (2016)
6. Berthold, M.R., Cebron, N., Dill, F., Gabriel, T.R., Kötter, T., Meinl, T., Wiswedel, B.: KNIME-the Konstanz information miner: version 2.0 and beyond. ACM SIGKDD Explor. Newslett. **11**(1), 26–31 (2009)
7. Cadenas, J., Garrido, M., Martínez, R., Muñoz, E., Bonissone, P.: A fuzzy K-nearest neighbor classifier to deal with imperfect data. Soft Comput. **22**(10), 3313–3330 (2018)
8. Ceruto, T., Lapeira, O., Rosete, A.: Quality measures for fuzzy predicates in conjunctive and disjunctive normal form. Ingeniería e Investigación **3**(4), 63–69 (2014)
9. Cumps, B., Martens, D., De Backer, M., Haesen, R., Viaene, S., Dedene, G., Snoeck, M.: Inferring comprehensible business/ICT alignment rules. Inf. Manag. **46**(2), 116–124 (2009)
10. Doong, S., Ho, S.: The impact of ICT development on the global digital divide. Electr. Commerce Res. Appl. **11**(5), 518–533 (2012)
11. Fernández, A., Lopez, V., del Jesús, M.J., Herrera, F.: Revisiting evolutionary fuzzy systems: taxonomy, applications, new trends and challenges. Knowl.-Based Syst. **80**, 109–121 (2015)
12. Fernández, A., Carmona, C.J., del Jesús, M.J., Herrera, F.: A view on fuzzy systems for big data: progress and opportunities. Int. J. Comput. Intell. Syst. **9**(1), 69–80 (2016)
13. Global Innovation Index (GII). Cornell University, INSEAD, WIPO (2019). Available https://www.globalinnovationindex.org/Home.Cited10Jan2019
14. Gosain, A., Sonika, D.: Performance analysis of various fuzzy clustering algorithms: a review. Proc. Comput. Sci. **79**, 100–111 (2016)
15. Hong, T., Lan, G., Lin, Y., Pan, S.: An effective gradual data-reduction strategy for fuzzy itemset mining. Int. J. Fuzzy Syst. **15**, 170–181 (2013)
16. Huarng, K.: A comparative study to classify ICT developments by economies. J. Bus. Res. **64**(11), 1174–1177 (2011)
17. Hullermeier, E.: Does machine learning need fuzzy logic? Fuzzy Sets Syst. **281**, 292–299 (2015)
18. ICT Development Index: IDI International Telecommunication Union (ITU) (2019). Available https://www.itu.int/net4/ITU-D/idi/2017/index.html.Cited25Jan2019

19. Internet Usage Statistics: Miniwatts Marketing Group (2019). Available https://www. internetworldstats.com/stats.htm.Cited21Jan2019
20. Kapila, D., Chopra, V.: A survey on different fuzzy association rule mining techniques. Int. J. Techno. Res. Eng. **2**(9), 2001–2007 (2015)
21. Kim, C., Lee, H., Seol, H., Lee, C.: Identifying core technologies based on technological cross-impacts: An association rule mining (ARM) and analytic network process (ANP) approach. Expert Syst. Appl. **38**(10), 12,559–12,564 (2011)
22. Kim, E., Kim, J., Koh, J.: Convergence in information and communication technology (ICT) using patent analysis. J. Inf. Syst. Technol. Manag. **11**(1), 53–64 (2014)
23. Kononova, K.: Some aspects of ICT measurement: comparative analysis of E-indexes. In: International Conference on Information & Communication Technologies in Agriculture, Food and Environment (HAICTA), pp. 938–945 (2015)
24. Lechman, E., Marszk, A.: ICT technologies and financial innovations: The case of exchange traded funds in Brazil, Japan, Mexico, South Korea and the United States. Technol. Forecast. Social Change **99**, 355–376 (2015)
25. Lin, J., Li, T., Fournier-Viger, P., Hong, T.: A fast algorithm for mining fuzzy frequent itemsets. J. Intel. Fuzzy Syst. **29**(6), 2373–2379 (2015)
26. Measuring the Information Society Report Pub series/76a34020-en. International Telecommunication Union (ITU) (2019). Available https://www.itu-ilibrary.org/science-and-technology. Cited23Jan2019
27. Nayak, J., Naik, B., Behera, H.S.: Fuzzy C-means (FCM) clustering algorithm: a decade review from 2000 to 2014. Comput. Intel. Data Mining **2**, 133–149 (2015)
28. Ogechi, A., Olaniyi, E.: Digital health: ICT and health in Africa. Actual Probl. Econ. **10**, 66–83 (2018)
29. Pradhan, R., Mallik, G., Bagchi, T.: Information communication technology (ICT) infrastructure and economic growth: a causality evinced by cross-country panel data. IIMB Manag. Rev. **30**(1), 91–103 (2018)
30. QS World University Rankings: Quacquarelli Symonds (2019). Available https://www. topuniversities.com/university-rankings.Cited14Jan2019
31. SCImago Journal & Country Rank: SCImago (2019). Available https://www.scimagojr.com. Cited22Jan2019
32. SCImago Institutions Ranking: SCImago (2019). Available http://www.scimagoir.com. Cited22Jan2019
33. Segatori, A., Marcelloni, F., Pedrycz, W.: On distributed fuzzy decision trees for big data. IEEE Trans. Fuzzy Syst. **26**(1), 174–192 (2018)
34. Sepehrdoust, H.: Impact of information and communication technology and financial development on economic growth of OPEC developing economies. Kasetsart J. Soc. Sci. 1–6 (2018). https://doi.org/10.1016/j.kjss.2018.01.008
35. Skryabin, M., Zhang, J., Liu, L., Zhang, D.: How the ICT development level and usage influence student achievement in reading, mathematics, and science. Comput. Educ. **85**, 49–58 (2015)
36. Strohmaier, E., Meuer, H., Dongarra, J., Simon, H.: The top500 list and progress in high-performance computing. Computer **48**(11), 42–49 (2015)
37. TOP500list: NERSC/Lawrence Berkeley National Laboratory, University of Tennessee, University of Mannheim (2019). Available https://www.top500.org/lists/.Cited24Jan2019
38. Witten, I., Frank, E., Hall, M., Pal, C.: Data Mining: Practical Machine Learning Tools and Techniques, 4th edn. Morgan Kaufmann, Burlington (2016)
39. Zadeh, L.: Fuzzy sets. Inf. Control **8**(3), 338–353 (1965)
40. Zimmermann, H.: Fuzzy Set Theory-and Its Applications, 4th edn. Springer, Berlin (2011)

Index

A

Acoustic levitation
  advantageous, 2
  near-field, 2–5
  principle, 2
  standing waves, 2
  type, 3
Aerospace CPS (ACPS), 77
Airborne laser scanning (ALS), 42
ALS, *see* Airborne laser scanning (ALS)
The Angle between vectors, 105
Applications
  Hilbertian approach
    first dynamical system, 22–24
    infinite dimensional optimization,
      25–26
    second dynamical system, 24–25
  multiobjective optimization, 35–36
  proposed methodology, 50
  variational approach
    first dynamical system, 28–29
    second dynamical system, 30, 31
Automated CPS, 77
Automated machine learning (AutoML)
  BA, 197
  Bayesian optimisation method, 194
  CASH, 194
  data-sets, 196–198
  execution time *vs.* performance, 198
  experimental set-up, 196–197
  freeway data, 195
  Holm post-hoc test, 200
  meta-learning approach, 194
  ML, 200

  non-parametric statistical tests, 198
  RF/NN, 198
  RMSE, 198, 199
  TF, 189, 192–194
  transportation area, 189
  urban data, 195
Automated verification techniques, 171
Autonomous navigation
  CVS, 230–232
  INS signal, 229
  NExt-PF, 235–237
  VO, 229–230
Auto-WEKA, 189, 192–195, 197–202
  See also Automated machine learning
    (AutoML)

B

Bacterial Evolutionary Algorithm (BEA),
  243
Baseline algorithms (BAs), 189, 193, 197–200,
  202
Battery, 158–160
Bayesian filters
  Kalman filter, 138
  MCMC, 138, 139
  online applications, 138
  SIR (*see* Sampling importance resampling
    (SIR))
Bayesian optimisation method, 194
Best-of-generation (BOG), 219
Binary support (BS), 258, 267
Biomimetic methodologies, 167
Boltzmann distribution, 112

© Springer Nature Switzerland AG 2020
O. Llanes Santiago et al. (eds.), *Computational Intelligence in Emerging Technologies for Engineering Applications*, Studies in Computational Intelligence 872, https://doi.org/10.1007/978-3-030-34409-2

Printed in the United States
By Bookmasters